T0291072

Introduction to Sustainability for Engineers

Introduction to Sustainability for Engineers

Toolseeram Ramjeawon

CRC Press
Taylor & Francis Group
Boca Raton London New York

CRC Press is an imprint of the
Taylor & Francis Group, an **informa** business

CRC Press
Taylor & Francis Group
6000 Broken Sound Parkway NW, Suite 300
Boca Raton, FL 33487-2742

Printed on acid-free paper

International Standard Book Number-13: 978-0-367-25445-2 (Hardback)

Library of Congress Cataloging-in-Publication Data

Names: Ramjeawon, Toolseeram, author.
Title: Introduction to sustainability for engineers/Toolseeram Ramjeawon.
Description: Boca Raton: CRC Press, [2020] | Includes bibliographical references and index.
Identifiers: LCCN 2019042492 (print) | LCCN 2019042493 (ebook) | ISBN 9780367254452 (hardback; acid-free paper) | ISBN 9780429287855 (ebook)
Subjects: LCSH: Environmental engineering. | Sustainable engineering. | Engineering design.
Classification: LCC TA170 .R36 2020 (print) | LCC TA170 (ebook) | DDC 620.0028/6–dc23
LC record available at https://lccn.loc.gov/2019042492
LC ebook record available at https://lccn.loc.gov/2019042493

Visit the Taylor & Francis Web site at
www.taylorandfrancis.com

and the CRC Press Web site at
www.crcpress.com

To my family – for always being there for

me – and to my students – for the

motivation to inspire a sustainable world

Contents

Preface

Over the past 15 years or more, sustainability and sustainable design have emerged as important features of engineering higher education degrees. Since the beginning of the 21st century, graduates from engineering programs have been expected to have a specific set of skills indicative of an appropriate level of practice, referred to as graduate attributes. These attributes are mandated, regulated, and updated by national accreditation bodies and they direct institutions toward the expected outcomes for their respective engineering curricula. There are now sustainability expectations for engineering graduates. For example, engineering graduates from all programs accredited under the International Engineering Alliance (IEA) Washington Accord are expected to have an understanding of sustainability in the context of engineering practice in their field. All engineering students should be taught about sustainability.

Current engineering curricula tend to devote little time to subjects outside the core curricula. If sustainability themes are addressed in universities, the coverage is likely to be brief. This book has the aim to offer new engineers knowledge to help them navigate the myriad of terms and concepts related to "sustainability." These various terms are carelessly thrown around without any real common understanding of what they mean, leaving them open to individual interpretation. Engineering graduates need a holistic understanding of sustainable development and the role that engineering can play in meeting important social and economic goals. While many higher education institutions have recognized this need and made moves to address it, it is still a challenge to include sustainability teaching within degree programs as a contextual focus for engineering design and problem solving. This book aims to address this challenge, help engineering educators incorporate sustainability into their curricula without adding separate courses, and provide key reading and learning materials on sustainability of which our newly minted engineers should be aware. A key feature of the book is case studies that are drawn from a range of sectors to demonstrate the various tools and approaches while at the same time recognizing the different needs of the different branches of engineering. Exercises are included at the end of every chapter to test the grasp of the key concepts and ideas being developed in the text. Useful websites on sustainable engineering are given throughout the text.

This introductory textbook aims to make the concepts of sustainable engineering especially accessible to senior undergraduate students of engineering. The students are encouraged to use this book as a starting point to explore how the principles of sustainable engineering are relevant to their chosen branch of study and future professional practice.

There are five key benefits of this book.

(1) It is a comprehensive textbook for engineering students to develop competency in sustainability.
(2) It presents a framework for engineers to put sustainability into practice.
(3) It presents the link between sustainability and the design process. It shows the application of a sustainable engineering design process for putting sustainability into practice.
(4) There are well-woven case studies and links to websites for learning in various engineering disciplines.
(5) It includes challenging exercises at the end of each chapter, which will inspire students and stimulate discussion in the class.

Acknowledgments

I am grateful to many people for their help during the preparation of this book. The most important guidance has come from my undergraduate, postgraduate, and PhD students over the years at the University of Mauritius whose interactions in class as well as their research contributed directly and indirectly to this book.

I wish to thank the contributing authors (Drs K. Elahee and R. Foolmaun) for their timely and fine contributions. Generous permission to use or adapt text or case studies or figures on sustainable engineering design was provided by the following:

- The United Nations Environment Programme (UNEP) for reproducing the content of a few case studies from their publications on Global Outlook on Sustainable Consumption and Production Policies and on Design for Sustainability and for the key messages of the Global Environment Outlook (GEO) 6 report.
- The International Energy Agency for the Executive Summary of the World Energy Outlook 2018
- The International Resource Panel for the reproduction of one of their fact sheet on Green Energy.
- The David Publishing Company for reproducing a figure in one of their publication.
- Institution of Civil Engineers (ICE) Publishing for reproducing a figure in one of their publications.
- The Water Environment Federation for permission to feature a case study.
- The BRE group for the use of a case study on BREEAM on their website.
- Switch-Asia for reproducing figures from one of their publication.
- Drs Inga Gurauskienė and Visvaldas Varžinskas from Kaunas University of Technology in Lithuania for the reproduction of the contents of their article as a case study on sustainable electrical engineering design.
- Dr Anders Andrae from Huawei Technologies Sweden for the reproduction of the contents of an article as a case study on sustainable electronic engineering design.
- The Ellen MacArthur Foundation for the reproduction of a case study on their website.

- Drs R. Santos and A. Costa from the University of Lisbon for the reproduction of the contents of an article as a case study on sustainable building design.

I would like to thank the outside reviewers of this text for their very valuable contributions. I would also like to thank the entire team at CRC Press for their great support and assistance with this book.

A special thanks to Mr Seeven Ramen and Mrs Natacha Boodhun, staff at the University of Mauritius, for helping in the typing of the manuscript and the preparation of drawings.

And most importantly, thank you to my family-my mother Seeyamwatee Devi, my sister Surekha Devi, my wife Poonam and my son Rushil -for their patience, understanding, and encouragement and for putting up with all the time spent on the preparation of the book.

About the Author

Professor **Toolseeram Ramjeawon** has more than 30 years of experience working for academic institutions, government agencies, industries, and international organizations. He is a professor of environmental engineering in the Department of Civil Engineering of the Faculty of Engineering at the University of Mauritius. He holds a Diplome D'Ingenieur and a PhD in environmental engineering and has been teaching undergraduate and postgraduate students at the University of Mauritius in environmental management and environmental engineering. His research skillsets are broad based as evidenced from the list of publications, ranging from environmental pollution, water resources management, solid waste management, treatment of domestic and industrial wastewaters, environmental modeling, and cleaner production in industry and life cycle assessments. In addition to his academic experience, he has acted as a technical adviser on environmental management to the Ministry of Environment of the Government of Mauritius and has been fortunate to undertake several consultancy projects for national and global clientele including the Government of Mauritius, the African Development Bank (ADB), the United Nations Environment Programme (UNEP), the United Nations Development Programme (UNDP), Government of Seychelles, the United Nations Economic Commission for Africa (UNECA), and the African Roundtable on Sustainable Consumption and Production(ARSCP). He works closely with UNEP on the areas of green economy, sustainable consumption and production (SCP), and life cycle assessment (LCA).

Contributors

Dr Rajendra Kumar FOOLMAUN is presently working as divisional environment officer at the Ministry of Social Security, National Solidarity and Environment and Sustainable Development (Environment and Sustainable Development Division), where he is heading the Information and Education Division. He has a PhD from the University of Mauritius (2013) in environment systems analysis. His research work was based on "Life Cycle Sustainability Assessment (LCSA) of four disposal scenarios for post-consumer Polyethylene Terephthalate (PET) bottles in Mauritius." He has also been a part-time lecturer at the University of Mauritius. Moreover, he was a resource person for UNEP for providing training on water footprinting.

Dr Mohammad Khalil ELAHEE is associate professor at the University of Mauritius in the Faculty of Engineering. He holds BA (Hons) and MA from Cambridge, United Kingdom, and PhD from the University of Mauritius. He has a background in engineering but has moved to economic, environmental, and ethical dimensions related to energy management. Climate change in the context of small island developing states (SIDS) is his critical concern, with his specialization being on mitigation particularly. While being an engaged and independent scholar with focus on sustainable energy management issues, he had the opportunity to collaborate with both public and private bodies, locally and at international level. He was the founding chairperson of the Energy Efficiency Management Office in Mauritius and currently heading a sub-committee of the Standards Bureau looking into GHG standards.

Mr Pravesh RAMLOGUN is an engineer working for a multinational company. His specialization is industrial systems engineering. He is also involved in design research.

1

Sustainable Development and Role of Engineers

> You can resist an invading army, you cannot resist an idea whose time has come.
>
> (Victor Hugo)

1.1 Introduction

There are many different definitions of engineering, but one which suits the purpose of this book is as follows: *The profession in which knowledge of the mathematical and natural sciences is applied with judgement to develop ways to utilize, economically, the materials and forces of nature for the benefit of mankind* (Accreditation Board for Engineering and Technology). The definition makes reference to the use of "judgement," which is applied for the "benefit of mankind." It also includes a context which is one of constraints, including those imposed by the law, finance, ethics, and people. The professional conditions of engineering and the societal problems engineering should solve changed in the course of time. Today, the context is one of resource scarcity, poverty for a significant fraction of the world's population who cannot meet their basic needs, and our exceedance of the capacity of our planet to accommodate our emissions. This new context of recognizing the need to live within ecological constraints and to ensure more fairness in access to limited resources lies at the heart of the concept of sustainable development and which is now influencing engineering practice. Engineers tend to struggle with the concept that looks abstract, as there are no equations to optimize or no widely agreed upon standards to which they can adhere to. This chapter has the objective of making the concept clear and show its importance in sustainable engineering practice.

1.2 Why and What Is Sustainable Development?

Development is some set of desirable goals or objectives for society. It includes the basic aim of securing a rising level of income per capita or economic growth (traditionally regarded as the "standard of living"). But it is much more than just rising incomes and there is emphasis also on the "quality of life," which includes health, educational standards, and environmental quality.

Appendix A provides the key messages about the state of the world environment in 2018 as per the latest United Nations Global Environment Outlook (GEO) report. Although millions of people have been lifted out of poverty through economic development over the past decades, the negative environmental and social impacts that have been created threaten to undermine, even reverse the development that has been achieved to date. Since the 1950s, the extraction of natural resources and related environmental impacts have greatly accelerated worldwide (Steffen et al., 2015a), and there are increasing pressures on the global environment. The world has to deal with issues such as climate change, deforestation, biodiversity losses, and decline in air and water quality. The planetary boundaries (PBs) concept is a recent scientific framework that considers global environmental limits. It identifies a set of nine physical and biological limits of the Earth system that should be respected in order to maintain conditions favorable to further human development. Crossing the suggested limits would lead to drastic changes in human society by disrupting some of the ecological bases that underlie the current socioeconomic system (Rockström et al., 2009). The nine PBs are as follows: climate change, rate of biodiversity loss (terrestrial and marine), interference with the nitrogen and phosphorus cycles, stratospheric ozone depletion, ocean acidification, global freshwater use, change in land use, chemical pollution, and atmospheric aerosol loading, which are described in Box 1.1. The concept was updated by Steffen et al. (2015b), and the PBs that define the safe operating space for humanity are illustrated in Figure 1.1.

BOX 1.1 THE NINE PLANETARY BOUNDARIES

Stratospheric ozone depletion

The stratospheric ozone layer in the atmosphere filters out ultraviolet (UV) radiation from the sun. If this layer decreases, increasing amounts of UV radiation will reach ground level. This can cause a higher incidence of skin cancer in humans as well as damage to terrestrial and marine biological systems. The appearance of the Antarctic ozone hole was proof that increased concentrations of anthropogenic ozone-depleting chemical substances, interacting with polar stratospheric clouds, had passed a threshold and moved the Antarctic stratosphere into a new regime. Fortunately, because of the actions taken as a result of the Montreal Protocol, we appear to be on the path that will allow us to stay within this boundary.

Loss of biosphere integrity (biodiversity loss and extinctions)

The Millennium Ecosystem Assessment of 2005 concluded that changes to ecosystems due to human activities were more rapid in the past 50 years than at any time in human history, increasing the risks of abrupt and irreversible changes. The main drivers of change are the demand for food, water, and natural resources, causing severe biodiversity loss and leading to changes in ecosystem services. These drivers are either steady, showing no evidence of declining over time, or are increasing in intensity. The current high rates of ecosystem damage and extinction can be slowed by efforts to protect the integrity of living systems (the biosphere), enhancing habitat, and improving connectivity between ecosystems while maintaining the high agricultural productivity that humanity needs. Further research is underway to improve the availability of reliable data for use as the "control variables" for this boundary.

Chemical pollution and the release of novel entities

Emissions of toxic and long-lived substances such as synthetic organic pollutants, heavy metal compounds, and radioactive materials represent some of the key human-driven changes to the planetary environment. These compounds can have potentially irreversible effects on living organisms and on the physical environment (by affecting atmospheric processes and climate). Even when the uptake and bioaccumulation of chemical pollution is at sub-lethal levels for organisms, the effects of reduced fertility and the potential of permanent genetic damage can have severe effects on ecosystems far removed from the source of the pollution. For example, persistent organic compounds have caused dramatic reductions in bird populations and impaired reproduction and development in marine mammals. There are many examples of additive and synergic effects from these compounds, but these are still poorly understood scientifically. At present, we are unable to quantify a single chemical pollution boundary, although the risk of crossing Earth system thresholds is considered sufficiently well-defined for it to be included in the list as a priority for precautionary action and for further research.

Climate change

Recent evidence suggests that Earth, now passing 390 ppmv CO_2 in the atmosphere, has already transgressed the planetary boundary and is approaching several Earth system thresholds. We have reached a point at which the loss of summer polar sea-ice is almost certainly irreversible. This is one example of a well-defined threshold above which rapid physical feedback mechanisms can drive the Earth system into a much warmer state with sea levels meters higher than present. The weakening or reversal of terrestrial carbon sinks, for example, through the ongoing destruction of the world's rainforests, is another potential tipping point, where climate-carbon cycle feedbacks accelerate Earth's warming and intensify the climate impacts. A major question is how long we can remain over this boundary before large, irreversible changes become unavoidable.

Ocean acidification

Around a quarter of the CO_2 that humanity emits into the atmosphere is ultimately dissolved in the oceans. Here it forms carbonic acid, altering ocean chemistry and decreasing the pH of the surface water. This increased acidity reduces the amount of available carbonate ions, an essential "building block" used by many marine species for shell and skeleton formation. Beyond a threshold concentration, this rising acidity makes it hard for organisms such as corals and some shellfish and plankton species to grow and survive. Losses of these species would change the structure and dynamics of ocean ecosystems and could potentially lead to drastic reductions in fish stocks. Compared to pre-industrial times, surface ocean acidity has already increased by 30%. Unlike most other human impacts on the marine environment, which are often local in scale, the ocean acidification boundary has ramifications for the whole planet. It is also an example of how tightly interconnected the boundaries are since atmospheric CO_2 concentration is the underlying controlling variable for both the climate and the ocean acidification boundaries, although they are defined in terms of different Earth system thresholds.

Freshwater consumption and the global hydrological cycle

The freshwater cycle is strongly affected by climate change and its boundary is closely linked to the climate boundary, yet human pressure is now the dominant driving force determining the functioning and distribution of global freshwater systems. The consequences of human modification of water bodies include both global-scale river flow changes and shifts in vapor flows arising from land-use change. These shifts in the hydrological system can be abrupt and irreversible. Water is becoming increasingly scarce – by 2050, about half a billion people are likely to be subject to water-stress, increasing the pressure to intervene in water systems. A water boundary related to consumptive freshwater use and environmental flow requirements has been proposed to maintain the overall resilience of the Earth system and to avoid the risk of "cascading" local and regional thresholds.

Land system change

Land is converted to human use all over the planet. Forests, grasslands, wetlands, and other vegetation types have primarily been converted to agricultural land. This land-use change is one driving force behind the serious reductions in biodiversity, and it has impacts on water flows and on the biogeochemical cycling of carbon, nitrogen, phosphorus, and other important elements. While each incident of land cover change occurs on a local scale, the aggregated impacts can have consequences for Earth system processes on a global scale. A boundary for human changes to land system needs to reflect not just the absolute quantity of land, but also its function, quality, and spatial distribution. Forests play a particularly important role in controlling the linked dynamics of land use and climate and are the focus of the boundary for land system change.

Nitrogen and phosphorus flows to the biosphere and oceans

The biogeochemical cycles of nitrogen and phosphorus have been radically changed by humans as a result of many industrial and agricultural processes. Nitrogen and phosphorus are both essential elements for plant growth, so fertilizer production and application is the main concern. Human activities now convert more atmospheric nitrogen into reactive forms than all of the Earth's terrestrial processes combined. Much of this new reactive nitrogen is emitted to the atmosphere in various forms rather than taken up by crops. When it is rained out, it pollutes waterways and coastal zones or accumulates in the terrestrial biosphere. Similarly, a relatively small proportion of phosphorus fertilizers applied to food production systems is taken up by plants; much of the phosphorus mobilized by humans also ends up in aquatic systems. These can become oxygen-starved, as bacteria consume the blooms of algae that grow in response to the high nutrient supply. A significant fraction of the applied nitrogen and phosphorus makes its way to the sea, and can push marine and aquatic systems across ecological thresholds of their own. One regional-scale example of this effect is the decline in the shrimp catch in the Gulf of Mexico's "dead zone" caused by fertilizer transported in rivers from the US Midwest.

Atmospheric aerosol loading

An atmospheric aerosol planetary boundary was proposed primarily because of the influence of aerosols on Earth's climate system. Through their interaction with water vapor, aerosols play a critically important role in the hydrological cycle, affecting cloud formation and global-scale and regional patterns of atmospheric circulation, such as the monsoon systems in tropical regions. They also have a direct effect on climate, by changing how much solar radiation is reflected or absorbed in the atmosphere. Humans change the aerosol loading by emitting atmospheric pollution (many pollutant gases condense into droplets and particles), and also through land-use change that increases the release of dust and smoke into the air. Shifts in climate regimes and monsoon systems have already been seen in highly polluted environments, giving a quantifiable regional measure for an aerosol boundary. A further reason for an aerosol boundary is that aerosols have adverse effects on many living organisms. Inhaling highly polluted air causes roughly 800,000 people to die prematurely each year. The toxicological and ecological effects of aerosols may thus relate to other Earth system thresholds. However, the behavior of aerosols in the atmosphere is extremely complex, depending on their chemical composition and their geographical location and height in the atmosphere. While many relationships among aerosols, climate, and ecosystems are well established, many causal links are yet to be determined.

Retrieved from www.stockholmresilience.org/research/planetary-bound aries/planetary-boundaries/about-the-research/the-nine-planetary-bound aries.html

As Figure 1.1 shows, the boundaries in three systems (rate of biodiversity loss, climate change, and human interference with the nitrogen cycle) are already exceeded. Evidence that humankind has already overshot these ecological thresholds is covered in the results of the UN Millennium Ecosystem Assessment, of the Intergovernmental Science-Policy Platform on Biodiversity and Ecosystem Services (IPBES), of the International Panel of Climate Change (IPCC) and of the International Resource Panel (IRP).

The UN Millennium Ecosystem Assessment Report (MEA, 2005) shows that close to two-thirds of the world's ecosystems – tropical forests, fisheries, soils, freshwater resources and so on – are now in serious decline. There are irreversible losses of ecosystems because of habitat change, nitrogen and phosphorus pollution, climate change, invasive species, and overexploitation (with fisheries and forestry). The IPBES 2019 Global Assessment Report on Biodiversity and Ecosystem Services concludes that one million plant and animal species are on the verge of extinction, with alarming implications for human survival.

Figure 1.2 shows yearly temperature anomalies from 1880 to 2014 as recorded by four agencies.

Though there are minor variations from year to year, all four records show rapid warming in the past few decades, and all show the last decade as the warmest. The world is nearly 1 °C above pre-industrial

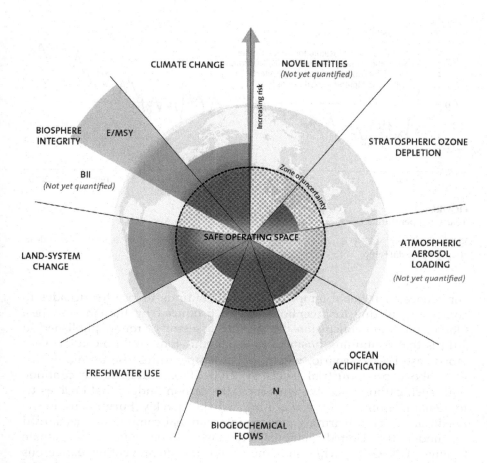

FIGURE 1.1
The nine planetary boundaries that we must work within to maintain the essential natural support systems of the planet.

(Steffen et al., 2015b). Retrieved from www.stockholmresilience.org/research/planetary-boundaries.html

level and nine of the ten warmest years have occurred since 2005, with the last five years (2014–2018) ranking as the five warmest years on record (National Oceanographic and Atmospheric Administration (NOAA); www.ncdc.noaa.gov/sotc/global/201813). The quickly rising temperatures over the past two decades follow a much longer warming trend that corresponds with the scientific consensus among researchers that climate change is caused by human activity. Published every six years, the IPCC Assessment Reports are the most comprehensive international assessments on the state of climate science. Scientists have high

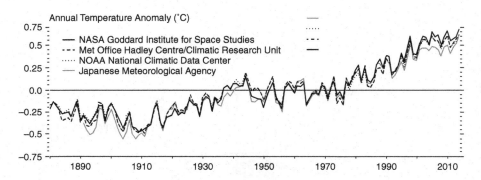

FIGURE 1.2

Yearly temperature anomalies from 1880 to 2014.

Retrieved from NASA Earth Observatory. https://earthobservatory.nasa.gov/world-of-change/DecadalTemp

confidence that global temperatures will continue to rise for decades to come, largely due to greenhouse gases produced by human activities. One of the main conclusions of the fifth assessment report published in 2014 is that continuing business-as-usual emissions could mean the very worst-case IPCC scenario, where global temperatures rise by more than 4 °C above pre-industrial levels by 2100 (IPCC, 2014). If we continue with such carbon-intensive emissions, the carbon budget that enables us to remain within 2 °C of warming will be quickly burnt. This is the maximum level of warming that the international community has agreed to, under the United National Framework Convention on Climate Change (UNFCCC), which has the stated aim of preventing dangerous human interference with the climate system and associated impacts. Appendix B provides a summary of the International Energy Agency World Energy Outlook in 2018 while Appendix C explains the relationship between energy and climate change.

As per the Global Resources Outlook 2019 published by the International Resource Panel (IRP, 2019b), global resource use has more than tripled since 1970 to reach 92 billion tonnes in 2017. Other key facts from the report are summarized in Box 1.2. During the 20th century, the annual extraction of ores and minerals grew by a factor of 27, construction materials by a factor of 34, fossil fuels by a factor of 12, and biomass by a factor of 3.6. In total, material extraction increased by a factor of about 8. The average person in high-income countries relied on close to 10 tonnes of primary materials sourced from elsewhere in the world in 2017. This reliance has been increasing at 1.6% per year since 2000. Figure 1.3 shows the exponential growth path for the extraction of many metals since the beginning of the 20th century. Other reports have illustrated that the use of

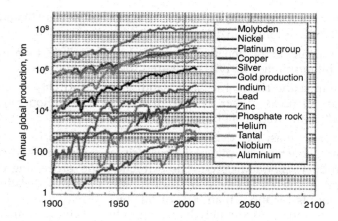

FIGURE 1.3
Extraction of many metals grew exponentially since the year 1900 (the ordinate on the picture being logarithmic).

(Source: Sverdrup et al., 2013)

some natural resources essential to prosperity – including freshwater, land and soils, and fish – have similarly increased, in many cases beyond sustainable levels.

Resource extraction and processing cause over 90% of global biodiversity loss and water stress, and more than half of global climate change impacts. Environmental impacts of material consumption are three to six times greater in high-income countries than in low-income countries. Without action, resource use would more than double from current levels to 190 billion tonnes by 2060. Related impacts would exceed the PBs and endanger human well-being.

BOX 1.2 GLOBAL RESOURCES CONSUMPTION

The *Global Resources Outlook 2019: Natural Resources for the Future We Want* (*www.resourcepanel.org/reports/global-resources-outlook*) prepared by the International Resource Panel examines the trends in natural resources and their corresponding consumption patterns since the 1970s to support policymakers in strategic decision-making and transitioning to a sustainable economy. Key facts from the report are reproduced as follows:

- Biomass extraction has increased 2.7-fold from 1970 to 2017 (9 billion tonnes in 1970 compared to 24 billion tonnes in 2017).

- Metal extraction has increased 3.5 times between 1970 and 2017 (2.6 billion tonnes in 1970 to 9.1 billion tonnes in 2017).

- Fossil fuel extraction was 2.5 times higher in 2017 than in 1970 (6 billion tonnes in 1970 compared to 15 billion tonnes).

- Nonmetallic mineral (mainly sand, gravel, and clay) extraction was 4.9 times higher in 2017 than in 1970 (9 billion tonnes in 1970 compared to 44 billion tonnes in 2017).

- Over the past five decades, our global population has doubled, the extraction of materials has tripled, and gross domestic product has quadrupled. The extraction and processing of natural resources has accelerated over the past two decades and accounts for more than 90% of our biodiversity loss and water stress and approximately half of our climate change impacts.

- On a per capita basis, high-income countries are reliant on 9.8 tonnes of primary materials mobilized elsewhere in the world. This is 60% higher than the upper-middle income group, and 13 times the level of the low-income groups.

- A scenario developed by the International Resource Panel on *Historical Trends* shows that unless a fundamental change toward decoupling drives natural resource use away from the status quo, resource use will continue to grow to 190 billion tonnes and over 18 tonnes per capita by 2060. Moreover, greenhouse gas emissions increase by 43% from 2015 to 2060, industrial water withdrawals increase by up to 100% from 2010 levels, and the area of agricultural land increases by more than 20% in that time, reducing forests by over 10%.

- Resource efficiency is essential, though not enough on its own. We need to move from linear to circular flows through a combination of extended product life cycles, intelligent product design and standardization and reuse, recycling and remanufacturing.

Climate Change, biodiversity loss, and resources scarcity will define the global agenda for generations to come. Another way to look at the scale of the challenges ahead is to consider our ecological footprint. The latter is a condensed way of understanding the relationship between our consumption of resources and the capacity of the planet to provide them and absorb the emissions generated. It compares human demand with our planet's ecological capacity to regenerate. It specifically measures the amount of biologically productive land and water area required to produce the resources an individual, population, or activity consume, and to absorb the waste it generates, given prevailing technology and resource management. This area is expressed as global hectares, hectares with world-average biological productivity. Using this assessment, it is possible to estimate how much of Earth (or how many planet Earths) it would take to support humanity if everybody lived a given lifestyle (Wackernagel and Rees, 1996). The Global Footprint network (www.footprintnetwork.org/our-work/ecological footprint/) estimated humanity's total ecological footprint in 2018 at 1.7 planet Earths – in other words, they estimate that humans are using natural resources 1.7 times faster than what ecosystems can regenerate – or, put another way, consuming 1.7 Earths. This means it now takes Earth one year and eight months to regenerate what we use in a year. We are using more ecological resources and services than nature can regenerate through overfishing, overharvesting forests, and emitting more carbon dioxide into the atmosphere than forests can sequester. And worse, the human population is expected to use the equivalent of 2 Earths of renewable resources per year by 2050. Simply put, this is not a sustainable path for our planet's future.

On the social dimension, while global poverty rates have been cut by more than half since 2000, one in ten people in developing regions are still living with their families on less than the international poverty line of USD1.90 a day, and there are millions more who make little more than this daily amount (www.un.org/sustainabledevelopment/poverty/). Up to 42% of the population in Sub-Saharan Africa continues to live below the poverty line. Poverty, besides the lack of income and resources to ensure a sustainable livelihood, is also manifested by hunger and malnutrition, limited access to education and other basic services, social discrimination and exclusion as well as the lack of participation in decision-making. Globally, inequality has also increased, with 62 of the world's richest people owning as much wealth as half of the world's population in 2015. In 2016, 22% of global income was received by the top 1% compared with 10% of income for the bottom 50%. In 1980, the top 1% had 16% of global income and the bottom 50% had 8% of income. Further, large disparities remain in access to basic services such as education, healthcare, and sanitation. Certain groups are disproportionately affected by such disparities, including migrants, women, children, persons with disabilities, ethnic and religious minorities, and indigenous peoples (www.undp.org/content/undp/en/home/sustainable-development-goals/goal-10-reduced-inequalities.html).

The concept of sustainable development emerged out of the environmental movement of the 1950s and 1960s, which was concerned that human activity was having severe negative impacts on the planet, and that patterns of growth and development would be unsustainable if they continued as they were. Key works that highlight this thinking included Rachel Carson's "Silent Spring" (1962), Garret Hardin's "Tragedy of the Commons" (1968), and the Club of Rome's "The Limits to Growth" report

(1972). The sustainability idea, as we know it today, emerged in a series of meetings and reports during the 1970s and 1980s. It received its first major international recognition in 1972 at the UN Conference on the Human Environment held in Stockholm. The term was not referred to explicitly, but nevertheless the international community agreed that both development and the environment, until then addressed as separate issues, should be considered together. In 1987, the UN-sponsored Brundtland Commission released "Our Common Future," a report that captured widespread concerns about the environment and poverty in many parts of the world (Brundtland, 1987). According to the Brundtland report, economic development is not to stop, but it must change course to fit within the planet's ecological limits. It also popularized the term "sustainable development," which it defined as "development that meets the needs of the present without compromising the ability of future generations to meet their own needs." Three concepts are inherent in this definition: (1) the concept of "need," and more particularly the basic needs of the most deprived, who should be given the highest priority, (2) the idea of the limitations that the state of our techniques and our social organization imposes on the capacity of the environment to meet current and future needs, and (3) the concept of "intra-generational and intergenerational equity" (i.e., the fair distribution of, and access to resources within the same generation, and between succeeding generations).

One must understand sustainable development as striving to *simultaneously* address three developmental dimensions: the social, the environmental, and the economic (also sometimes called "People, Planet and Profit"). These three dimensions when considered separately usually pull society in different directions. The prevailing tendency has been to privilege the economic facet of development to the detriment of social and the environmental dimensions. Sustainable development is thus not simply a matter of trading off positive impacts in one area against negative impacts in another. It requires the *simultaneous* achievement of economic success, social benefit, and high environmental quality together. Measuring success against all three factors at the same time is often referred to as the "triple bottom line" approach.

It is popular to represent the three dimensions of sustainable development in the form of a simple Venn diagram as illustrated in Figure 1.4.

Sustainability is the region in the middle of the diagram where human activities are compatible with all three sets of constraints. Sustainable development, therefore, is the process of finding our way toward that central sustainable overlap region. This simple Venn diagram implies that the economy, society, and environment are of equal priority. However, within the Earth's single planet limit, the environment supports our human society, which has invented the economy to serve its needs. The climate change issue reminds us that it is the laws of nature that are unchangeable while economic rules and society's behavior can be

FIGURE 1.4
The concept of sustainable development.

changed. Ecosystem services – such as the production of atmospheric gases or the purification of soil, water, and air or the storage and cycling of freshwater and nutrients or the regulation of the chemistry of the atmosphere and oceans or supporting biodiversity and maintenance of habitats for wildlife – are not substitutable by technological innovation. So the actual relationship between the three dimensions is more like the social and economic dimensions being nested in the environment dimension as shown in Figure 1.5. We can thus think of sustainable development as the process of moving the circles together so that they almost completely overlap but with the societal and economic circles enclosed within the environmental circle, at which point all human activity is sustainable.

It is important to remember that what is to be sustained is not a predetermined environmental future but a process of development that implies continuous improvement with multiple benefits – environmental ones but also economic benefits and social benefits. Sustainable development is a journey or direction for proceeding toward sustainability. According to Chambers et al. (2000), "sustainable development" and "sustainability" differ, in that sustainability is the goal, while sustainable development is the process or journey for achieving that goal.

Engineering is indirectly linked to the three components of sustainability. Resources used in engineering, whether fuels, minerals, or water, are required in virtually all economic sectors. They are obtained from the environment and the latter also acts as a sink for all the wastes generated from engineering processes. Finally, the services provided by engineering

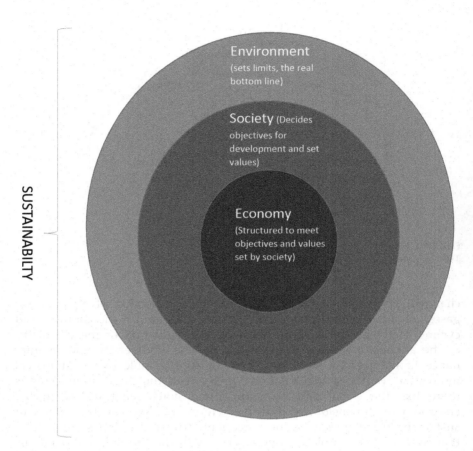

FIGURE 1.5
Environment is the real bottom line.

allow for good living standards, and often support social stability as well as cultural and social development. Achieving of sustainability in engineering is thus a critical aspect of achieving sustainable development, in individual countries and globally.

Another model that has been developed to provide another way of understanding sustainability, based on the balancing of social, economic, and social factors is the "five capital model," which looks at different kinds of capital from which we derive the goods and services we need to improve the quality of our lives:

• *Natural capital* is any stock or flow of energy and material that produces goods and services and also creates amenity and beauty.

- *Human capital* consists of people's health, knowledge, skills, and motivation.

- *Social capital* refers to the stable relationships between individuals and groups within society and concerns the institutions that help us maintain and develop human capital in partnership with others, for example, families, businesses, trade unions, voluntary organizations, and educational institutions.

- *Manufactured capital* comprises material goods or fixed assets, which contribute to the production process rather than being the output itself, for example, machines, tools, buildings, and infrastructure.

- *Financial capital* plays an important role in our economy, enabling the other types of capital to be owned and traded. However, unlike the other types, it has no real value itself but is representative of natural, human, social, or manufactured capital, for example, shares, bonds, and bank notes.

Sustainability poses the challenge of determining whether we can hope to see the current level of well-being at least maintained in the future, and whether these levels of well-being can be sustained over time depends on whether the above stocks of capital that matter for our lives are passed on to future generations. Sustainable development is seeking to preserve and enhance the five capitals and then "living off the interest."

One reason for postponing actions to tackle sustainability issues and to behave in a reactive mode is that future costs are less burdensome than current costs. This reflects what economists call "time preference" or "discounting the future." In contrast, in an anticipatory mode, we try to anticipate problems and incur the costs now before the problem occurs in the future. Cost escalation in the future, the possible irreversibility of damages, and the scientific uncertainty about how ecosystems function favor the anticipatory mode. The concept of sustainable development therefore favors strongly an anticipatory approach to sustainability policies.

1.3 The SDGs, Paris Agreement, and Role of Engineering

Global efforts to address sustainability challenges have significantly increased during the last few decades. For UN and intergovernmental cooperation, 2015 was a major success, which included not only the global agreement on the Sustainable Development Goals (SDGs) but also the Paris Agreement on Climate Change.

Nations met in September 2015 at the UN in New York and committed to the SDGs – 17 global goals (Figure 1.6) with 169 targets – to be met by 2030 (UN, 2015) (for information on the targets please consult

FIGURE 1.6
The 17 SDGs to "end poverty, protect the planet, and ensure prosperity for all as part of a new sustainable development agenda".

(Retrieved with permission from www.un.org/sustainabledevelopment/news/communications-material/)

https://www.un.org/sustainabledevelopment/). These goals apply to *every* nation and every sector. While the SDGs' predecessor, the Millennium Development Goals (MDGs), was mainly aimed at governments and into developing countries, the SDGs are global and aimed at government, business, and nongovernmental organizations alike. Cities, businesses, schools, organizations, *all* are challenged to act. It must also be recognized that each goal is important in itself and they are all connected. Achieving these goals involves making a transformational change in how we live on Earth. We have to move on all the SDGs at the same time, but we have to move forward very urgently on climate change (SDG 13) otherwise we will not get the resources to achieve the others. The Paris Agreement was reached by the parties to the United Nations Framework Convention on Climate Change (UNFCCC) on 12 December 2015 in Paris and symbolizes a fundamentally new course in the global fight against climate change. It marked an unprecedented political recognition of the risks of climate change. As the first ever universal and legally binding global climate agreement, it sends a clear signal to all stakeholders that resources have to shift away from fossil fuels. It also shifted from a top-down to a bottom up process with the nationally determined contributions (NDCs). All countries have a legally binding obligation to prepare, communicate, and maintain an NDC that consists of the plan of a particular country to achieve the objective of the convention.

Both the 2030 Agenda and the Paris Agreement are highly significant in that they commit industrialized and emerging economies and other developing countries to join forces to eradicate poverty, and protect Earth's resources and ecosystems for the benefit of present and future generations. Achieving the SDGs and the Paris Agreement will require engineering expertise to plan and implement cost-effective, feasible solutions in collaboration with scientists, policy-, and decision-makers as well as international and national financial institutions. Goals 6 (clean water and sanitation), 7 (affordable and clean energy), 9 (industry, innovation, and infrastructure), and 11 (sustainable cities and communities) have obvious engineering requirements associated with them. However, every one of the SDGs requires engineering involvement as discussed subsequently:

GOAL 1: END POVERTY IN ALL ITS FORMS EVERYWHERE

Eliminating world poverty in all its forms and dimensions by 2030 will require an enormous effort by governments, civil society, and the private sector. Engineering contributes to the eradication of poverty both indirectly and directly. The provision of services such as planning and implementing new infrastructure to provide clean water and sanitation contributes indirectly to lift people out of poverty. The provision of urban systems by engineers makes economic development possible to eradicate poverty.

GOAL 2: END HUNGER, ACHIEVE FOOD SECURITY AND IMPROVED NUTRITION, AND PROMOTE SUSTAINABLE AGRICULTURE

Ending hunger and all forms of malnutrition by 2030 is a tough global challenge. Approximately 10% of the world population are estimated to be chronically undernourished and about 90 million children under the age of 5 are dangerously underweight. Fulfillment of the goal calls for action on various agendas from establishing sustainable food production systems and increased agricultural productivity to increased investment in rural infrastructure, agricultural research and extension services, technology development, sustainable irrigation, sustainable transportation, and so on. The production of more and more nutritious food will be aided by factors that include pest outbreak and climate modeling, safe and sustainable herbicides and fertilizers, and the development of different strains of crops and animals. All of this will require expertise from the relevant engineering disciplines.

GOAL 3: ENSURE HEALTHY LIVES AND PROMOTE WELL-BEING FOR ALL AT ALL AGES

Ensuring healthy lives and promoting well-being for all are essential to sustainable development. Access to good health and well-being is a human right. Engineers contribute to good health and well-being in numerous ways – directly as well as indirectly. For instance, they work with and assist local governments with capacity building to ensure that relevant policies and guidelines are in place to mitigate the impact of infrastructural development upon human beings and health. They prevent and reduce deadly diseases and epidemics from spreading when assisting governments and local authorities to implement water and sanitation programs. By engineering better medicine, advancing health informatics, providing clean air, clean water, and safe food production, developing bioengineering and nanotechnology life improvement devices and materials, engineers contribute significantly to this goal.

GOAL 4: ENSURE INCLUSIVE AND EQUITABLE QUALITY EDUCATION AND PROMOTE LIFELONG LEARNING OPPORTUNITIES FOR ALL

Achieving quality education for all will not come easy and will require increasing efforts. Education is one of the most powerful and proven accelerators for sustainable development and will therefore contribute to the fulfillment of the other 16 SDGs. Engineering contributes to this goal or instance when assisting national and local authorities with capacity building and implementation of employment policies or advancing personalized learning;

GOAL 5: ACHIEVE GENDER EQUALITY AND EMPOWER ALL WOMEN AND GIRLS

Providing women and girls with equal access to education, healthcare, decent work, and representation in political and economic decision-making processes will fuel sustainable economies and benefit societies and humanity at large. Engineers seek to ensure that gender equality remain central to all projects that they implement in partnership with other stakeholders. Engineers associations work to increase participation of women in all engineering fields.

GOAL 6: ENSURE AVAILABILITY AND SUSTAINABLE MANAGEMENT OF WATER AND SANITATION FOR ALL

Access to clean water and sanitation and comprehensive freshwater management are essential to life, health, environment, and economic prosperity in the developing countries. With more than 2 billion people globally living in countries with excess water stress and some 2.4 billion people with inadequate or no access to sanitation, this is undoubtedly one of the most important ones for engineers. They design water and sanitation facilities; develop education programs for hygiene, sanitation, and water usage; and provide capacity building to governments and local authorities in water and sanitation.

GOAL 7: ENSURE ACCESS TO AFFORDABLE, RELIABLE, SUSTAINABLE, AND MODERN ENERGY FOR ALL

Energy is essential to keep life, production, growth, and development going. However, being the main contributor to climate change, energy also constitutes one of the biggest challenges that the world and humanity face today. This calls for new green and sustainable energy solutions and behavioral change. Engineers design and implement stand-alone energy systems, provide technical assistance to local authorities, assist governments with energy efficiency strategies, or design integrated renewable energy solutions, including wind, solar, biomass, and hydro so as to make them affordable.

GOAL 8: PROMOTE SUSTAINED, INCLUSIVE, AND SUSTAINABLE ECO-NOMIC GROWTH, FULL AND PRODUCTIVE EMPLOYMENT, AND DECENT WORK FOR ALL

Unemployment and low wages are significant global challenges. Around half the world's population lives on the equivalent of about USD 2, and having a job does not guarantee the ability to escape that situation. It requires sustainable economic growth and conditions that allow people to have quality jobs that stimulate the economy while not harming the environment. Engineers contribute to economic growth, productive employment, and decent work when assisting authorities, with plans to address resource efficiency in consumption and production, and when working with development institutions with a specific focus upon initiatives to boost economic development and job creation.

GOAL 9: BUILD RESILIENT INFRASTRUCTURE, PROMOTE INCLUSIVE AND SUSTAINABLE INDUSTRIALIZATION, AND FOSTER INNOVATION

Insufficient infrastructure constitutes a major challenge in many developing countries. A well-developed infrastructure is key to industrialization, market access, job creation, information, education, and training. Engineers develop quality, reliable and sustainable railways, harbors, roads, bridges, and so on to support economic development and human well-being. Engineers also assist both national and local governments with infrastructure-related capacity building, for instance, addressing road maintenance and road safety. Engineers play a major role also in sustainable manufacturing.

GOAL 10: REDUCE INEQUALITY WITHIN AND AMONG COUNTRIES

Reducing inequality within and among countries is an issue that calls for global solutions. Today, the richest 10% earns around 40% of total global income. Fighting income inequality is one of the goals where engineering companies can contribute directly such as ensuring a living wage for workers, or indirectly such as strengthening land rights for some of the poorest in the world.

GOAL 11: MAKE CITIES AND HUMAN SETTLEMENTS INCLUSIVE, SAFE, RESILIENT, AND SUSTAINABLE

By 2030, almost 60% of the world's population will live in urban areas. Common urban challenges, as a result, include congestion, lack of funds to provide basic services, shortage of adequate housing, pollution, and declining infrastructure. Engineers work with planners to develop solutions that will match the specific needs of a city while simultaneously strengthening the capacity of city planners and administrators to control urbanization. More specifically, engineers for instance assist to build sustainable and climate-resilient buildings utilizing local materials, helping create carbon-neutral cities, and mitigating the environmental impacts, for example, air pollution, waste management of growing cities, and expanding infrastructure. Bridging the digital divide, where most of the world does not have access to the internet, is also crucial. Engineers will need to leverage existing and widely deployed technologies and future developments in Information and Communication Technologies (ICT) – including next-generation mobile broadband, the Internet of Things (IoT), artificial intelligence, 3D printing, and others – to provide the tools for integrated solutions for sustainable development.

GOAL 12: ENSURE SUSTAINABLE CONSUMPTION AND PRODUCTION PATTERNS

Sustainable growth and development require minimization of the natural resources and toxic materials used and of the waste and pollutants generated throughout the entire production and consumption process. We need to do more with less, pollute less, and we need to recycle and upcycle what we yesterday called waste. Engineers work with various stakeholders to promote the circular economy and the sustainable management and efficient use of natural resources, including food, water, and energy consumption.

GOAL 13: TAKE URGENT ACTION TO COMBAT CLIMATE CHANGE AND ITS IMPACTS

Climate change is now affecting every country on every continent, and mitigating climate change and its impacts will require global commitment, enormous resources, and innovative ideas and processes. Engineers contribute to the fight against climate change when working to adapt to and mitigate the significant impacts of climate change, including changing weather patterns, rising sea levels, and extreme weather events. They work to strengthen resilience and adaptive capacity to climate-related hazards and natural disasters, to integrate climate change measures into national policies, strategies and planning, and to institutional strengthening and capacity building on climate adaptation, mitigation, impact reduction, and so on. They develop carbon sequestration method and make sustainable energy affordable for all as well as make the built environment adaptable and resilient to climate change.

GOAL 14: CONSERVE AND SUSTAINABLY USE MARINE RESOURCES FOR SUSTAINABLE DEVELOPMENT

Oceans contain nearly 200,000 identified species. Oceans serve as the world's largest source of protein, with more than 3 billion people depending on the oceans as their primary source of protein. However, as much as 40% of the world's oceans are badly affected by human activities such as pollution, depleted fisheries, and loss of coastal habitats. Engineers work in partnership with various stakeholders to prevent and reduce marine pollution of all kinds, to sustainably manage and protect marine and coastal ecosystems, to develop conservation plans in line with national and international policies and conventions, and to provide institutional strengthening and capacity building to relevant authorities in all aspects of marine management and conservation. They also improve monitoring and work toward better provision of data for resources management.

GOAL 15: PROTECT, RESTORE, AND PROMOTE SUSTAINABLE USE OF TERRESTRIAL ECOSYSTEMS, SUSTAINABLY MANAGE FORESTS, COMBAT DESERTIFICATION, AND HALT AND REVERSE LAND DEGRADATION AND HALT BIODIVERSITY LOSS

The preservation of our forests and enrichment of the globe's biodiversity are both central themes to sustainable development. Engineers work to conserve and restore inland freshwater ecosystems to promote the sustainable use of forests, develop conservation plans for differing ecosystems to integrate ecosystem and biodiversity values into national and local planning, to provide training and capacity building, and to work with national and local authorities to implement international laws and conventions. They work to restore organic matter to degraded soils and to better integrate biofuels.

GOAL 16: PROMOTE PEACEFUL AND INCLUSIVE SOCIETIES FOR SUSTAINABLE DEVELOPMENT, PROVIDE ACCESS TO JUSTICE FOR ALL AND BUILD EFFECTIVE, ACCOUNTABLE, AND INCLUSIVE INSTITUTIONS AT ALL LEVELS

Peace, stability, human rights, and effective governance, based on the rule of law, are essential for the hope of sustainable development. Engineers work to secure the cyberspace and nuclear energy as well as assisting local and national authorities with capacity building.

GOAL 17: STRENGTHEN THE MEANS OF IMPLEMENTATION AND REVITALIZE THE GLOBAL PARTNERSHIP FOR SUSTAINABLE DEVELOPMENT

A successful sustainable development agenda requires partnerships between governments, the private sector, and civil society. Engineers place a significant emphasis upon developing inclusive partnerships based upon shared principles and values. They play an active role in the global partnership as practitioners of sustainable development.

1.4 Sustainable Development and the Engineering Profession

Engineering and sustainable development are closely linked, with many aspects of sustainable development depending directly and significantly on appropriate and timely actions by engineers. Products and services that engineers design provide the interface between humans and environment so that all engineering projects either contribute to sustainability or not. It requires engineers asking new questions at the right time at each stage of project delivery. The International Federation of Consulting Engineers (FIDIC) notes that

> These changes are beginning to fundamentally shift the way engineering project performance is judged, and they add invisible design criteria that will ultimately affect every engineering project, whether for products, processes, facilities or infrastructure. The effect of sustainable development will be to bring broad resources, ecological and social issues into the mainstream of engineering design and it has become critically important that engineers understand these issues and look for ways to incorporate these considerations in all that they do.
>
> (FIDIC (2013))

This needs new engineering skills and an important one is that of systems thinking and dealing with complexity. The natural and man-made world is highly interconnected, so that actions in one sphere can have unintended and unforeseen consequences elsewhere. Engineering solutions have to operate within highly complex systems and engineers need to have a habit of mind that allows them to see the "big picture" and to cope with change and uncertainty. Engineers also work in multidisciplinary teams where they are involved with other specialists and this means that they may or may not have control of, or be solely responsible for, a particular project. However, they should be engaged for preparing and presenting clear justifications to implement more sustainable solutions that serve the public interest. They should strive to understand and manage the environmental aspects of projects that they are involved with and need to learn to engage better with communities. The Declaration of Barcelona presents a list of engineers' preferable attributes for sustainable development (EESD, 2004), which includes the following abilities:

- to "understand how their (engineers') work interacts with society and the environment, locally and globally, in order to identify potential challenges, risks, and impacts";
- to "understand the contribution of their work in different cultural, social, and political contexts and take those differences into account";

- to "work in multidisciplinary teams, in order to adapt current technology to the demands imposed by sustainable lifestyles, resource efficiency, pollution prevention, and waste management";
- to "apply a holistic and systemic approach to solving problems and the ability to move beyond the tradition of breaking reality down into disconnected parts";
- to "participate actively in the discussion and definition of economic, social, and technological policies to help redirect society towards more sustainable development";
- to "apply professional knowledge according to deontological principles and universal values and ethics"; and
- to "listen closely to the demands of citizens and other stakeholders and let them have a say in the development of new technologies and infrastructures."

A professional person such as an engineer is one who engages in an activity that requires a specialized and comprehensive education and is motivated by a strong desire to serve humanity. Professionalism is a way of life. True professionals are those who pursue their learned art in a spirit of public service. One of the common characteristics of a professional is that it has, as a result of specialized expertise, significant power to affect individual clients and wider society. As a result of the power their skills bring society places great trust in professionals to exercise those skills wisely. Thus, common to all professions is a commitment to use expertise in pursuit of the public good. This creates a critical role for ethics to guide decision-making. It is clearly no longer possible to be a professional engineer and ignore the challenges and opportunities that arise from the need to achieve sustainable development. Often it will be – and should be – engineers who lead processes of making decisions about the use of material, energy, and water resources, the development of infrastructure, the design of new products, and so on. It is an engineer's responsibility to understand the consequences of actions on projects in respect of their environmental and societal implications. Professional Societies and Codes of Ethics provide a set of guidelines of how engineers should behave with respect to clients, the profession, the public, and the law. The World Federation of Engineering Organization (WFEO) has developed a model code of ethics for engineers globally (WFEO, 2016). Sustainability is explicitly included in the WFEO Model Code of Ethics under Canon 4 "Protect the Natural and Built Environment." Many engineering societies now mention in their code of ethics that engineers shall consider environmental impact and sustainable development in the performance of their professional duties.

1.5 Key Attributes of the Graduate Engineer

Developing a professional frame of mind on sustainability begins with engineering education. According to the WFEO, it is critical that engineering graduates are equipped with the relevant knowledge and skills to effectively address such challenges in society. *Graduate attributes* form a set of individually assessable outcomes that are the components indicative of the graduate's potential to acquire competence to practise at the appropriate level. The International Engineering Alliance (IEA) Graduate Attributes and Competences are the foundation for the accreditation of engineering programs under the Washington Accord (WA) (IEA, 2013). To be recognized under the WA, the accreditation process must provide assurance that the attributes of graduates of a signatory's programs are substantially equivalent to the IEA graduate exemplars.

Graduates are expected to be able to demonstrate both knowledge and competencies. In terms of knowledge profile, the expected level of knowledge is WK7: "Comprehension of the role of engineering in society and identified issues in engineering practice in the discipline: ethics and the professional responsibility of an engineer to public safety; the impacts of engineering activity: economic, social, cultural, environmental and sustainability." Thus, engineering graduates from all programs accredited under the IEA WA can be expected to have an understanding of sustainability in the context of engineering practice in their field. Linked to this knowledge profile, the graduate attribute or competency is "Understand and evaluate the sustainability and impact of professional engineering work in the solution of complex engineering problems in societal and environmental contexts (WA7)." Engineering students should develop this competency through all projects undertaken during their studies.

Sustainable development redefines the contexts within which engineering skills must be deployed. It is a new integrative principle and it is therefore very important for graduate engineers to gain a feel for sustainable development during their studies, so that, as with concepts such as justice, they can recognize it as a guiding principle to be interpreted for each instance in which the principle is needed in their future professional career. Concepts such as life cycle thinking, industrial ecology, circular economy, sustainable consumption and production, systems thinking, and stakeholder engagement are now important elements in the education of the modern engineer.

1.6 Summary

Sustainable development refers to systemic conditions where both social and economic development takes place within the carrying capacity of the environment without compromising the ability of future generations to meet their own needs (i.e., maintaining the natural capital which will be passed on to future generations) and ensuring everyone has the same access to global natural resources. This chapter explained how the concept affects the way in which engineering must now be practiced and what should be the key attributes of the engineer for sustainable or good engineering. Achieving the SDGs and the Paris Agreement on climate change requires strong engineering involvement. It has become critically important that engineers understand sustainability issues and look for ways to incorporate these considerations in all that they do. It requires ethics and a set of new skills. It is critical that engineering graduates are equipped with the relevant knowledge and skills to effectively address such challenges in society during their professional career.

SUSTAINABLE ENGINEERING IN FOCUS: AN ADAPTING COOKING TECHNOLOGY POWERED BY RECYCLED MATERIALS: SUSTAINABLE OPPORTUNITY FOR GREEN ENERGY AND WOMEN LED MICRO ECONOMIC INDUSTRY

The retained heat cooker (RHC) is an energy efficient nonelectric cooking material using principle of thermal insulation. In fabrication process, two layers of fabric are sewn according to the desired shape. Recycled materials such as wool blankets, old sweaters, and newspapers can be used as good insulation materials. RHC saves up to 50% of the required cooking energy securing green energy consumption.

Approximately 3 billion people rely on biomass fuels to meet household energy demand. Household air pollution (HAP) associated with burning biomass fuels is the seventh-largest risk factor for global burden of disease; HAP contains harmful pollutants including fine particulate matter (PM2.5), carbon monoxide (CO), and polycyclic aromatic hydrocarbons (PAH). Unsustainable harvesting of biomass and use of inefficient technologies have adverse consequences for the environment, economic development, and climate. Also to be noticed, the various economic slabs and pressure points in different areas, pressurizing the urban poor and women to use such neo-conventional methods of using energy for household cooking and facing unsustainable economic development. Mostly women and children are the victims of the hazardous physical stability caused by the HAP. In developing countries, household energy use has 10% share in world's primary energy consumption, a total of 1090 million tonnes of oil equivalent. The main use of energy in the households in developing countries is for cooking followed by lighting and heating. Household use of biomass in these countries accounts for almost 7%

of world primary energy demand. The genesis of the project is in the city of Kolkata that has the highest gross CO_2 emission footprints in India, of which city's municipal solid waste and landfill accounts for nearly 45% of the total emissions. The future of the radically unequal and explosively unstable urban world that makes "citizens of dirt" and waste recycling were concerns of South Asian Forum for Environment (SAFE) behind creating this entrepreneurial opportunity called RESOLVE for the urban poor. They initiated this innovative adaptive economic model of RESOLVE with the slum dwellers, since every slum portrays a vast humanity warehoused in shanty towns ostracized from the formal world economy. Social audits conducted by SAFE showed that the urban slums adjacent to municipal garbage hills increasing landfill emission, spreading pollution, and diseases, whereas the solid wastes recreate resources for dwellers to survive. This unorganized sector has created opportunity for handful of recyclers to make huge profit by exploiting the underprivileged in the lure of a tiny amount.

At the project's opening day, there were three groups of employee volunteers with four to six members in each group, moved toward the locality of Dhapa-Chaynabhi separately. The three groups covering more than 1500 m discovered the problems and lags of local households while using any kind of energy for cooking food. LPG and kerosene as fuel for cooking are costly options for marginal women in urban/slums and clusters in Kolkata, and wood works as an alternative fuel for the women in the area of Chainavi.

After the Employee Volunteer Programme (EVP), there was a demonstration of the RHC or the magic cooker by the members. RHC is a stand-alone, nonelectric insulated bag designed to reduce the amount of fuel required to cook food. Instead of being placed on a stove for the entire duration, food is heated to a boiling temperature and transferred to the cooker. It uses the principle of thermal insulation to continue the cooking process without requiring any additional heat. RHC provides the possibility of smooth cooking by retaining the heat within the enclosed space using thermal insulation. It reduces the emission of CO_2 and fuel consumption that will be creating a positive impact in our environment. In fabrication process, 180–190 GSM polyester microfiber-made fabric with air porosity properties is used. Here, two layers of fabric are sewn according to the desired shape. For getting thermal insulation properties, polystyrene beads from EPS are used. By the simple and easy cooking process, RHC saves up to 50% of the required cooking energy, keeps food hot or cold for up to 10 hours. RHC is portable, cost-effective, and safe, able to reduce stove time by two-thirds. Recycled materials such as wool blankets, old sweaters, and newspapers can be used as good insulation materials for fabricating this bag. The bag can be made within one hour by any semi-expert women using the waste cloths. With an investment of less than 100 INR, The selling price of the RHC could be 400–500 INR. The RHC is promoting green energy while also contributing to the income of the women of urban and semi-urban slums.

Retrieved with permission from One Planet Network (www.oneplanetnet work.org/).

References

Brundtland, G. (Ed.). 1987. *Our Common Future: The World Commission on Environment and Development*, Oxford University Press, Oxford.

Carson, R., Darling, L. and Darling, L. 1962. *Silent Spring*, Houghton Mifflin, Boston and Cambridge, MA.

Chambers, N., Simmons, C. and Wackernagel, M. 2000. *Sharing Nature's Interest: Ecological Footprint as an Indicator of Sustainability*, Earthscan Publications Ltd, London and Sterling, VA.

EESD (Engineering Education in Sustainable Development). 2004. Declaration of Barcelona. http://eesd15.engineering.ubc.ca/declaration-of-barcelona/ (accessed 15th July 2019).

FIDIC. 2013. *Project Sustainability Management – Applications Manual*, 2nd ed. International Federation of Consulting Engineers (FIDIC), Geneva.

Hardin, G. 1968. The Tragedy of the Commons. *Science*, 162, 1243–1248.

IEA. 2013. Graduate Attributes and Professional Competencies. www.ieagreements. org/IEA-Grad-AttrProf-Competencies.pdf.

IPCC. 2014. Summary for Policymakers. In O. Edenhofer, R. Pichs-Madruga, Y. Sokona, E. Farahani, S. Kadner, K. Seyboth, A. Adler, I. Baum, S. Brunner, P. Eickemeier, B. Kriemann, J. Savolainen, S. Schlömer, C. von Stechow, T. Zwickel and J.C. Minx (Eds.), *Climate Change 2014:Mitigation of Climate Change. Contribution of Working Group III to the Fifth Assessment Report of the Intergovernmental Panel on Climate Change*, Cambridge University Press, Cambridge.

IRP. 2019a. Assessing Global Resource Use: A Systems Approach to Resource Efficiency and Pollution Reduction. In S. Bringezu, A. Ramaswami, H. Schandl, M. O'Brien, R. Pelton, J. Acquatella, E. Ayuk, A. Chiu, R. Flanegin, J. Fry, S. Giljum, S. Hashimoto, S. Hellweg, K. Hosking, Y. Hu, M. Lenzen, M. Lieber, S. Lutter, A. Miatto, A. Singh Nagpure, M. Obersteiner, L. van Oers, S. Pfister, P. Pichler, A. Russell, L. Spini, H. Tanikawa, E. van der Voet, H. Weisz, J. West, A. Wiijkman, B. Zhu and R. Zivy. A Report of the International Resource Panel. United Nations Environment Programme. Nairobi, Kenya.

IRP. 2019b. Global Resources Outlook 2019: Natural Resources for the Future We Want. A Report of the International Resource Panel. United Nations Environment Programme, Nairobi.

MEA (Millennium Ecosystem Assessment). 2005. *Ecosystems and Human Well-being: Synthesis*, Washington, DC, Island Press.

Meadows, D.H., Meadows, D.L., Randers, J., Behrens III, W.W. 1972. *The Limits to Growth. A Report for the Club of Rome's Project on the Predicament of Mankind*, Universe Books, New York.

Rockström, J., Steffen, W., Noone, K., Persson, Å., Chapin, F.S., Lambin, E.F., Lenton, T.M., Scheffer, M., Folke, C., Schellnhuber, H.J., Nykvist, B., de Wit, C.A., Hughes, T., van der Leeuw, S., Rodhe, H., Sörlin, S., Snyder, P.K., Costanza, R., Svedin, U., Falkenmark, M., Karlberg, L., Corell, R.W., Fabry, V.J., Hansen, J., Walker, B., Liverman, D., Richardson, K., Crutzen, P. and Foley, J.A. 2009. A Safe Operating Space for Humanity. *Nature*, 461, 472–475. doi:10.1038/461472a.

Steffen, W., Broadgate, W., Deutsch, L., Gaffney, O. and Ludwig, C. 2015a. The Trajectory of the Anthropocene: The Great Acceleration. *The Anthropocene Review*, 2, 81–98.

Steffen, W., Richardson, K., Rockström, J., Cornell, S.E., Fetzer, I., Bennett, E.M., Biggs, R., Carpenter, S.R., De Vries, W., De Wit, C.A., Folke, C., Gerten, D., Heinke, J., Mace, G.M., Persson, L.M., Ramanathan, V., Reyers, B. and Sörlin, S. 2015b. Planetary Boundaries: Guiding Human Development on a Changing Planet. *Science*, 347, 736. 1259855.

Sverdrup, H.U., Koca, D. and Ragnarsdóttir, K.V. 2013. Peak Metals, Minerals, Energy, Wealth, Food and Population: Urgent Policy Considerations for a Sustainable Society. *Journal of Environmental Science and Engineering*, 2(B), 189–222.

UN. 2015. *Transforming Our World: the 2030 Agenda for Sustainable Development*, United Nations, New York.

UNEP. 2012. *Global Environmental Outlook 5: Environment for the future we want: United Nations Environment Programme*, Nairobi, Kenya.

UNEP. 2016. *Resource Efficiency: Potential and Economic Implications*, A report of the International Resource Panel. P. Ekins, N. Hughes et al. United Nations Environment Programme, Paris.

United Nations. 1987. *Our Common Future*, Reprinted. Oxford University Press, Oxford and New York.

Wackernagel, M. and Rees, W. 1996. *Our Ecological Footprint*, New Society Publishers, Gabriola Island, BC.

WFEO. 2013. World Federation of Engineering Organizations – Model Code of Practice for Sustainable Development and Environmental Stewardship. October 2013.

WFEO. 2016. World Federation of Engineering Organizations. Code of Ethics. www.wfeo.org/ethics/.

Exercises

1. The planetary boundaries concept states that when planetary boundaries are crossed, irreversible and abrupt environmental change will occur. Identify which of the planetary boundary(s) will be negatively impacted by the combustion of fossil fuels.

2. Environment, Economy, and Population/Society are three variables that interact with one another. Explain, in a maximum of 1000 words, giving relevant examples from your country, how does the following interaction occur:

   - how economic development impact on the environment?
   - how can environment degradation undermine the goals of economic development?
   - how can the population/society impact on the environment?
   - how can the environment impact on population/society?
   - how can the population impact on the economy?
   - how can the economy impact on the population?

3. Search the internet for an example of a major collapse of a culture in the past due to unsustainable behavior.

4. What is the relationship between the five capitals model and the three dimensions of the sustainable development concept? A manufacturing organization decides to develop a sustainable development strategy based around the five capitals model. Identify the capital under which the organization's health and safety strategy would sit.

5. Weak sustainability is maintaining total capital intact without regard to the partitioning of that capital among the five kinds. Strong sustainability asserts that different kinds of capital cannot be easily substituted. This implies maintaining the natural capital. Ecosystem services that nature provides for free are not cost effectively replaceable or substitutable by technological innovation. These services complement and are depended on by life on our planet. There are numerous natural ecosystem processes that are complementary to human capital and inventions, but are not substitutable. Brainstorm as many ecosystem services as you can and then compare them to page 42 of the book "The Natural Advantage of Nations".

6. The World Bank as part of the conditions around funding a transport infrastructure project requires that the latter be appraised with regards to the UN Sustainable Development Goals. One of the major benefits of the project as per the appraisal is job creation during both the construction and operational phases. Which of the SDGs does the project address?

7. Through a literature search, explain in a maximum of 200 words the concept of ecological footprint. There are a number of ecological footprint calculators available on the web (see for example: http://footprint.wwf.org.uk/). Find out your ecological footprint; discover your biggest areas of resource consumption (among food, housing, purchasing or transportation habits). In a maximum of 500 words,

suggest changes in your lifestyle that can help to reduce the footprints by 25%. You may wish to consult the following website: www.youthx change.net

8. Write an essay of 500 to 1000 words in which you explain what climate change means for the country in which you live. You can refer to the specific Nationally Determined Contribution (NDC) of your country.

9. What is a carbon footprint? What are the six Green House Gas (GHG) emissions it accounts for? What is the unit of measurement? A variety of tools are available to help us estimate our carbon footprint and to enable us to take responsibility. A good carbon footprint calculator can be found at: www.carbonfootprint.com/calculator.aspx. (or at www.nature.org/greenliving/carboncalcula tor/). Please go to one of these sites and estimate your carbon footprint for the past year. How does your annual carbon footprint compare to the global average per capita carbon footprint? What would be some reasonable ways for you to reduce your own carbon footprint by 25%?

10. Choose a specific field of engineering (civil or chemical or mechanical or electrical/electronic, etc.) and discuss how it can help meet the SDGs.

11. Consider an engineer who is a director in an environmental engineering firm and is requested by a developer client to prepare an impact assessment of a piece of property adjacent to a wetlands area for potential development as a residential estate. During the impact assessment, one of the engineering firm's biologists reports to the engineer that in his opinion, the project could threaten a bird species that inhabits the adjacent protected wetlands area. The bird species is not an "endangered species," but it is considered a "threatened species" by regulators. In subsequent discussions with the developer client, the engineer mentions the concern, but he does not include the information in a written report that will be submitted to a public authority that is considering the developer's proposal. Was it ethical for the engineer not to include the information about the threat to the bird species in a written report that will be submitted to a public authority that is considering the developer's proposal? You can refer to the WFEO Code of Ethics.

12. Read an article from a journal, or from a popular magazine, or from the Internet, on some environmental issue of interest to you. Summarize the article in a short memorandum format, addressed to your instructor. In the body of the memorandum, limit the length to one page of single-spaced text, including graphics/tables (if needed). Use of headings is appropriate and be sure to reference information

sources. Structure the memorandum in this way: a. Introduction and motivation; b. A description of the issue; c. A description of what engineers are doing, have done, or are going to do to address the challenge

13. Describe your vision of a sustainable world, including a description of the lifestyle you would expect in such a world.

Recommended Reading and Websites

- Hargroves, K. and Smith, M.H. (2005) The Natural Advantage of Nations: Business Opportunities, Innovation and Governance in the twenty-first Century, Earthscan, London.
- IPCC (2014) Climate Change 2014: Synthesis Report of the IPCC Fifth Assessment Report (AR5) Available at www.ipcc.ch/report/ar5/syr/. Accessed on 7 August 2019.
- Meadows, D.H., Meadows, D.L., Randers, J. and Behrens III, W.W. (1972). The Limits to Growth; A Report for the Club of Rome's Project on the Predicament of Mankind Universe Books: New York.
- United Nations (1987) Report of the World Commission on Environment and Development: Our Common Future, Chapter 2: Towards Sustainable Development. Geneva, 3/20/1987.
- United Nations (2005) Millennium Ecosystem Assessment. Available at www.millenniumassessment.org/en/index.aspx. (Accessed 5 January 2007).
- United Nations. Global Resources Outlook. www.resourcepanel.org/reports/global-resources-outlook
- United Nations: About the Sustainable Development Goals www.un.org/sustainabledevelopment/sustainable-development-goals/
- WFEO Model Code of Ethics www.wfeo.org/wp-content/uploads/code_of_ethics/WFEO_MODEL_CODE_OF_ETHICS.pdf
- United Nations. Transforming our world: the 2030 Agenda for Sustainable Developmenthttps://sustainabledevelopment.un.org/post2015/transformingourworld
- The WWF Living Planet Index Report is a periodic update on the state of the world's ecosystems (www.wwf.panda.org)

2

Sustainable Engineering: Concepts, Principles, and Frameworks

2.1 Introduction

There is no universal agreement on a definition of sustainable engineering. However, in some specific areas, such as energy, some engineering-related sustainability definitions have been formulated, which emphasize some of the essential points of sustainable engineering. Rosen (2009) defines energy sustainability as the provision of energy services for all people in a sustainable manner, that is, in ways that, now and in the future, are sufficient to provide basic necessities, affordable, not detrimental to the environment, and acceptable to communities and people. Sustainable engineering may thus be defined as

> the provision of engineering services in a sustainable manner, which in turn necessitates that engineering services be provided for all people in ways that, now and in the future, are sufficient to provide basic necessities, affordable, not detrimental to the environment, and accep-table to communities and people.
>
> (Rosen, 2012)

The aim of this chapter is to present a framework for sustainable engineering, which builds on concepts and principles. A concept is the formulation of an idea how to achieve sustainability prior to actually creating it. Principles provide the rules to keep in mind when solving an engineering problem and give more precise guidance for action. We need to be able to set our choices and engineering decisions against these guiding principles for sustainability. Concepts and principles are built upon to construct frameworks. A framework, by definition, is a little loose. It is an approach to solving a problem that provides a rough outline of the methodology that will achieve a specific goal, but that does not identify the specific tools that may be used. A framework is powerful because it provides guidance while being flexible enough to adapt to changing conditions.

2.2 Key Concepts for Sustainable Engineering

2.2.1 Factor 4 and Factor 10: the Goals of Sustainability

A useful way to think about the sustainability challenges ahead is to examine the driving forces behind those challenges. One of the more famous expressions of these driving forces is provided by the Ehrlich and Holdrens (1971) Master Equation:

$$\text{Environmental Impact} = (\text{Population}(\times(\text{GDP/Person})$$
$$\times (\text{Environmental Impact/Unit of GDP})$$

where GDP is a country's gross domestic product, a measure of economic activity. This equation has traditionally been called the IPAT equation where I = Environmental Impact, P = Population, A = Affluence or material standard of living (GDP/person), and T = Technology (environmental impact/unit of GDP). The third term in the master equation – T or the environmental impact per unit of GDP – is an expression of the degree to which technology is available to permit development without serious environmental impact. It is a technological term that offers the greatest hope for a transition to sustainable development and it is modifying this term that is among the central tenets of sustainable engineering. The IPAT equation can be used conceptually to suggest goals for technology and society. If our aim is to maintain the environmental impact of humanity to its present level, we need to look at the probable trends in the three terms of the equation. The first will likely increase by a factor of 1.5 over the next 50 years, while the second is predicted to likely increase by a factor of 2 to 3 over the same time period. Accordingly, to just maintain the level of environmental impact to what it is today, the third term must decrease by 65–80%. The technology factor must decrease 4–10 times to counterbalance the expected growth in population and material consumption. This is the inspiration for calls for factor 4 or factor 10 reductions in environmental impact per unit of economic activity. Factor 4 encapsulates the idea that natural resources can be used four times more efficiently in all spheres of daily life, either by generating more goods, services, and quality of life from the available resources, or by using fewer resources to maintain the same standard. The concept was introduced in 1998, in a book of the same name written by L. Hunter Lovins and Amory Lovins of the Rocky Mountain Institute, and Ernst von Weizsäcker, founder of the Wuppertal Institute for Climate, Environment & Energy. The book illustrates 50 examples of technologies that can help to deliver the necessary improvement in resource efficiency, including energy efficient cars or buildings. The concept of factor 10 evolved from the concept of factor 4 taking into consideration that in the long term resource use in developed countries, which are currently overconsuming resources, needs to be decreased by a factor of 10 to approach sustainability.

2.2.2 Systems Thinking

A system is a set of things interacting in a way that produces something greater than the sum of its parts. Systems thinking is concerned with expanding our awareness to see the relationships between parts and wholes rather than looking at just discrete, isolated parts and it is particularly useful in addressing complex situations. The latter cannot be solved by any one actor (businesses, people, or government) and one can understand a complex system only by looking at it from different perspectives. Systems thinking helps us see the big picture and enables us to identify multiple leverage points that can be addressed to solve the problem. It also helps us see the connectivity between elements in the situation, so as to support joined-up actions. At early stages of an engineering project, a systems thinking approach is needed for the integration of sustainability in the planning and design as it enables to look at an issue in a broader context. Rather than just trying to address the immediate symptoms, a systems approach looks at the underlying causes of an identified problem. It helps to avoid any burden shifting that may occur. For example, while it has been shown that certain kinds of biofuels can reduce emissions of greenhouse gases if they replace fossil fuels, it has also been shown that large-scale expansion of biofuels could lead to reduced supply of food crops and feed crops and thereby result in increasing food prices. In order to promote sustainable development, it is thus important that such side effects are identified at an early stage in the decision-making process.

2.2.3 Life Cycle Thinking

Life cycle thinking (LCT) extends the boundaries of the production site and manufacturing processes to include environmental, social, and economic impacts of a product over its entire life cycle. With this kind of thinking, we aim to reduce a product's resource use and emissions to the environment as well as improve its socioeconomic performance through its life cycle. Figure 2.1 shows in a schematic manner the different stages of a product life cycle. Through an LCT approach, negative impacts are minimized while avoiding burden shifting between life cycle stages or between environmental impacts. When applied to product design, production processes, and as a decision-making tool, LCT is an important concept for devising and implementing effective sustainability strategies.

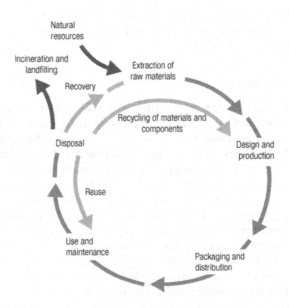

FIGURE 2.1
Typical product life cycle stages.

(Source: www.lifecycleinitiative.org/starting-life-cycle-thinking/what-is-life-cycle-thinking/)

2.2.4 The Circular Economy

An alternative to a traditional linear economy (make, use, dispose) is a circular economy, in which we keep resources in use for as long as possible, extract the maximum value from them whilst in use, then recover and regenerate products and materials at the end of each service life (Figure 2.2). The concept of "circular economy" is one that is "restorative and regenerative by design and aims to keep products, components, and materials at their highest utility and value at all times, distinguishing between technical and biological cycles" (Ellen MacArthur Foundation, 2016). One key aspect of the circular economy is the reusability of products through activities to extend the lifespan along the value chain, such as repairing, reusing, refurbishing, reconditioning, and remanufacturing as well as recyclability of raw materials. In such an economy, production strategies and business models need to be redesigned by manufacturers to allow for more durable, reparable, and recyclable products. In addition, the awareness and engagement of consumers are raised to change consumption patterns and to increase demand for reused products and services. Circular economy needs system thinking. All actors (businesses, people, and organisms) are part of a network in

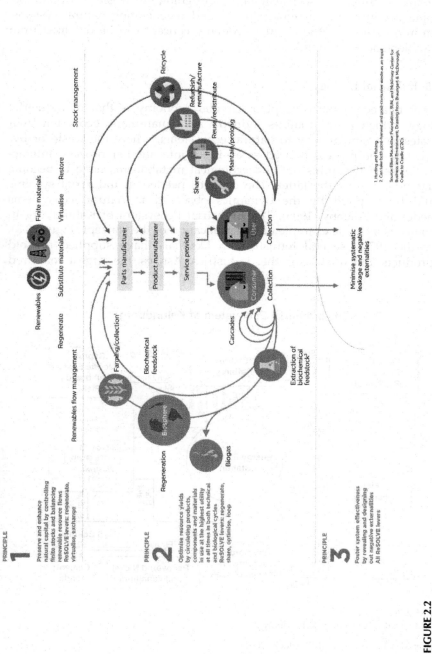

FIGURE 2.2

Outline of a circular economy.

(Reprinted with Permission from www.ellenmacarthurfoundation.org/circular-economy/infographic)

which the actions of one actor impact other actors. Supported by a transition to renewable energy sources, the circular economy builds economic, natural, and social capital by designing out waste and pollution, keeping products and materials in use and regenerating natural systems. More information on the circular economy concept can be obtained from www.ellenmacarthurfoundation.org.

2.2.5 Industrial Ecology

Closely related to the concept of circular economy is that of industrial ecology, which conceptualizes industry as a man-made ecosystem that operates in a similar way to natural ecosystems, where the waste or by-product of one process is used as an input into another process. Interactions between the industrial and ecological metabolisms have to become benign to preserve the latter. The changes needed in industrial systems should be inspired by the dynamics observed in natural ecosystems (Graedel and Allenby, 1995). The Industrial Ecosystem of Kalundborg in Denmark is a classic example of the application of the concept (Figure 2.3). Since the 1970s, several industries in Denmark have supplied or sold by-products and wastes to other industries. Asnaes, the largest coal-fired

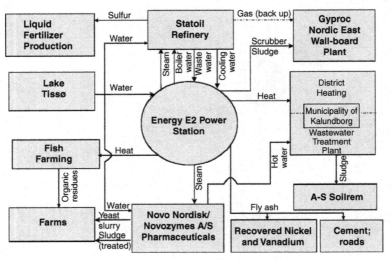

FIGURE 2.3

The industrial ecosystem of kalundborg.

(Source: M. Chertow and M. Portlock (2002))

power plant in Denmark, sold processed steam to Statoil (an oil refinery) and Novo Nordisk (a pharmaceutical plant). Some of Asnaes' surplus heat was supplied to the town's heating scheme, reducing the number of domestic oil burning systems in use. Surplus heat was also used to heat the water of Asnaes' commercial fish farm. Local farmers used sludge from the fish farm as fertilizer. By treating some of its waste, Novo Nordisk sold high-nutrient liquid sludge to farmers. Statoil supplied cooling and purified wastewater to Asnaes, which reduced Asnaes' freshwater extraction. In addition, Statoil removed sulfur from its surplus gas and sold all of its cleaned surplus gas to Asnaes and Gyproc (a plasterboard factory). The removed sulfur was sold to Kemira (a sulfuric acid producer). By desulfurizing its smoke, Asnaes sold the resulting calcium sulfate to Gyproc as an alternative to mined gypsum, which was being imported. These partnerships were formed voluntarily and negotiated independently. Initially for purely economic reasons, some of the later deals were made for environmental reasons. Industrial ecology is best implemented within a reasonable transport distance between industries.

2.2.6 Green Economy and Low-Carbon Economy

The United Nations Environment Programme (UNEP) defines green economy (GE) as an economy "that results in improved human well-being and social equity, while significantly reducing environmental risks and ecological scarcities." This definition emphasizes the importance of "getting the economy right" as a precondition for achieving sustainability (UNEP, 2011). A low-carbon economy or decarbonized economy is an economy based on low carbon power sources, which therefore has a minimal output of greenhouse gas (GHG) emissions into the biosphere, but specifically refers to the greenhouse gas carbon dioxide. Many countries around the world are designing and implementing low emission development strategies. These strategies seek to achieve social, economic, and environmental development goals while reducing long-term GHG emissions and increasing resilience to climate change impacts. The concepts of green economy and low carbon economy complement the concept of sustainable development by emphasizing the importance of the economy, and especially of innovations, for achieving sustainability (Olsen, 2012).

2.2.7 The Natural Step

Developed by Swedish oncologist Karl-Henrik Robert in1989, the Natural Step considers the effects of material selection on human health. He suggested that many human health problems result from materials we use in our daily lives. The Natural Step articulates the four system conditions that should be followed to eliminate the effects of material practices on our health. (https://thenaturalstep.org/approach/). The four system conditions are as follows:

1. In order for a society to be sustainable, nature's functions and diversity are not systematically subject to increasing concentrations of substances extracted from the Earth's crust.

2. In order for a society to be sustainable, nature's functions and diversity are not systematically subject to increasing concentrations of substances produced by society.

3. In order for a society to be sustainable, nature's functions and diversity are not systematically impoverished by overharvesting or other forms of ecosystem manipulation

4. In a sustainable society, resources are used fairly and efficiently in order to meet basic human needs globally.

2.2.8 Resources Efficiency and Decoupling

With a business as usual consumption rate by the world population, annual global resource extraction could triple from 2000 levels to 140 billion tons in 2050. This scenario would seem to represent an unsustainable future in terms of resource use, emissions, and environmental impacts. Resource efficiency enhances the means to meet human needs while respecting the ecological carrying capacity of the earth by producing more well-being with less material consumption. The term "resource efficiency" encompasses a number of ideas: the technical efficiency of resource use (measured by the useful energy or material output per unit of energy or material input); the resource productivity, or extent to which economic value is added to a given quantity of resources (measured by useful output or value added per unit of resource input); and the extent to which resource extraction or use has negative impacts on the environment (increased resource efficiency implies reducing the environmental pressures that cause such impacts). The rate of resource productivity (doing more with less) needs to be improved faster than the economic growth rate. This is the notion behind "decoupling," which is the idea that economic growth does not actually have to be accompanied by ever-greater use of natural resources. Through decoupling, economies can continue to grow without overusing their natural resource base and degrading the environment. The UNEP International Resource Panel (2011) defines two aspects of this: *resource decoupling* means reducing the rate of use of resources per unit of economic activity and *impact decoupling* means maintaining economic output while reducing the negative environmental impact of the underlying economic activities (Figure 2.4).

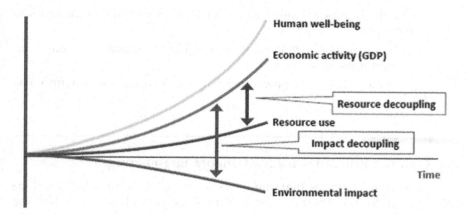

FIGURE 2.4
Decoupling of resource use and environmental impacts from GDP growth.

(Source: UNEP, 2011)

2.2.9 Eco-efficiency

Originated by the World Business Council for Sustainable Development (WBCSD) in 1992, the concept of eco-efficiency includes environmental impacts and costs as a factor in calculating business efficiency (WBCSD, 2000). Linking environmental and economic performance, eco-efficiency is primarily a management concept. It differs from sustainability, in that eco-efficiency does not measure social aspects. Businesses using eco-efficiency principles are more profitable and competitive as they use fewer resources, generate less waste and pollution, improve production methods, develop new products or services, use or recycle existing materials, extend product durability, and increase the service intensity of goods and services. Eco-efficiency applies to all business aspects, from purchasing and production to marketing and distribution.

2.2.10 Triple Bottom Line

A company's bottom line, from the financial point of view, means its profit margins. In 1994, John Elkington, a famed British management consultant, coined the phrase "triple bottom line" (TBL) as his way of measuring performance in corporate America. Elkington's TBL framework advances the goal of sustainability in business practices, in which companies look beyond profits to include social and environmental issues to measure the full cost of doing business. According to the TBL theory (Elkington, 1997), companies should be working simultaneously on these three bottom lines:

1. Profit: the traditional measure of corporate profit – the profit and loss (P&L) account.
2. People: measures how socially responsible an organization has been throughout its operations.
3. The Planet: measures how environmentally responsible a firm has been.

2.3 Guiding Principles for Sustainable Engineering

It is of vital importance that engineers incorporate sustainable development principles into their professional practice. It is much more than the narrow discipline-specific activity of "protection of the environment" and engineers are required to take a wider perspective including goals such as poverty alleviation and social justice. There are a number of academic and professional institutions who have formulated sustainable engineering principles. The overarching goal is to generate a *balanced solution* to any engineering problem. If an engineering project benefits only one of the three sustainability dimension but ignores the others, it will be unsustainable in the long term. Some of the aspects that differentiate the traditional and sustainable approaches in engineering are given in Table 2.1.

Engineers understand the concept of physical principles such as conservation of mass, energy, or momentum. They provide the ideas, rules, or concepts to keep in mind when solving an engineering problem. Similarly, we need to be able to set our choices and engineering decisions against guiding principles for sustainability. The WFEO Model Code of Practice for Sustainable Development and Environmental Stewardship links the

TABLE 2.1

Sustainability Approaches in Engineering

Traditional Engineering	Sustainable Engineering
Considers the object or process	Considers the whole system in which the object or process will be used
Focuses on technical issues	Considers both technical and nontechnical issues synergistically
Solves the immediate problem	Strives to solve the problem for infinite future
Considers the local context	Considers the global context
Assumes others will deal with political, ethical, and societal issues	Acknowledges the need to interact with the experts in other disciplines related to the problem

Code of Ethics with professional practice. It defines and explains a set of ten principles that guide engineering practice (WFEO, 2013):

1. Maintain and continuously improve awareness and understanding of environmental stewardship, sustainability principles, and issues related to your field of practice.

2. Use expertise of others in the areas where your own knowledge is not adequate to address environmental and sustainability issues.

3. Incorporate global, regional, and local societal values applicable to your work, including local and community concerns, quality of life, and other social concerns related to environmental impact along with traditional and cultural values.

4. Implement sustainability outcomes at the earliest possible stage employing applicable standards and criteria related to sustainability and the environment.

5. Assess the costs and benefits of environmental protection, ecosystem components, and sustainability in evaluating the economic viability of the work, with proper consideration of climate change and extreme events.

6. Integrate environmental stewardship and sustainability planning into the life-cycle planning and management of activities that impact the environment, and implement efficient, sustainable solutions.

7. Seek innovations that achieve a balance between environmental, social, and economic factors while contributing to healthy surroundings in both the built and natural environment.

8. Develop locally appropriate engagement processes for stakeholders, both external and internal, to solicit their input in an open and transparent manner, and respond to all concerns – economic, social, and environmental in a timely fashion in ways that are consistent with the scope of your assignment. Disclose information necessary to protect public safety to the appropriate authorities.

9. Ensure that projects comply with regulatory and legal requirements and endeavour to exceed or better them by the application of best available, economically viable technologies and procedures.

10. Where there are threats of serious or irreversible damage but a lack of scientific certainty, implement risk mitigation measures in time to minimize environmental degradation.

The Sandestin Sustainable Engineering Principles, developed as an outcome of a multidisciplinary engineering conference, emphasize the need for holistic thinking and the use of environmental impact and integrative analysis tools such as life-cycle assessment (Abraham and Nguyen, 2004). The principles are as follows:

Principle 1: Engineer processes and products holistically, use system analysis, and integrate environmental impact assessment tools.

Principle 2: Conserve and improve natural ecosystems while protecting human health and well-being.

Principle 3: Use life cycle thinking in all engineering activities.

Principle 4: Ensure that all material and energy inputs and outputs are as inherently safe and benign as possible.

Principle 5: Minimize depletion of natural resources.

Principle 6: Strive to prevent waste.

Principle 7: Develop and apply engineering solutions, while being cognizant of local geography, aspirations, and cultures.

Principle 8: Create engineering solutions beyond current or dominant technologies; improve, innovate and invent technologies to achieve sustainability.

Principle 9: Actively engage communities and stakeholders in development of engineering solutions.

Similarly, the Royal Academy of Engineering of UK (2005) defines the following 12 guiding principles on the basis of case studies illustrating the issues connected to sustainable development in engineering:

1. **Look beyond your own locality and the immediate future:** In considering the effects of our decisions on the wider world we need to: (i) identify the potential positive and negative impacts of our proposed actions, not only locally and soon but also outside our immediate local environment, organization and context, and into the future; (ii) seek to minimize the negative, while maximizing the positive, both locally and more widely, and into the future.

2. **Innovate and be creative:** A sustainable development approach is creative, innovative and broad, and thus does not mean following a specific set of rules. It requires an approach to decision-making that strikes a balance between environmental, social, and economic factors.

3. **Seek a balanced solution:** Approaches like the "three pillars" and the "five capitals" seek to deliver economic, social, and environmental success all at the same time, and so seek to avoid any product, process, or project that yields an unbalanced solution.

4. **Seek engagement from all stakeholders:** Society will ultimately say what is needed or wanted for any development, sustainable, or otherwise.

5. **Make sure you know the needs and wants:** Effective decision-making in engineering for sustainable development is only possible when we know what is needed or wanted – the framework of the problem, issue, or challenge to be tackled. This should be identified as clearly as possible, including identifying any legal requirements and constraints. We should use teamwork and assistance of immediate colleagues to improve problem definition.

6. **Plan and manage effectively:** In planning our engineering projects, we need to express our aims in sufficiently open-ended terms so as not to preclude the potential for innovative solutions as the project develops.

7. **Give sustainability the benefit of any doubt:** This encapsulates the "precautionary principle" and, to be implemented, forces us to address the future impacts of today's decisions. Essentially, the precautionary approach is seen as a "better safe than sorry" principle that focuses on the benefits of prevention rather than cure, particularly in the context of potentially irreversible environmental damage. It emphasizes that we have to take anticipatory action to prevent harm in the face of scientific uncertainty.

8. **If polluters must pollute ... then they must pay as well:** The adverse, polluting effects of any decision should, in some way, be paid for or compensated for by the proponent of an engineering project, scheme, or development; they should not be transferred to others without fair compensation.

9. **Adopt a holistic, "cradle-to-grave" approach:** To deliver this approach, the effects on sustainability throughout the whole life cycle of a product or infrastructure scheme should be systematically evaluated.

10. **Do things right, having decided on the right thing to do:** Adhering to the above should ensure that right decisions from a sustainability point of view have been made in relation to the circumstances that apply. The implementation of these right decisions must then pay full regard to doing things right, again from a sustainability point of view. To deliver this principle, we need to retain the sustainability focus on the intended outcome right through the implementation of the solution.

11. **Beware cost reductions that masquerade as value engineering:** We are unlikely to arrive at our best decisions first time every time we need to challenge ourselves and refine those decisions, whilst remaining focused on the intended outcome. We therefore need to avoid sacrificing the sustainability desires incorporated in a design when seeking cost reductions and include any adverse effects on sustainability in the "value equation" and value engineering.

12. **Practice what you preach.** Be prepared to be accountable for your design and engineering and uphold by example the beliefs it reflects. Change yourself before you seek to change others.

Figure 2.5 presents a consolidated framework of the many sustainable engineering principles available in the literature. It is mainly adapted from the work of Gagnon et al. (2009). It lists the various principles of sustainable engineering versus environmental, social, and economic poles. Some of these principles lie toward one of the corners of this triangle and thus address particular societal, environmental, or economic concern. But some others, which are placed along the sides of the triangle, have connections to two of the poles of the diagram and address both societal and economic, or both economic and environmental concerns in some proportion. Those principles placed in the center of the diagram combine all three aspects of sustainability to a certain degree and hence their implementation would benefit all societal, environmental, and economic stakeholders.

The above principles can be viewed as guidelines to fit any engineering project. They can be rephrased into questions to guide the engineers' thinking concerning a project's sustainability from the moment it is initiated – for example, "how to involve stakeholders and other professionals in the project if holistic solutions are sought?" The principles must be contextualized for a given project and broken down into a more specific set of criteria and indicators along which design options can be compared.

2.4 Frameworks for Sustainable Engineering

The above engineering principles establish a framework for sustainable engineering practice. Sustainable engineering can be regarded as the operational arm of the above concepts and principles. Frameworks are needed to integrate all the above concepts and principles for their applications in sustainable engineering practice. Three useful frameworks are presented below.

A first sustainability framework is presented in Figure 2.6, which helps to identify the types of interaction between engineering projects and the physical or social systems associated with them. The upper part of the diagram shows the following overlapping physical systems: the biosphere (which includes the aquatic and terrestrial ecosystems), the anthroposphere (which encompasses the agricultural, industrial, and urban systems); both of these are positioned at the triple boundary between the atmosphere, lithosphere, and hydrosphere of the earth. The anthroposphere box can be expanded to indicate the significant role played by engineering projects as human activity. Design and engineering increase

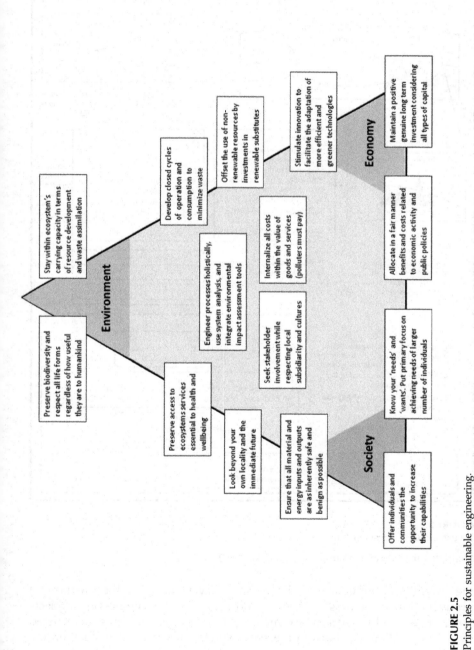

FIGURE 2.5
Principles for sustainable engineering.

(Reprinted with Permission from Gagnon et al., 2009)

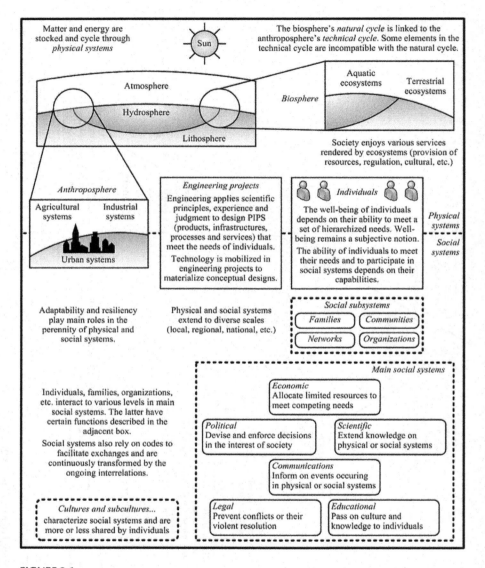

FIGURE 2.6
A sustainability framework for engineering.

(Reprinted with Permission from Gagnon et al., 2009)

the extent and influence of the technical cycle through the creation of products, infrastructures, processes, and services. The Engineering Projects box is linked to Individuals box, meaning that technical progress is incurred by and benefits individuals in the social sphere. There is a feedback on the natural systems due to the expansion of the anthropogenic technical systems, which can undermine the benefits human beings and societies derive from the environment. Individuals are given a central place in the framework since physical and social systems both contribute toward their well-being. The Social Systems box at the bottom of the diagram provides more detailed representation of those benefits, which include economic, political, scientific, legal, educational values, and communications. Social subsystems, such as Families, Communities, Networks, and Organizations interact to a various degree with the main social systems and fulfill certain functions. This framework diagram is very broad, and each box could be presented as a separate sub-system with its own internal connections. But this is the big picture, which allows us to visualize the diversity of the factors that contribute to the sustainability of human society. Technology fits anywhere at the connections of the anthropogenic spheres with the physical systems. For example, through technologies, society can utilize natural resources. Technologies can reconcile the processes and matter flows between the anthroposphere and environmental spheres as well as create conflicts between them. Thus, engineering projects undermining the resilience or adaptability of ecosystems, social systems, or individuals might bring benefits in the short term but are likely to have long-term negative outcomes. What we call *sustainable* technologies are designed to render the feedbacks and connections mutually harmless or mutually beneficial in the best-case scenario. To assess technologies from this angle, we need to learn to recognize the feedbacks and effects they create on a large scale. This framework can be used to define problems and also to set up design teams and to evaluate projects.

The second framework, the Sustainable Consumption and Production (SCP) framework (Figure 2.7), reflects the incorporation of life cycle thinking in decision-making. Consumption and production patterns are main drivers behind depletion of natural resources, environmental degradation, and growing amount of wastes. "Sustainable Consumption and Production (SCP) is a holistic approach to minimizing the negative environmental impacts from consumption and production systems while promoting quality of life for all" (UNEP, 2015). SCP aims to minimize the negative environmental impacts from consumption and production systems, considering all stages of the life cycle of products and services while promoting quality of life for all. SCP is a systems approach that has at its core a life cycle perspective. It avoids shifting problems between stages of consumption and production, geographic areas or impact categories in the life cycle. It is a broad framework that encompasses many operational solutions that are key for designing and implementing policies and measures to achieve economic, social, and environmental sustainability. These solutions from a life cycle thinking perspective, as Figure 2.7 illustrates, are sustainable resource management, design for sustainability, cleaner production, sustainable mobility, eco-labeling and certification, sustainable procurement,

FIGURE 2.7
The SCP framework.

(Source: UNEP, 2010)

sustainable marketing, sustainable lifestyles, waste management, and resource efficiency along the value chain.

As per this framework, engineers can contribute to sustainable development along the entire chain of modern production and consumption, including the following:

- Extracting and developing natural resources
- Processing and modifying resources
- Designing and production
- Transportation
- Meeting the needs of consumers
- Recovering and reusing resources

Hauschild et al. (2017) present a Life Cycle Engineering (LCE) framework useful to position manufacturing in the context of absolute environmental sustainability. LCE is a systematic "cradle to grave" approach that "provides the most complete environmental profile of goods and services." It focuses on design and manufacture of products, optimizing the product life cycle, and minimizing pollution and waste while at the same time encouraging economic progress. LCE tools are life cycle oriented. Considering the challenges to achieve sustainable manufacturing and the central role of life cycle engineering in addressing these challenges, there is a need for a systematic framework of life cycle engineering that organizes engineering activities throughout the life cycle of a product or technology and position them according to the leverage in terms of sustainable production systems. Consider Figure 2.8. In order to develop SCP patterns, engineering activities and in particular LCE must consider the planetary boundaries from the scope of both environmental and temporal concern.

The space of the framework is defined by the temporal scope and the scope of environmental concern. Following a top-down approach, sustainability describes the wider space defined by the Earth's life support system and the span of our civilization. One level down, at the edge of societies/economies and with a time scope between the generation life-time and the civilization span, sustainable development is positioned expressing the continuous process of change. One further level down, industrial ecology looks at industry and its surrounding systems. Life cycle management covers not only all activities but also structures and behavior of an organization. The framework sees the environmental dimension as the basis and boundary for

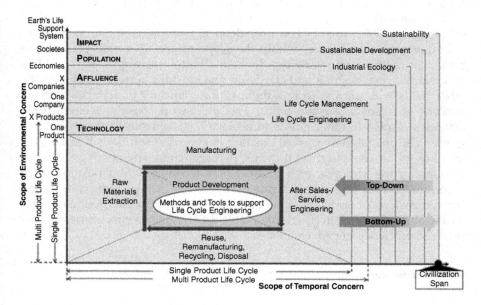

FIGURE 2.8
Life cycle engineering framework.

(Reprinted with Permission from Hauschild et al., 2017)

economic and social sustainability. It defines the scope of LCE as looking at products from a multitude of different products to a single product over all the stages of the life cycle and structures LCE with regard to the main activities and life cycle stages. In the framework, the scope of environmental concern deals with the type and scale of environmental impact both in terms of spatial or geographical scale and in terms of organizational scale ranging from the manufacturing system over the society's entire economy and the overall impact of society to the Earth's life support system, as considered in the setting of planetary boundaries. The scale of temporal concern deals with the time scale ranging from single- and multi-product life cycles to the lifetime of a generation until the civilization span. Along the two axes, the increase in temporal and environmental scope is linked to each of the four factors in the IPAT equation. The technology factor T has the scope of one or multiple products in their life cycle and is determined by the decisions made in product development. The affluence factor A has the scope of the economy and is influenced by the governance of and political management of the latter. The population factor P has society's scope of environmental concern and the temporal scope of generations, while the impact factor I representing our interference with the environment has the scope of Earth's life support systems and the temporal scope of civilizations. When interpreting the framework bottom up, LCE is at the center. Product development plays a central role for LCE, as a large share of the later environmental impact is already decided at this stage. LCE methods and tools should help to support decision-making toward the upper scopes of concern, orienting it toward absolute sustainability. With this, LCE is now defined as sustainability-oriented product development activities within the scope of one to several product life cycles. Using this framework, top-down and bottom-up approaches can thus be identified, establishing and clarifying relationships between LCE and the other fields with an aim to link them to the planetary boundaries and the concept of absolute sustainability. The framework positions engineering activities relative to other efforts to achieve a sustainable development and guide LCE practitioners toward creating engineering solutions that are sustainable in absolute terms. It helps in organizing engineering activities throughout the life cycle of a product or a technology and position them according to their leverage in terms of promoting SCP patterns considering the dimensions of scale and time and introducing absolute boundaries for sustainability. It helps in developing tools that target eco-effectiveness rather than eco-efficiency improvements.

2.5 Summary

A unique vocabulary is emerging to describe concepts related to sustainable engineering. The engineer should be familiar with concepts such as the TBL, factor 4 and factor 10, systems and life cycle thinking, and circular economy. Principles provide the ideas, rules, or concepts to keep in mind when solving an engineering problem. Engineers need to be able to set their choices and engineering decisions against guiding principles for sustainability. The three frameworks presented in this chapter integrate all the concepts and principles for their applications in sustainable engineering practice. The contribution of a project toward sustainable development can be analyzed with a wide variety of tools and the tools selected to analyze a particular project need to be positioned on the frameworks to clearly identify which component or interaction they intend to evaluate. Any selected tool should fit in a logical manner within the framework.

SUSTAINABLE ENGINEERING IN FOCUS: CONSERVING MATERIALS FOR THE NEXT MOBILITY REVOLUTION

GEM Co. Ltd (GEMChina) is a publicly listed urban mining, resource recycling, and WEEE (*Waste Electrical and Electronic Equipment*) recycling business founded in 2001 in Shenzhen. GEMChina are internationally recognized for their comprehensive industrial chains for resource collection and recycling across five categories: (i) batteries; (ii) cobalt, nickel, tungsten, and carbide; (iii) electronics; (iv) scrap automobile parts; and (v) waste residues, mud, and wastewater. In 2016, GEMChina processed a total of three million tonnes of waste, recycled 37 resource categories, and avoided 7.5 million tonnes of carbon emissions, saving 14 million barrels of fossil fuel and 18 million acres of forests.

The growing importance of battery recycling: The Chinese government has set the target by 2020 to increase the number of electric vehicles (EVs) by five million. The country is on its way to reaching this ambitious target. In 2015, Didi, the biggest Chinese ride-hailing platform, had 90,000 registered EVs. By 2020, it plans to expand its electric fleet by more than tenfold to one million. Beijing is also set to invest CNY 9 billion (USD 1.5 billion) to replace its entire fleet of 70,000 taxis with EVs. Shenzhen has already reached its target to fully electrify its entire bus fleet and have expanded its ambitions to include all taxis by 2020. These huge planned shifts in urban mobility systems create urgency around the issue of used EV batteries. In 2015, China's output of lithium-ion automotive batteries was 16.9 GWh, with demand projected to reach 125 GWh by 2020. Used batteries, if not properly managed, can pose huge problems to the environment, as they can leach heavy metals and other toxic residues that could contaminate water and soil. The government has been clamping down on private handlers who do not have stringent enough pollution control in place. At the same time, companies that invest in technological innovations and centralized treatments are encouraged to take over, presenting an opportunity for remanufacturers and recyclers at scale. GEMChina has the highest used battery recycling capacity in China, processing more than 10% of the total waste battery or about 300,000 tonnes of waste battery per year. Their technology enables the recycling of scrapped lithium batteries from EVs, extracting the nickel, cobalt, and other important resources, transforming them into materials used by battery producers such as Samsung SDI and Ecopro Co Ltd.

Putting e-waste back in circulation, China is the world's largest producer of consumer electronics and household appliances, accounting for almost 40% of global output. Meanwhile, China is also now one of the world's largest markets for these products, due to its growing affluence and urbanization. The high turnover of electronic goods has unwelcome consequences. Every year, some six million tonnes of electronic products are discarded, representing a huge loss in economic value as well as a risk to the environment.

GEMChina has invested CNY 2 billion (USD 300 million) to build eight treatment centers around China, with a combined annual processing capacity of 1.2 million tonnes of used household appliances (15% of China's total); 30,000 tonnes of circuit board (20% of China's total); and 10,000 tonnes of waste plastics. Speciality disassembly lines are in place for televisions, washing machines, LCD screens, small appliances such as telephones and circuit boards.

Future prospects for recycling in China: The government has set a target of two million annual EV sales by 2020 and seven million by 2025. Such growth could put China in a dominant position not only in the global EV industry, but also in EV-related businesses such as battery recycling. It has been estimated that the battery recycling market could be worth CNY 31 billion (USD 5 billion) by 2023. GEMChina has taken the leading position in the high-tech recycling market in China. However, while recycling is important as a way of capturing materials, a higher value-conserving strategy, would focus at the design stage of electronics products including batteries, so that old products can be easily disassembled into components that can be easily reused. Recognizing this, the Chinese Ministry of Industry has urged the sector to introduce standardized designs that could facilitate this approach. Companies such as GEMChina could certainly play a role in this future prospect by feeding in expertise and experiences in dealing with end-of-life products.

Source: Ellen MacArthur Foundation, "Conserving materials for the next mobility revolution", www.ellenmacarthurfoundation.org/case-studies/avoiding-3-million-tonnes-of-waste

References

Abraham, M. and Nguyen, N. 2004. "Green Engineering: Defining Principles" – Results from the Sandestin Conference. *Environmental Progress*, 22, 233–236. doi:10.1002/ep.670220410t.

Chertow, M. and Portlock, M. 2002. *Developing Industrial Ecosystems: Approaches, Cases, and Tools*, Yale University, New Haven, CT.

Ehrlich, P. and Holdren, J. 1971. Impact of Population Growth. *Science*, 171, 1212–1217.

Elkington, J. 1997. *Cannibals with Forks. The Triple Bottom Line of 21st Century*, John Wiley and Sons, London.

Ellen MacArthur Foundation. 2016. Circular Economy. Retrieved 15th July 2019, from www.ellenmacarthurfoundation.org/circular-economy.

Gagnon, B., Leduc, R. and Savard, L. 2009. Sustainable Development in Engineering: A Review of Principles and Definition of a Conceptual Framework. *Environmental Engineering Science*, 26(10), 1459–1472. doi:10.1089/ees.2008.0345.

Graedel, T.E. and Allenby, B.R. 1995. *Industrial Ecology*, Prentice Hall, Englewood Cliffs.

Hauschild, M., Herrmann, C. and Kara, S. 2017. An Integrated Framework for Life Cycle Engineering. *Procedia CIRP*, 61, 2–9.

IRP. 2011. *International Resource Panel. Decoupling Natural Resource Use and Environmental Impacts from Economic Growth*, United Nations Environment Programme, Paris.

Olsen, L.S. (Ed.). 2012. Integrating Green Growth in Regional Development Strategies. *Nordregio Report*, 2012, 6.

Rosen, M.A. 2009. Energy Sustainability: A Pragmatic Approach and Illustrations. *Sustainability*, 1, 55–80.

Rosen, M.A. 2012. Engineering Sustainability: A Technical Approach to Sustainability. *Sustainability*, 4, 2270–2292.

Royal Academy of Engineering. 2005. *Engineering for Sustainable Development: Guiding Principles*, R. Dodds and R. Venables (Eds.), The Royal Academy of Engineering, London.

UNEP. 2010. *ABC of SCP. Clarifying Concepts on Sustainable Consumption and Production*, United Nations Environment Programme, Paris.

UNEP. 2011. *Towards a Green Economy: Pathways to Sustainable Development and Poverty Eradication*, United Nations Environment Programme, Paris.

UNEP. 2015. *Sustainable Consumption and Production: A Handbook for Policymakers Global Edition*, United Nations Environment Programme, Paris.

von Weizsäcker, E., Lovins, A.B. and Lovins, H.L. 1997. *Factor Four: Doubling Wealth, Halving Resource Use*, Earthscan, London.

WBCSD. 2000. *Eco-Efficiency: Creating More Value with Less Impact*, World Business Council for Sustainable Development, Geneva.

WFEO. 2013. WFEO Model Code of Practice for Sustainable Development and Environmental Stewardship. Think Global and Act Local. www.wfeo.org/wp-content/uploads/code-of-practice/WFEOModelCodePractice_SusDevEnvStewardship_Interpretive_Guide_Publication_Draft_en_oct_2013.pdf (Accessed 15 November 2019).

Exercises

1. Explain briefly the meaning of the following concepts: Resource Efficiency; Decoupling, Industrial Ecology, Eco-efficiency, Factor 4, Circular Economy. Life Cycle Thinking.

2. Derive a definition of an eco-industrial park from existing literature. From that definition, create a set of at least four evaluation criteria to determine whether an industrial park could be called an eco-industrial park.

3. In order to reach a circular or closed-loop economy, companies need to move away from the usual linear business model to closed-loop business models.

    (a) Explain the differences between those two types of business models.

    (b) Give a rough outline of what a closed-loop business model would look like for: (i) a building, (ii) a mobile phone, and (iii) a (nonelectric) razor.

    (c) For one of the three above examples, discuss the major bottlenecks to implement such a closed-loop business model and possible measures that might facilitate a transition

4. A life cycle can begin with extracting raw materials from the ground and generating energy. Materials and energy are then part of manufacturing, transportation, use (wearing and washing the t-shirt, for instance) and eventually recycling, reuse, or disposal. A life cycle approach means we recognize how our choices influence what happens at each of these points, so we can balance trade-offs (e.g., regional or organic) and positively impact the economy, the environment and society. A life cycle approach is a way of thinking, which helps us recognize how our selections and decisions – such as buying electricity or a new t-shirt – are one part of a whole system of events. Considering a t-shirt, the different life cycle phases are as follows:

    (1) *Raw materials*: This includes the growing and harvesting of cotton and the production of the raw fiber.

    (2) *Production*: This phase comprises the transport of the raw materials, the whole production of the garment, including yarn manufacturing, knitting, pre-treatment, dyeing, finishing, and making-up.

    (3) *Distribution and retail*: This encompasses the transport of the product and the selling/retail of the clothing. In this specific analysis, the retailers are assumed to be big supermarkets/retailers or clothing stores/chains.

(4) *Consumption*: This includes the transport to the place of purchase and also the washing, drying, and/or ironing of the clothing.

(5) *End of life*: This encompasses the disposal or possible recycling of the product.

Put on "life cycle glasses" and analyze a t-shirt in a structured way on issues concerning its environmental and social impact. Hot spots are areas where there is a significant social or environmental issue. Depending on the environment and social issues (listed in the chart) state which phase of the life cycle of a t-shirt has a high or low impact on environmental and social issues. For a low impact, give it a small dot; for high impact, give it a large dot; for very little or no impact, leave it empty (e.g., during the life cycle of a cotton t-shirt, a lot of water is being used for the growing of the cotton, so "Raw Materials/Water" gets a large dot (representing a very important hot spot). You may wish to discuss with students in other disciplines to fill this chart. Justify your answers.

	Raw Materials	Production	Distribution and Retail	Consumption	End of Life
Energy					
Water					
Air					
Land Use and Biodiversity					
Human Rights					
Worker's Health and Safety					

5. "The framework presented in Figure 2.6 would also make it easier for professors to justify the relevance of non-technical courses (ethics, management and economics, legal issues, environmental management, etc.) since engineers need the related aptitudes to contribute more actively towards sustainable development in the future". Discuss this statement.

6. Provide examples of "risk shifting," a concept introduced in Sandestin Green Engineering Principle 1.

7. Bagasse, a waste product of sugarcane processing, is the woody stalk of the cane plant remaining after sugar is extracted. In Brazil or Mauritius, for example, bagasse is often used as a fuel for power

generation at the locations of cane processing. Provide one or more examples of wastes from key product systems that have found beneficial uses through engineering designs (Sandestin Green Engineering Principle 6).

8. Investigate and report on how engineers became involved in community engagement in your chosen field of study. What kind of community engagement activities were the engineers involved in (Sandestin Green Engineering Principle 9)?

9. Reading assignment: Engineering for Sustainable Development: Guiding Principles, Royal Academy of Engineering, Dodds, R., and Venables, R., Eds., 2005. pp. 12–24. www.raeng.org.uk/publications/other/engineering-for-sustainable-development.

Read the seven case studies. Choose one case study according to your field and show which principles in Figure 2.5 are related to it.

Recommended Reading and Websites

- Meadows, D.H., Thinking in Systems, Chelsea Green (2008). This book is great reading on systems philosophy; it will really help you grasp the idea of how complex systems work.
- Hargroves, K. and Smith, M.H. (2005). The Natural Advantage of Nations: Business Opportunities, Innovation and Governance in the 21st Century, Earthscan, London.
- Hawken, P., Lovins, A.B. and Lovins, L.H. (1999). Natural Capitalism: the Next Industrial Revolution, Earthscan, London. Chapters of the publication are freely downloadable from www.natcap.org
- McDonough and Braungart, M. (2002). Cradle to Cradle. Remaking the way we make things. North Point Press, NY 2002. Numerous examples and stories on sustainable design. The authors explain how products can be redesigned in such a way that after their service life, they do not become waste, but nourishment for something new.
- von Weizsäcker, E., Lovins, A.B. and Lovins, H.L. (1997). Factor Four: Doubling Wealth, Halving Resource Use, Earthscan, London.
- von Weizsacker and E.U et al. (2009). Factor Five: Transforming the Global Economy through 80% Improvements in Resource Productivity, The Natural Edge Project, Earthscan Publication, UK. The Natural Edge Project is an important contribution to a growing corpus of work regarding energy and resource efficiency. It provides a coherent framework and synthesis of the crucial issues of resource-use efficiency and decoupling of production from material and energy

throughput. There are numerous examples of resource productivity improvements from the most relevant sector.

- Gagnon, B., Leduc, R. and Savard, L., Sustainable development in engineering: a review of principles and definition of a conceptual framework. Cahier de recherche/Working Paper 08-18, 2008. This chapter was largely based on this paper.

- McDonough, W. and Braungart, M. (2001). "The Next Industrial Revolution", in Allen, P. (Ed.) Metaphors for Change, Greenleaf Books, London.

- Ellen MacArthur Foundation website on circular economy with several case studies:

www.ellenmacarthurfoundation.org/circular-economy.

3

Tools for Sustainability Assessment

3.1 Introduction

Measuring the level of sustainability – or a sustainability assessment (SA) – is required if we want to know whether we are moving toward it following an engineering decision. According to Waas et al. (2014), SA is any process that aims to:

- contribute to a better understanding of the meaning of sustainability and its contextual interpretation (interpretation challenge);
- integrate sustainability issues into decision-making by identifying and assessing (past and/or future) sustainability impacts (information-structuring challenge);
- foster sustainability objectives (influence challenge).

SA tools are operational methods supporting the concepts and principles discussed in the previous chapter. Figure 3.1 depicts a framework that captures the various concepts of sustainability and the tools that can be used for an SA. Analytical tools provide technical information as to the consequences of a choice, while procedural tools focus on the procedures to guide the way to reach a decision (Wrisberg et al., 2002). Analytical tools model the system in a quantitative or qualitative way aiming at providing technical information for a better decision. A further distinction between analytical tools is made based on physical metrics and nonphysical metrics. Analytical tools can be used within the framework of procedural tools (Finnveden and Moberg, 2005). All types of tools are supported by technical elements such as mass balance models and evaluation models, and the technical elements themselves are supported by data.

This chapter describes the main tools that are specifically aimed at measuring sustainability that feed the decision-making process in the engineering problem-solving methodology. They are described in a general way. A detailed explanation of the tools can be found elsewhere and sources are given for guidance. Some of the key tools will be further expanded in the

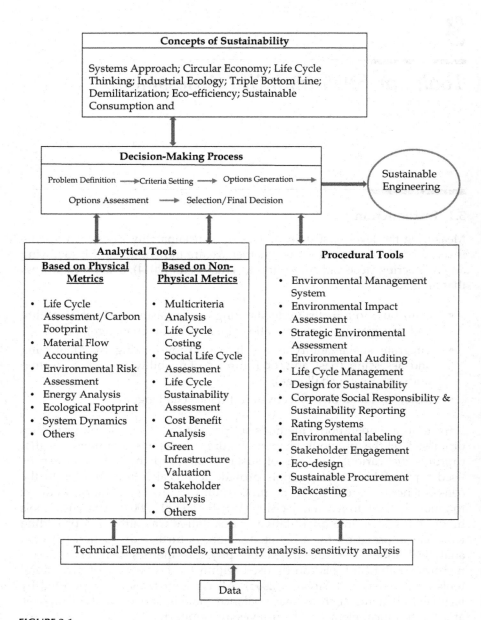

FIGURE 3.1
Sustainable Engineering Decision-Making Tools Framework.

Reprinted with Permission and Adapted from Wrisberg et al. (2002).

forthcoming chapters to move beyond the rhetoric and to enable the actual realization of sustainable development in engineering projects. The tools selected for an SA need to be positioned on the sustainability frameworks presented in the previous chapter to clearly identify which component or interaction they intend to evaluate.

3.2 Procedural Tools

3.2.1 Environmental Management System

An environmental management system (EMS) is a powerful tool for managing the adverse impacts of activities of an organization on the environment. An EMS is built around the way an enterprise operates, and it specifies the way it may organize its environmental efforts. It is the part of the overall management system that includes organizational structure, planning activities, responsibilities, practices, procedures, processes, and resources for developing, implementing, achieving, reviewing, and maintaining an environmental policy. An EMS is integrated into the overall management system of an enterprise. Like an overall management system, it represents a process of continual analysis, planning, and implementation; it requires that top management commits and organizes such resources as people, money, and equipment to achieve enterprise objectives, and it requires that resources be committed to support the management system itself.

The following are the basic elements of an EMS:

1. Establishing an overall policy (broad goals, aims, mission, and values) to guide everything that follows (this can be considered part of the planning activity or a step that comes before it).
2. Assessing the current situation.
3. Determining exactly what is to be achieved (setting explicit goals, objectives, targets, and performance standards).
4. Examining different ways of achieving it.
5. Working out in detail what seems like the best course of action (type of program, project, plan, action plan, and initiative).
6. Carrying out the plan (implementation).
7. Monitoring how things are going.
8. Making corrections as needed to stay on course.

The ISO 14000 family of standards provides practical tools for companies and organizations of all kinds looking to manage their environmental responsibilities, such as auditing, environmental labeling, and life cycle assessment (LCA). The ISO 14001 is the international standard for EMSs and provides systematic guidance on how to identify, evaluate, manage, and improve the environmental impacts of an organization's activities, products, and services. The basic steps in the ISO 14001 EMS are (a) environmental policy, (b) planning, (c) implementation and operation, (d) checking and corrective action, and (e) management review (see Figure 3.2). The first standards were published in 1996, and ISO 14001:2015 is the latest, improved version. ISO 14001 is a generic standard and is the only standard in the ISO 14000 family that can be used for certification. Generic means that the same standards can be applied to any organization, large or small, whatever its product or service, in any sector of activity, and whether it is a business enterprise, a public administration, or a government department. An EMS will ensure a continuous cycle of improvement, learning from success and failures. The UNEP/ICC/FIDIC Environmental Management System Training Resource Kit (2001) offers companies a systematic approach to integrate environmental considerations into all aspects of their activities. Additional guidance for the implementation of EMS can be obtained from *Environmental Management Systems: A Step-by-Step Guide to Implementation and Maintenance.*[*]

3.2.2 Environmental Auditing

Environmental auditing is a management tool used in companies for evaluation of their environmental performance. It is defined as

> a management tool comprising systematic, documented, periodic and objective evaluation of how well environmental organization management and equipment are performing with the aim of helping to safeguard the environment by facilitating management control of practices and assessing compliance with company policies, which would include regulatory requirements and standards applicable.
>
> (ICC, 1989)

It is usually an integral part of EMS, but can be conducted independently of these. It is a series of activities initiated by management to evaluate environmental performance, to check compliance with legislation, and to assess whether the management systems in place are effective. Conducting an environmental audit is no longer an option but a sound precaution and a proactive measure in today's heavily regulated environment. It helps to trigger new awareness and new priorities in policies and practices. The ICC has published a guide to environmental auditing (ICC, 1991).

---

[*] Sheldon, C. and Yoxon, M. 2006. *Environmental Management Systems: A Step-by-Step Guide to Implementation and Maintenance*, 3rd ed, Earthscan, London.

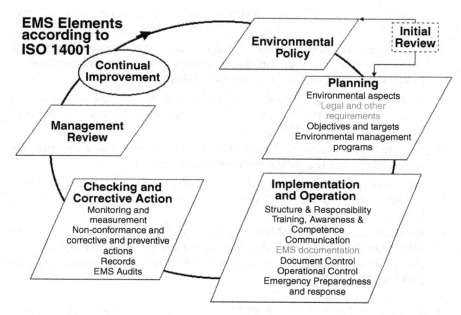

FIGURE 3.2
Basic Elements of an EMS

Source: UNEP (2001).

3.2.3 Cleaner Production Assessment

Cleaner production (CP) is a strategy to constantly improve environmental performance by reducing pollution at source. It has been defined as "the continuous application of an integrated preventive environmental strategy to processes, products, and services so as to increase efficiency and reduce risks to humans and the environment" (www.unep.fr/scp/cp/). CP is essentially similar in approach to related concepts such as waste minimization, pollution prevention, and eco-efficiency. It is implemented through the following strategies:

1. Good housekeeping: Take appropriate *managerial and operational* actions to prevent leaks, spills, and to enforce existing operational instructions.

2. Input material change: Substitute input materials by less toxic or by renewable materials or by adjunct materials that have a longer service lifetime in production.

3. Better process control: Modify operational procedures, equipment instructions, and process record keeping in order to run the

processes more efficiently and at lower waste and emission generation rates.

4. Equipment modification: Modify the existing production equipment and utilities in order to run the processes at higher efficiency and to lower waste and emission generation rates.

5. Technology change: Replace existing technology, processing sequence, and/or synthesis pathway in order to minimize the rates of waste and emission generation during production.

6. On-site recovery/reuse: Reuse wasted materials in the same process for another useful application within the company.

7. Production of useful by-products: Consider transforming waste into a useful by-product, to be sold as input for companies in different business sectors; and

8. Product modification: Modify product characteristics in order to minimize the environmental impacts of the product during or after its use (disposal) or to minimize the environmental impacts of its production.

A cleaner production assessment (CPA) is a systematic approach to identifying opportunities through the above-named strategies in a company. A CPA can be described as consisting of four basic steps (see Figure 3.3):

1. Planning and organization, which consist of the following activities:
   a. Obtain (further) management commitment.
   b. Identify assessment focus area. Often a limited assessment focus is useful to apply for each assessment (e.g., a section within the factory, or a raw material group, energy, or water).
   c. Organize the project team, with members from all concerned parts of the company. Do not forget accounting!
   d. Identify possible barriers and solutions by a (time limited) preassessment.

2. Assessment: This step consists of a more thorough analysis of material flows and material balances within individual steps of the manufacturing process. It is through this control of reality versus theory that most sources for waste generation are revealed.
   a. Identify sources (where).
   b. Identify and analyze causes (why).
   c. Generate possible options (how).

3. Feasibility analysis
   a. Screen options (technical, economic, and environmental).
   b. Prioritize and select best options.

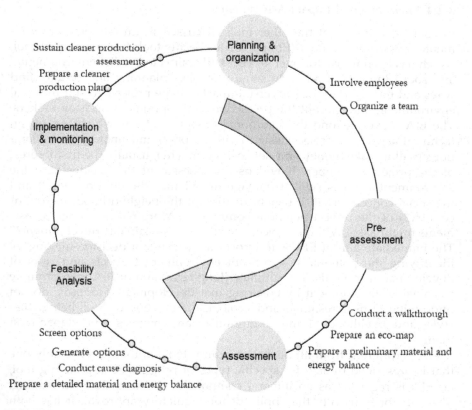

FIGURE 3.3
How to Conduct Cleaner Production Assessments?

Source: www.unep.fr/shared/publications/other/WEBx0072xPA/manual_cdrom/Guidance%20Manual/PDF%20versions/Part4_Sec%202.pdf

4. Implementation
   a. Option implementation.
   b. Monitoring and evaluation.
   c. Sustain and continue.

An EMS provides the framework to ensure the incorporation of CP initiatives. As such an EMS and a CP strategy can be mutually reinforcing ways of improving environmental performance in companies. A useful document to learn on how to conduct CPA can be retrieved from: www.unep.fr/shared/publications/other/WEBx0072xPA/manual_cdrom/Guidance%20Manual/PDF%20versions/Part4_Sec%202.pdf

3.2.4 Environmental Impact Assessment

In contrast to many of the other tools discussed in this chapter, environmental impact assessment (EIA) is a site-specific tool. It is a common tool, which has been in use for over 40 years. EIA aims to predict environmental and social impacts at an early stage in project planning and design, find ways and means to reduce adverse impacts, shape projects to suit the local environment, and present the predictions and options to decision-makers. The EIA moves beyond the traditional feasibility study, which focuses on technical aspects of project design (will it work?) and financial considerations (will it make money and not cost too much?), usually the first interest of the project developer. It widens the assessment to consider both the environmental issues (will it damage or enhance the environment?) and the social concerns (will it benefit or disrupt the neighborhood, region, or country?) of the project neighbor community. More recently, such assessments might ask: "Will the project contribute to sustainable development?" The immediate aim of EIA is to inform the process of decision-making by identifying the potentially significant environmental effects and risks of development proposals. The ultimate (long-term) aim of EIA is to promote sustainable development by ensuring that development proposals do not undermine critical resource and ecological functions or well-being, lifestyle, and livelihood of the communities and peoples who depend on them.

EIA is different from all other procedural tools in that it is the only one that is prescribed by law for specific types of projects. As a planning tool, an EIA is regarded as an integral component of sound decision-making. The initiator is usually the applicant for regulatory approval. It has both information gathering and decision-making components, which provides the decision-maker (such as local planning authorities or ministries of environment) with an objective basis to either grant or deny approval for a proposed development. Following the conclusions of an EIA project planners and engineers can shape the project, so that its benefits can be achieved and sustained without causing significant impacts. The EIA process is shown in Figure 3.4, and its key elements are as follows:

- Scoping: Identify key issues and concerns of interested parties.
- Screening: Decide whether an in-depth EIA is required based on initial information collected.
- Identifying and evaluating alternatives: List alternative sites and techniques and the impacts of each.
- Mitigating measures dealing with uncertainly: Review proposed action to prevent or minimize the potential adverse effects of the project.
- Issuing environmental statements: Report the findings of the EIA.

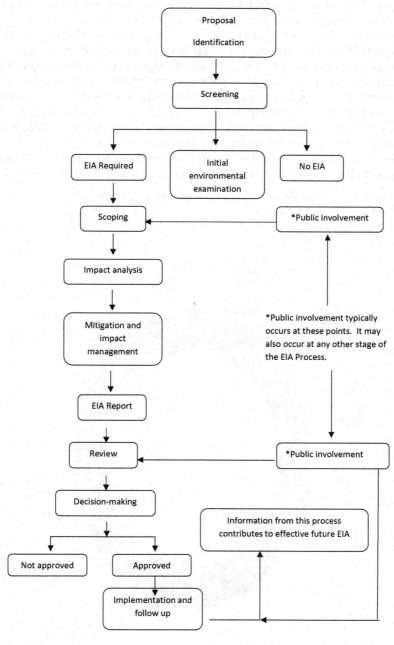

FIGURE 3.4

Generalized EIA Process Flowchart.

Source: UNEP (2002).

Figure 3.5 shows a generalized project cycle, presenting when and how an EIA can contribute positively to the cycle's progress.

An EIA should be integrated within the project cycle. The key lies in the management of the process by designing it so that it provides useful information to decision-makers at just the right time in the project cycle. In addition to an EIA, a social impact assessment (SIA) may also be carried out. Even when there is no specific requirement to consider social impacts, there may be a range of regulations that apply to social issues, such as employment conditions, ambient noise levels, protection of heritage sites, residential zoning requirements, and sanitation standards.

The *Environmental Impact Assessment Training Resource Manual* is intended to support capacity development in Environmental Impact Assessment (EIA):

UNEP. 2002. *Environmental Impact Assessment Training Resource Manual*, 2nd ed, UNEP, Geneva.

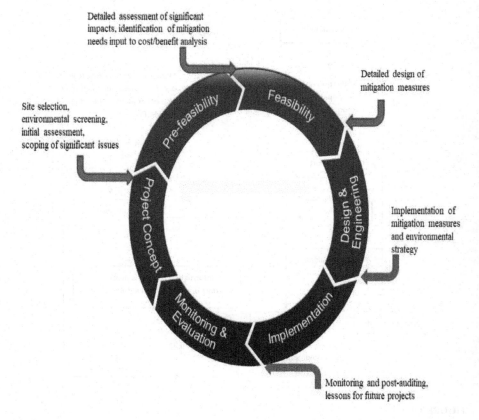

FIGURE 3.5
EIA in the Project Cycle.

Another suggested reading is Glasson, J., Therivel, R. and Chadwick, C. 2012. *Introduction to Environmental Impact Assessment*, Routledge, New York.

3.2.5 Strategic Environmental Assessment

Strategic environmental assessment (SEA) can be seen as an improvement of and a complement to the EIA. While EIAs are used at the project level looking at a physical object, SEAs are intended to provide the tool for influencing decision-making at an earlier stage when plans and programs – which give rise to individual projects – are being developed. They consider a greater scale and longer time interval than project EIAs. Impact predictions are subject to much greater uncertainty, but the time for data gathering is often longer and the degree of detail required is much less than that for project evaluations. SEAs take the impact assessment "upstream" into planning rather than outline design and broaden is scope in space and time. SEAs can be useful and effective for a number of applications where SEAs are not formally required. For example, SEA could be a useful tool for energy or waste management policies developed at national level. While first generation SEAs were conducted as extended EIAs, following the same logic and structure with the generation of a report rather than mainstreaming, second-generation SEAs are more process-oriented and are geared toward mainstreaming sustainability issues involving all relevant stakeholders as well as capacity development. The following report provides information and guidance on EIA and SEA good practice: UNEP. 2004. *Environmental Impact Assessment and Strategic Environmental Assessment: Towards an Integrated Approach.*

Another suggested reading is: Therivel, R. 2010. *Strategic Environmental Assessment in Action.* 2nd ed, Routledge, New York.

3.2.6 Life Cycle Management

Life cycle management (LCM) is in a nutshell about the application of life cycle thinking to a product management system, aiming to minimize environmental and socioeconomic impacts associated with the product during its entire life cycle and value chain. It implies that everyone in the whole chain of the product's life cycle has a responsibility and a role to play. LCM is not a single tool or methodology but a management system collecting, structuring, and disseminating product-related information from the various programs, concepts, and tools incorporating environmental, economic, and social aspects of products, across their life cycle. The organization must "go beyond its facility boundaries" and be willing to expand its scope of collaboration and communication to all stakeholders in its value chain (UNEP, 2007). LCM concerns all the departments within an organization, such as manufacturing, procurement, marketing, research and development (R&D), or environment, health, and safety (EHS) (Figure 3.6). It is often the company's department of environment or sustainability who initially suggests and coordinates the implementation of an LCM system. A useful guide to LCM: UNEP (2007). *Life Cycle Management. A Business Guide to Sustainability*. United Nations Environment Programme. Paris.

3.2.7 Design for Sustainability

It is only the designer who has the overview of the whole design and manufacture process. The greater demands and expectations placed on new products effectively require that designers must take a greater role in specifying and controlling the new product from its inception right through manufacture to packaging and even marketing. This is effectively design for sustainability (D4S). D4S is an umbrella term used to describe the techniques for incorporating sustainability considerations at the earliest

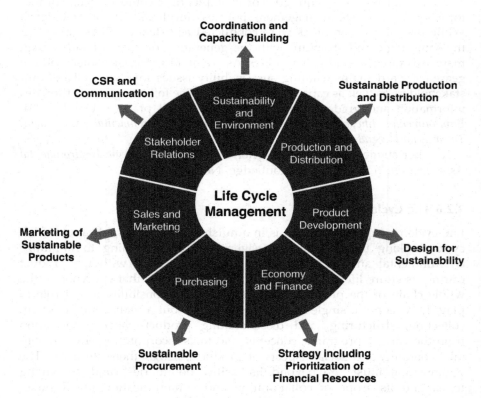

FIGURE 3.6
Where Does LCM Apply in the Organization?

Source: UNEP (2007).

possible stages of product development and design. D4S strategies may identify design measures that can be taken to reduce the sustainability impact in each of the following key phases of a product lifecycle:

- Raw materials: Designing in opportunities for resource conservation and low-impact raw material usage.
- Manufacturing: Providing for measures relating to cleaner production and eco-efficiency during the production phase.
- Product use: Ensuring provision in the product-use phase for considerations relating, for example, to improved energy and water efficiency, reduced material use, and increased durability.
- End-of-life: Key design considerations include design for disassembly, produce reuse, and design for recycling.

D4S approaches are categorized in different innovation levels. It includes the more limited concept of eco-design or design for the environment, which is at the product innovation level focusing on improving existing or developing completely new products. Guidelines and toolkits advocating Design for X or DfX (X standing for any of the "more preferable" attitudes in design from recycling to recyclability to ease of dismantling to reparability) have been developed. The life cycle approach of eco-design is usually supported by life cycle assessment methods. However, eco-design focuses solely on environmental performance and therefore disregards social dimensions of sustainability. The concept of D4S requires that the design process and resulting product take into account not only environmental concerns but also social and economic concerns as well. D4S goes beyond how to make a "green" product and embraces how to meet consumer needs in a more sustainable way. D4S is closely linked to wider concepts such as sustainable product–service systems and systems' innovations. The following two reports are a good guide for those looking to further their understanding of the concept:

UNEP. 2006. *Design for Sustainability: A Practical Approach for Developing Economies*, United Nations Environment Programme, Paris.
UNEP. 2009. *Design for Sustainability: A Step-by-Step Approach*, United Nations Environment Programme, Paris. A useful website is www.d4s-sbs.org

3.2.8 Sustainable Procurement

Procurement laws are generally based on the guiding principles of "best value for money" to ensure cost-efficiency through competition and "acting fairly" to ensure a level playing field for market participants. Sustainable procurement builds on the principles and good practices of "traditional" procurement and considers additional factors to maximize social, environmental, and economic benefits for the procuring organization, its supply chain and society as a whole.

Sustainable procurement is defined as

> a process whereby organizations meet their needs for goods, services, works and utilities in a way that achieves value for money on a whole life basis in terms of generating benefits not only to the organization, but also to society and the economy, whilst minimising damage to the environment.
>
> (UNEP, 2012)

It seeks to achieve the appropriate balance between the three pillars of sustainable development: economic, social, and environmental factors.

- Economic factors include the costs of products and services over their entire life cycle, such as acquisition, maintenance, operations, and end-of-life management costs (including waste disposal) in line with good financial management;
- Social factors include social justice and equity; safety and security; human rights and employment conditions;
- Environmental factors include emissions to air, land, and water; climate change; biodiversity; natural resource use and water scarcity over the whole product life cycle.

UNEP has a set of guidelines on sustainable procurement in the following publication: UNEP (2012). *Sustainable Public Procurement Implementation Guidelines*, United Nations Environment Programme, Paris.

When purchasing goods and services, the most inexpensive offers are often not the most economic ones. Inexpensive products might cause higher follow-up costs compared to more expensive alternatives. Such higher costs might arise, for example, from the consumption of auxiliary materials or energy during the usage phase, the installation and mainte-nance costs and the costs at the end of the useful life (in particular, costs for picking up, disposal, and recycling). This includes the costs arising from external effects of the environmental pollution, which are associated with the advertised capacity during the life cycle. Life cycle costing includes these factors in the calculation of the actual costs for a product. For that reason, the method can also be used to promote environmentally friendly products and thus contribute to sustainability.

3.2.9 Stakeholder Engagement

A stakeholder is any individual or group who is affected by, or can influence, the activities of another group. For a company, this typically extends to employees and trade unions, shareholders, customers, suppli-ers, business peers, communities, regulators, NGOs, and the media – individuals who have a legitimate interest in the activities of the company and to whom the company owes an account of its conduct. Stakeholder engagement is the process used by an organization to engage relevant stakeholders for a purpose to achieve accepted outcomes. It refers to the process of interaction between an organization and its stakeholders, beyond the one-way communication of data. Such engagement may be undertaken in order to gather information and ideas, build and strengthen relationships and trust, improve decision-making, and enhance the company's reputation. Important vehicles for stakeholders engagement and dialogue are sustainability reports, company websites (inter/intranet), company newsletters, product information (e.g., product declarations and eco-labels), focus groups, workshops, community panels, consumer surveys, and annual consultative stakeholder dialogue and meetings. Effective stakeholder engagement is valuable in fostering trust and developing social capital. Engineering projects should identify and effectively engage with stakeholders throughout the project cycle to ensure public support. Stakeholder engagement should be pursued through a clearly defined stakeholder engagement plan that includes provisions for soliciting stakeholder feedback and grievances.

A guide to stakeholder engagement is available on www.projectengineer.net/guide-to-stakeholder-engagement/

3.2.10 Corporate Social Responsibility and Sustainability Reporting

The treatment of stakeholders in an ethically and socially responsible manner is at the core of Corporate Social Responsibility (CSR). It is also now emphasized that it is not just about what a company does with its profits, but how it earns them too. Corporate responsibility goes beyond philanthropy and compliance to address the manner in which companies manage their economic, social, and environmental impacts, and their stakeholder relationships in all their spheres of influence: the workplace, the marketplace, the supply chain, the community, and the public policy realm. As a support to CSR, nonfinancial reporting on sustainability has expanded over the last 20 years. The Global Reporting Initiative (GRI) is a network-based organization that aims to mainstream a firm's disclosure on environmental, social, and governance performance. The GRI Sustainability Reporting Guidelines include a detailed set of criteria relating to a company's economic, social, and environmental performance. G4 is the latest generation of these guidelines. The guidelines consist of principles for defining report content and ensuring the quality of the information reported. The guidelines also include Standard Disclosures consisting of Performance Indicators and other disclosure items. Further information is available on their website:

www.globalreporting.org/information/sustainability-reporting/Pages/default.aspx

Many companies report against a set of core indicators (including those contained in the GRI guidelines) as well as against issues of concern identified by external stakeholders. Chapter 7 further elaborates on the links between business and sustainability.

3.2.11 Eco-Labeling and Environment Product Declaration

An eco-label is a market-driven environmental policy instrument with the aim of promoting environmentally preferable goods and services through consumers' purchasing power. The label affirms that the product or service complies with certain predetermined environmental – and sometimes also social – criteria. The International Organization for Standardization (ISO) has classified eco-labels into three typologies – Types I, II, and III – and has specified the preferential principles and procedures for each one of them:

Type I – Eco-labels (ISO 14024:1999) are designed to enable consumers to easily identify products that meet specific environmental or health standards. The criteria required be fulfilled by a particular product type are defined in advance by an independent issuing organization. Manufacturers may then apply to be issued the label. This requires submission of documentation that substantiates the manufacturer's compliance with the criteria that has been established. Depending on the program, evaluations may also include site inspections, sampling, and laboratory testing. If all issuance criteria have been met, the manufacturer will be permitted to use the label on the product that has been evaluated. This group of labels is most useful for procurement managers and also directly for the consumer.

Type II – Self-declared environmental claims (ISO 14021:1999). The labels belonging to this group do not share some of the usual characteristics of environmental labels, the main difference being that they are not awarded by an independent authority. These labels are developed internally by companies, and they can take the form of a declaration, a logo, a commercial, and so on, referring to one of the company products.

Type III – Environmental impact labels (ISO 14025:2006). Type III labels consist in qualified product information based on life cycle impacts. Environmental parameters are fixed by a qualified third party, then companies compile environmental information into the reporting format, and these data are independently verified. The environmental impacts are expressed in a way that makes it easy to compare different products and sets of parameters. Type III labels only shows the objective data, and their evaluation is left to the buyer. This group of labels is particularly used by procurement managers.

Two useful websites on an introduction to eco-labels are as follows:

1. https://globalecolabelling.net/assets/Uploads/intro-to-ecolabelling.pdf
2. www.ecospecifier.com.au/knowledge-green/technical-guides/technical-guide-9-introduction-to-ecolabels-and-environmental-product-declarations/

3.2.12 Buildings and Infrastructure Rating Systems

The shift to a sustainable built environment is being propelled by the growing evidence of accelerated destruction of our planetary ecosystems and the increasing demand for natural resources due to population growth and increased consumption. There is the need to assess performance of the built environment from a broader perspective, taking into account also the environmental, social, and economic impacts of constructions. This has prompted the creation of green building and infrastructure standards, certifications, and rating systems aimed at mitigating the impacts through the measurement and recognition of sustainability performance. Rating tools are design checklists and credit rating calculators developed to assist designers in identifying design criteria and documenting proposed design performance. A common objective of these rating systems is that projects awarded or certified within these programs are designed to reduce the overall impact of the built environment on human health and the natural environment. The use of sustainability rating tools in the built environment began in the 1990s and 2000s with the introduction of rating systems for buildings. BREEAM (Building Research Establishment Environmental Assessment Method) was launched in the United Kingdom in 1990, LEED (Leadership in Energy and Environmental Design) in the United States in 2000, and Green Star (Australia and New Zealand) in 2003 and 2007, respectively. The need to look beyond individual buildings or facilities and consider networks and neighborhoods has resulted in the introduction of neighborhood and community tools, for example, BREEAM Communities LEED-ND. While a greater emphasis of these sustainability rating tools was on the buildings, there was a lack of similar tools for the "horizontal" infrastructure beyond buildings. Recently, sustainable infrastructure rating systems have been developed with three of them (Envision in the United States, CEEQUAL in the United Kingdom, and the Infrastructure Sustainability (IS) Rating scheme in Australia being able to evaluate all types and sizes of civil infrastructures, including ports, airports, highways, dams, bridges, water, and wastewater treatment facilities, tunnels, and railways).

The essential purpose of most sustainability rating systems is a tool to encourage sustainability practices beyond the regulatory minimum and to communicate sustainability in a comprehensible manner. Building and infrastructure rating systems promote the development and diffusion of sustainable construction practices. Their use encourages building owners and the construction industry to strive for higher levels of sustainability and in turn elevate the ambition of government building codes and regulation, workforce training, and corporate strategies. Chapter 9 further details these rating systems, which can help engineers for more sustainable designs.

A recommended reading on the subject is Kubba, S. 2012. *Handbook of Green Building Design and Construction: LEED, BREEAM and Green Globes*, Butterworth-Heinemann, London.

3.2.13 Backcasting

Backcasting provides a way of connecting the future to the present (Weaver et al., 2000). Forecasting techniques, which are widely applied in strategic thinking, largely extrapolate current trends out into the future, often as a set of scenarios identifying potential future outcomes. In contrast, backcasting works "backward" from a preferred future state, allowing exploration of strategic steps forward to meet it from the current situation (Figure 3.7). Backcasting is typically well suited for sustainability issues as it works through envisioning and analyzing sustainable futures and then supporting strategic planning toward that preferred future. This can happen through "leapfrogging" rather than by incremental improvements from the current situation. For example, through forecasting techniques, a company may decide to invest in energy efficiency as a priority, whereas a backcasting approach, after recognizing that the current process

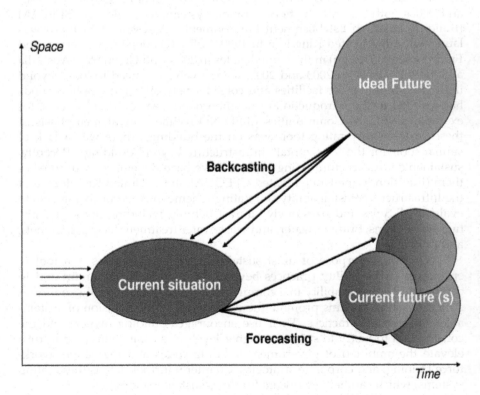

FIGURE 3.7
Difference between Backcasting and Forecasting.

Source: Dostal (2004).

is energy intensive and may not be sustainable in the long term, will encourage managers to look to alternative solutions, identifying novel products and processes rather than continue to invest into a nonsustainable business strategy.

The steps in a backcasting process involve analyzing needs, identifying options for improvement, creating a common future vision with stakeholders, developing pathways that could lead to this vision, and developing consensus on these pathways (Mulder, 2006). The main advantage of the backcasting technique is that it helps focus on a development trajectory that avoids merely extrapolating forward from within present constraints. The key to successful backcasting is that future scenarios that underpin it are desired outcomes, which have emerged from a collective envisioning process carried out by a diverse range of community and professional stakeholders. A good example of the application of the backcasting technique is given by Svenfelt et al. (2011), who applied the technique to examine how to decrease energy use in buildings by 50% by 2050.

A good guide to the backcasting tool is Dreborg, K.H. 1996. Essence of backcasting. *Futures*, 28(9), 813–828.

3.3 Analytical Tools

3.3.1 Based on Physical Metrics

3.3.1.1 Life Cycle Assessment

The most established and well-developed analytical tool for sustainable engineering is LCA. LCA is based on a "cradle-to-grave" approach to assessing the environmental impacts of a product or service throughout its life cycle. This is achieved by considering the primary resources consumed and the emissions and wastes generated during the extraction and processing of raw materials, manufacturing, transportation and distribution, use, reuse and maintenance, recycling, and final disposal. Figure 3.8 gives an example of the different stages considered in an LCA.

The LCA methodology is standardized by a series of ISO standards (ISO 14040: 2006). An LCA can be a powerful tool for effective engineering decision-making when comparing the relative environmental merits of two or more products or service categories. LCA activities often form an important component of product eco-labels and design for environment activities, as well as in identifying effective cleaner production possibilities. Given the importance and usefulness of the tool for the engineer, the next chapter details the methodology.

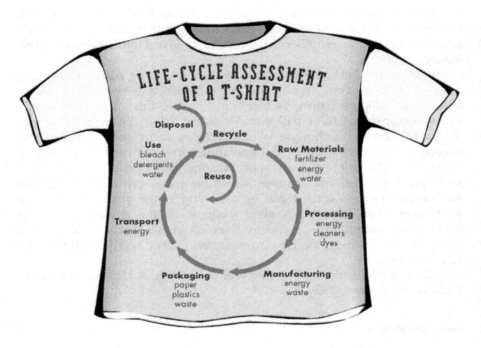

FIGURE 3.8
Stages in the Life Cycle Assessment of a T-Shirt.

Reprinted with permission from Worldwatch Institute (2003).

3.3.1.2 Carbon, Water, and Product Environmental Footprints

A Product Carbon Footprint (PCF) is a means for measuring, managing, and communicating greenhouse gas (GHG) emissions related to goods and services. It is based on an LCA but focuses on a single issue: global warming. In response to the need for transparency in the GHG emissions of products, several standards have been developed. There are three main PCF standards that are applied worldwide: PAS 2050, GHG Protocol, and ISO 14067. All three standards provide requirements and guidelines on the decisions to be made when conducting a carbon footprint study and they all build on existing LCA methods established through ISO 14040 and ISO 14044. ISO 14067:2018, *Greenhouse gases – Carbon footprint of products – Requirements and guidelines for quantification* is the most recent published international standard, providing globally agreed principles, requirements, and guidelines for the quantification and reporting of the carbon footprint of a product.

Carbon emissions are a global issue; emissions of carbon (and carbon equivalents) all have the same impact on the atmosphere, no matter where they are emitted. Although this creates great complexity in policies, it is relatively simple to calculate the effect of these emissions with LCA. Although a water footprint is similar, measuring water footprinting requires a different approach than carbon and it can be more complicated to calculate. Water is not an emission, but a resource, so both supply and regionality must be considered. While there is basically one accepted method for calculating a carbon footprint (IPCC GWP 2013), there are several methods for calculating a water footprint. ISO 14046: Water footprint: Principles, requirements, and Guidelines published in August 2014 is an international standard that defines the principles, requirements, and guidelines for conducting an LCA-based water footprint for products, processes, or organizations.

A product environmental footprint (PEF) is a methodology by the European Commission's Joint Research Center (JRC; https://ec.europa. eu/jrc/en), which is based on LCA. Its goal is to provide "a common way of measuring environmental performance" for companies within the EU wishing to market their product. Like a regular LCA, a PEF study measures all quantifiable environmental impacts over the life cycle of a product, including emissions to water, air, and soil; resource use; and depletion, and impacts from land and water use. However, it has more stringent rules than a regular LCA and may have product category-specific rules, all determined by the European Commission. Therefore, PEF studies are more robust, and can be compared if they consider the same product group.

Some useful websites and manuals on these footprinting tools are as follows:

www.carbontrust.com/resources/guides/carbon-footprinting-and-reporting/carbon-footprinting/

https://waterfootprint.org/media/downloads/TheWaterFootprintAssessmentManual_2.pdf

https://ec.europa.eu/environment/eussd/pdf/footprint/PEFmethodologyfinaldraft.pdf

Franchetti, M.J. and Apul, D. 2012. *Carbon Footprint Analysis: Concepts, Methods, Implementation and Case Studies*, CRC Press, Boca Raton, FL.

3.3.1.3 Material Flow Accounting

Material Flow Accounting (MFA) is a systematic assessment of the flows and stocks of materials within a system defined in space and time. It connects the sources, the pathways, and the intermediate and final sinks of a material. A simple material balance comparing all inputs, stocks, and outputs of a process can control the results of an MFA. It is this distinct characteristic of MFA that makes the method attractive as a decision-support tool in sustainable engineering. MFA is a family of different methods with distinct focus (Bringenzu et al. 1997). Two types are briefly mentioned here, which have applications in engineering practice: material intensity per unit service (MIPS) and substance flow analysis (SFA) also sometimes referred to as material flow analysis.

The MIPS tool uses a resource indicator to measure the environmental performance of a cradle-to-grave business activity. It is defined for the final good providing the service and calculations are made per unit of delivered "service" or function in the product during its entire life cycle (manufacturing, transport, package, use, reuse, recycling, new manufacturing from recycled material, and final disposal as waste). The whole material and energy input are indicated in kilogram or ton. It should be remembered that the more material is used in a production process, the more likely it has to have a bigger environmental impact. Closely linked to MIPS is the concept of the ecological rucksack. The latter is the total quantity (in kg) of materials moved from nature to create a product or service, minus the actual weight of the product. That is, ecological rucksacks look at hidden material flows. Extracted materials could be said to carry a rucksack or extraction burden. The rucksack factor is the quantity (in kg) of materials moved from nature to create 1 kg of the resource. For example, the rucksack factor for aluminum is 85:1 (85 kg materials moved for every 1 kg of aluminum obtained), for recycled aluminum it is 3.5:1, and for diamond, the rucksack factor is 53,000,000:1. One kilogram of aluminum from bauxite requires displacing 85 kg of materials compared to moving only 4 kg to produce 1 kg of recycled aluminum.

SFA is a type of MFA that focuses on tracing the flow of a selected chemical (or group of substances) through a defined system. SFA can be defined as a detailed level application of the basic MFA concept tracing the flow of selected chemical substances or compounds – for example, heavy metals (mercury (Hg), lead (Pb), etc.), nitrogen (N), phosphorous (P), persistent organic substances, such as polychlorinated biphenyls (PCBs) – through society. An SFA identifies these entry points and quantifies how much of and where the selected substance is released. Its general aim is to identify the most effective intervention points for policies of pollution prevention.

Two useful publications in this area are as follows:

Brunner, P.H. and Rechberger, H. 2004. *Practical Handbook of Material Flow Analysis.* CRC Press LLC, Boca Raton, FL.

Ritthoff, M., Rohn, H. and Liedtke, C. 2003. *Calculating MIPS – Resource Productivity of Products and Services.* Wuppertal Institute.

3.3.1.4 Environmental Risk Assessment

Environmental Risk Assessment (ERA) is a structured process for describing a hazard, identifying the potential for exposure to the hazard, estimating the risk or likelihood of a negative effect based on the hazard and exposures and considering uncertainties associated with the hazard. An ERA aims to answer three questions:

- What can go wrong?
- What is the likelihood and severity of any adverse occurrence?
- What can be done to manage any significant adverse occurrence, and who should be involved?

The risk assessment process provides the data and information for the risk management process. Risk management uses the results or risk assessment and takes into account economic, social, and political factors in the decision-making process. ERA typically consists of human health risk assessment and ecological risk assessment. ERA is conducted through several interactive steps. After having decided on the purpose and scope of the risk assessment, the process usually begins by collecting measurements that characterize the nature and extent of chemical contamination in the environment, as well as information needed to predict how the contaminants behave in the future. The risk assessor uses this information to evaluate the frequency and magnitude of human and ecological exposures that may occur as a consequence of contact with the contaminated medium, both now and in the future. This evaluation of exposure is then combined with information on the inherent toxicity of the chemical (that is, the expected response to a given level of exposure) to predict the probability, nature, and magnitude of the adverse health effects that may occur. Given the uncertainty on data, fate, and transport processes, the magnitude and frequency of human and ecological exposure, and the inherent toxicity of all of the chemicals all risk estimates are uncertain to some degree. For this reason, a key part of all good risk assessments is a fair and open presentation of the uncertainties in the calculations and a characterization of how reliable (or how unreliable) the resulting risk estimates really are. Risk managers then use this information to help them decide how to protect humans and the environment from contaminants. As stated earlier, a risk assessment is an iterative process, which involves researchers identifying and filling data gaps in order to develop a more refined assessment of the risk. Since risk is closely related to uncertainty, risk analysis cannot be separated from uncertainty analysis (Rotmans, 1998).

Two useful websites from the United States Environmental Protection Agency on ERA are as follows:

www.epa.gov/risk/conducting-human-health-risk-assessment
www.epa.gov/risk/conducting-ecological-risk-assessment

3.3.1.5 Green Infrastructure Valuation

Green infrastructure (GI) is the network of natural environmental components and green and blue spaces that lie in and around our nation's towns and cities. It provides the raw materials for our economy but more importantly it is our life support system, regulating the quality of our air, water, and soils. Similarly to other infrastructure, GI underpins our health and well-being. The valuation of GI can help to inform the project decision-making and design processes in engineering. It is important to value those things that are relevant both to an individual project and to the wider society affected by a particular project. Most benefits from GI are a combination of economic, social, and environmental values, and are often collectively termed as ecosystem services. For example, we can calculate the costs/benefits associated with a project by the following methods (source: http://whygreeneconomy.org):

- Climate change adaptation and mitigation, for example, carbon capture and storage, avoided carbon emissions and reduced energy consumption.
- Water management and flood alleviation, for example, reduced wastewater treatment costs, reduced energy, and carbon emissions.
- Areas of natural beauty, for example, willingness of people to pay for a view of natural beauty.
- Health and well-being, for example, health cost savings from reducing air pollution and green spaces for physical exercise.
- Land and property values, for example, impact of the environment on property prices.
- Tourism, for example, number of people visiting nature spaces.
- Recreation, for example, use of different natural environments by the local population.
- Biodiversity, for example, willingness to pay for protection/enhancement of biodiversity.
- Land management, for example, users calculate the annual estimate of the value of crops, timber, and other saleable produce at market prices.

The GI valuation toolkit provides a set of calculator tools to assess the value of green asset or a proposed green investment. Where possible, the benefits of GI are given an economic value. It is available at https://ecosystemsknowledge.net/green-infrastructure-valuation-toolkit-gi-val (accessed on 20 July 2019). Another useful resource is:

Green Infrastructure Valuation Network. 2010. *Building Natural Value for Sustainable Economic Development: The Green Infrastructure Valuation Toolkit User Guide*. Green Infrastructure Valuation Network.

3.3.2 Based on Nonphysical Metrics

3.3.2.1 Stakeholder Analysis

Stakeholders are all those who need to be considered in achieving project goals and whose participation and support are crucial to its success.

Stakeholder analysis is a way to identify a project's key stakeholders, assess their interests and needs, and clarify how these may affect the project's viability. Project managers can use the analysis to plan how to address these social and institutional aspects. The analysis can also contribute to project design by identifying the goals and roles of different stakeholder groups, and by helping to formulate appropriate forms of engagement with these groups. Stakeholder analysis can be usefully implemented at any stage of the project cycle, but its use should always be considered at the outset of a project. Participation of interested stakeholders is an area where many engineers will be called on to manage and participate with specialists who work in the social assessment area. Stakeholder analysis is an entry point to SIA as part of an EIA and participatory work. It addresses strategic questions such as who are the key stakeholders? What are their interests in the project? What are the power differentials between them? What relative influence do they have on the operation? This information helps to identify institutions and relationships which, if ignored, can have negative influence on proposals, or if considered can be built upon to strengthen them.

A useful guide on how to perform a stakeholder analysis is available at: www.projectengineer.net/how-to-perform-a-stakeholder-analysis/

3.3.2.2 Cost Benefit Analysis

Cost benefit analysis (CBA) is an economic tool used to determine whether or not the benefits of an investment or a policy outweigh its costs (Wrisberg et al., 2002). It aims at expressing all positive and negative effects of an activity in monetary units. It often includes those environmental and social costs and benefits that can be reasonably quantified. CBA is mainly applied to determine if an investment (or decision) is sound, ascertaining if – and by how much – its benefits outweigh its costs and to provide a basis for comparing investments (or decisions), comparing the total expected cost of each option with its total expected benefits. Benefits and costs in CBA are expressed in monetary terms and are adjusted for the time value of money; all flows of benefits and costs over time are expressed on a common basis in terms of their net present value, regardless of whether they are incurred at different time. As a technique, it is used most often at the start of a program or project when different options or courses of action are being appraised and compared, as an option for choosing the best approach.

A step-by-step guide on CBA is available at: www.projectmanager.com/blog/cost-benefit-analysis-for-projects-a-step-by-step-guide

3.3.2.3 Environmental Life Cycle Costing (E-LCC)

LCC is an economic approach that sums up "total costs of a product, process or activity discounted over its lifetime" (Gluch and Baumann, 2004). LCC summarizes all costs associated within the life cycle of a product that are directly covered by one, or more, of the actors in the product life cycle (e.g., supplier, producer, user/consumer, end-of-life actor). In principle, LCC is not associated with environmental costs, but costs in general. There are two different types of LCCs: conventional LCC and environmental LCC. The conventional LCC is, to a large extent, the historic and current practice in many governments and firms, and is based on a purely economic evaluation, considering various costs associated with a product which is born directly by a given actor. External costs are often neglected. Environmental LCC (E-LCC) summarizes all costs associated within the life cycle of a product that are directly covered by one, or more, of the actors involved in the products life cycle, including externalities that are anticipated to be internalized in the decision-relevant future. These costs must relate to real money flows. The environmental LCC is not a stand-alone technique but is seen as a complementary analysis to the environmental LCA. In environmental LCC, it is obligatory to include all life cycle stages, while the conventional LCC often does not take into consideration the end-of-life costs (Hunkeler et al., 2008). More information on the tool is given in the next chapter. A useful guide on the topic is

Hunkeler, D., Lichtenvort, K., Rebitzer, G., Ciroth, A., Huppers, G., Klopffer, W., Rudenauer, I., Steen, B. and Swarr, T. (2008) Environmental Life Cycle Costing, SETAC: New York, NY.

3.3.2.4 Social Life Cycle Assessment

A social life cycle assessment (SLCA) is a method that can be used to assess the social and sociological aspects of products, their actual and potential positive as well as negative impacts along the life cycle. SLCA makes use of generic and site-specific data, can be quantitative, semi-quantitative, and complements the environmental LCA and LCC. It can either be applied on its own or in combination with the other techniques. SLCA does not provide information on the question of whether a product should be produced or not – although information obtained from an SLCA may offer "food for thought" and can be helpful for taking a decision.

Although SLCA follows the ISO 14040 framework, some aspects differ, some are more common or some are amplified at each phase of the study. The UNEP Guidelines for Social Life Cycle Assessment of Products proposes a methodology to develop life cycle inventories. A life cycle inventory is elaborated for indicators (e.g., number of jobs created) linked to impact categories (e.g., local employment), which are related to five main stakeholder groups: (i) worker, (ii) consumer, (iii) local community, (iv) society, and (v) chain actors.

3.3.2.5 Life Cycle Sustainability Assessment

Life cycle sustainability assessment (LCSA) refers to the evaluation of all environmental, social, and economic negative impacts and benefits in decision-making processes toward more sustainable products throughout their life cycle.

$$LCSA = LCA + ELCC + SLCA$$

LCSA enables practitioners to organize complex environmental, economic, and social information and data in a structured form. LCSA helps in clarifying the trade-offs between the three sustainability pillars, life cycle stages and impacts, products and generations by providing a more comprehensive picture of the positive and negative impacts along the product life cycle. This tool is further elaborated in the next chapter. Useful guides on this tool that will gain importance in the years to come are as follows:

Finkbeiner, M., Schau, E., Lehmann, A. and Traverso, M. 2010. Towards Life Cycle Sustainability Assessment. *Sustainability* 2, 3309–3322.

Klöpffer, W. and Renner, I. 2008. Life-Cycle Based Sustainability Assessment of Products.

UNEP. 2011. *Towards a Life Cycle Sustainability Assessment; United Nations Environment Programme*, United Nations Environment Programme, Nairobi, Kenya.

3.3.2.6 Multi-criteria Analysis

For integration of sustainability in decision-making, engineers must adopt a wider range of criteria when judging the performance and outcomes of a project. These criteria should reflect a range of economic, social, and environmental considerations. To enable the handling of all this information with incommensurate units effectively, multi-criteria analysis (MCA) is a structured tool for ranking alternatives and making selections and decisions.

In an MCA, we consider how great an effect is (score) and how important it is (weight). A general outline of the MCA method is shown in Figure 3.9. MCA describes a system of assigning scores to individual effects (e.g., impact on traffic, human health risk reduction, and use of energy). These can then be combined into overall aggregates on the basis of the perceived importance (weighting) of each score, as illustrated in the figure. With MCA, ranking and decision-making processes can be made very transparent. This tool can be used to evaluate the relative importance of all criteria involved and reflect their importance in the final decision-making process. This approach has the advantage of incorporating both qualitative and quantitative data into the process (Wrisberg et al., 2002). Techniques are either qualitative or quantitative, depending on the information input they are able to handle. Qualitative techniques include frequency analysis or the analytical hierarchy process, whereas quantitative techniques include weighted summation and concordance discordance analysis. The analytic hierarchy process is an MCA technique quite often used to solve complex decision-making problems in engineering.

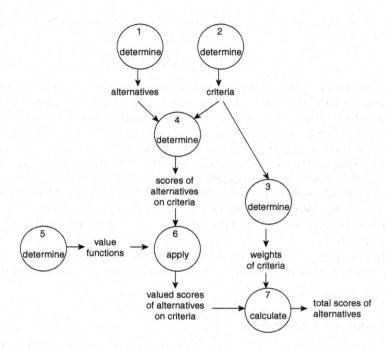

FIGURE 3.9
A General Outline of Multi-Criteria Analysis.

A suggested reading on the tool is
Linkov, I. and Moberg, E. (2011). *Multi-Criteria Decision Analysis: Environmental Applications and Case Studies.* CRC Press, Boca Raton, FL

3.3.2.7 System Dynamics

The field of systems thinking and the development of systems dynamics modeling were created to address the need to understand complex systems and their behavior over time and to enable decision-making. "System dynamics is designed to model the behaviour of constantly changing systems" (Forrester 1991). It deals with internal feedback loops and time delays that affect the behavior of the entire system. The methodology provides a framework for analyzing how actions and reactions cause and influence each other, and how and why elements and processes in the system change. In this way, it allows interested parties to understand how the system works and to predict how situations might develop over time (Forrester 1991). It is a method that permits the analyst to decompose a complex system into its constituent components and then integrate them into a whole that can be easily visualized and simulated. System dynamics modeling is an ideal tool for examining complex systems characterized by feedbacks and delayed effects, characteristics that underlie so many sustainability issues. It is an excellent tool to study problems that arise in closed-loop systems. Figure 3.10 shows an example of a causal loop diagram, which captures the interdependencies and feedbacks between various sub-systems of the water system in Singapore. A system dynamics model can then analyze the long-term impacts of various investment plans.

Recommended readings on this tool are as follows:

Hjorth, P. and Bagheri, A. (2006). Navigating towards Sustainable Development: A System Dynamics Approach. *Future*, 38(1), 74–92.

Lobontiu, N. 2018. *System Dynamics for Engineering Students. Concepts and Applications*, 2nd ed., Academic Press, Oxford, UK.

3.4 Summary

SA tools are operational methods supporting the concepts and principles for sustainable engineering. They contribute to a better understanding of

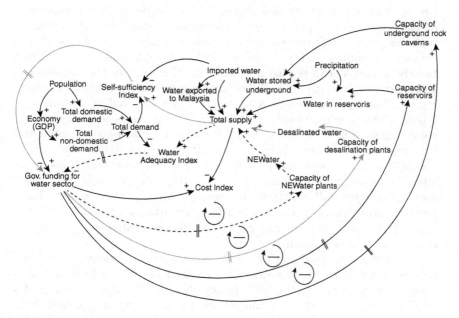

FIGURE 3.10
Causal Loop Diagram for the Water Systems in Singapore.

Reprinted with permission from Xi and Poh (2013)

the meaning of sustainability and its contextual interpretation, help to measure the level of sustainability, and integrate sustainability issues into decision-making. This chapter described the main tools that are specifically aimed at measuring sustainability that feed the decision-making process in the engineering problem-solving methodology. These tools are operational methods supporting sustainable engineering concepts and principles. Analytical tools provide technical information as to the consequences of a choice, while procedural tools focus on the procedures to guide the way to reach a decision. A further distinction between analytical tools is made based on physical metrics and nonphysical metrics. Analytical tools can be used within the framework of procedural tools.

SUSTAINABLE ENGINEERING IN FOCUS: APPLYING CIRCULAR ECONOMY PRACTICES TO OPTIMIZE ENVIRONMENTAL PERFORMANCE – A CASE STUDY OF JG AFRIKA'S OPERATIONAL MATERIALS MANAGEMENT PLAN FOR HOTEL VERDE, SOUTH AFRICA

Hotel Verde in Cape Town, South Africa, is the first hotel in Africa to offer carbon-neutral accommodations and conferencing facilities. Known as "Africa's Greenest Hotel," it has implemented a comprehensive range of interventions, including energy-saving heating systems, gray water recycling measures, and green building certifications to rubber-stamp it as a world-class sustainable tourism facility. Hotel Verde is also a pioneer in offering the Hotel Carbon Management Initiative (HCMI), carbon-offsetting program in the region. Some of the design features and systems that contribute to Hotel Verde's eco-friendly status are its power-generating gym, solar panels, and wind turbines, an intelligent building management system, efficient and intelligent heating, ventilation and air-conditioning systems, energy-efficient lighting, geothermal system, sub-soil drainage, and rainwater harvesting. Responsible procurement strategies fall squarely within JG Afrika's core belief that product and system redesign, waste avoidance, and minimization as well as resource efficiency are fundamental strategies in moving toward a circular economy.

JG Afrika, an engineering and environmental consultancy, in collaboration with Envirosense CC, assisted Hotel Verde to implement a "zero waste to landfill" policy through the development of an Operational Materials Management Plan, aiming to provide the Hotel control over the potential waste generated onsite through an informed procurement process. While there have been some changes in the Plan over the operational life of the hotel, the scorecard remains in place to assist with the selection of suppliers and products used onsite.

In 2013, Hotel Verde had gone the extra mile in its steadfast commitment to the principles of sustainability by applying responsible procurement as part of its hotel operations. The plan outlines its core procurement and operational strategies to evaluate products and select the best ones for the guests, community, planet, and hotelier, aiming for green procurement and waste elimination. It uses a scorecard approach as a guide in making responsible purchasing decisions and revises supplier behavior. The following four main targets are outlined in the scorecard:

(1) Better for the guest – evaluates the impact of supplier's operations, products, and services on guest health, focusing on efforts such as those that improve indoor air quality, help reduce allergens, and minimize product toxicity.

(2) Better for the community – assesses suppliers' economic, social, and environmental impacts as they relate to relationships in the workplace, the marketplace, the supply chain, the community, and public policy.

(3) Better for the planet – measures supplier's efforts to protect the environment and preserve resources through its operations, products, and services as reflected in its raw material use, carbon, energy, waste, and water footprints.

(4) Better for the hotelier – measures the positive contribution of suppliers' operations, products, and services to improving a hotel's sustainability, guest satisfaction, and business performance.

One of the successes of the plan was promoting better waste management among Hotel Verde's suppliers. Its milk and fruit juice provider supplied beverages in bulk returnable stainless steel containers. Previously, milk and juice had been delivered in 5 L, single-use plastic containers. The milk and juice are now decanted and provided in reusable glass bottles and jugs, eliminating carton and plastic bottle waste. Apart from the procurement system, Hotel Verde encourages guests to recycle and implement composting to reduce food waste.

Retrieved with permission from: One Planet Network. www.oneplanetnet work.org/

References

Bringezu, S., Fischer-Kowalski, M., Kleijn, R. and Palm, V. (Eds.). 1997. Analysis for Action: Support for Policy towards Sustainability by Material Flow Accounting. Proceedings of the ConAccount Conference, 11–12 Sept, 1997, Wuppertal, Germany, Wuppertal Institute, Wuppertal Special 6.

Dostal, E. 2004. *Biomatrix – A Systems Approach to organizational and Societal Change*, Sun Press, Stellenbosch.

Finnveden, G. and Moberg, A. (2005) Environmental Systems Analysis – An Overview. *Journal of Cleaner Production*, 13(12), 1165–1173.

Forrester, J.W. 1991. System Dynamics—Adding Structure and Relevance to Pre-College Education. In K.R. Manning (Ed.), *Shaping the Future*, pp. 118–131. MIT Press, Cambridge, MA.

Gluch, P. and Baumann, H. (2004) The Life Cycle Costing (LCC) Approach: A Conceptual Discussion of Its Usefulness for Environmental Decision-Making. *Building and Environment*, 39, 571–580.

ICC. 1989. *Environmental Auditing*, June 1989, ICC Publication No 468, International Chamber of Commerce (ICC), Paris.

ICC. 1991. *ICC Guide to Effective Environmental Auditing*, ICC Publication No 483, International Chamber of Commerce (ICC), Paris.

Mulder, K. 2006. *Sustainable Development for Engineers: A Handbook and Resource Guide*, Greenleaf Publishing, Sheffield.

Rotmans, J. 1998. Methods for IA: The Challenges and Opportunities Ahead. *Environmental Modeling & Assessment*, 3(3, Special issue: Challenges and Opportunities for Integrated Environmental Assessment, Rotmans, J. and Vellinga, P., eds.), 155–179.

Svenfelt, Å., Engström, R. and Svane, O. 2011. Decreasing Energy Use in Buildings by 50% by 2050 – A Backcasting Study Using Stakeholder Groups. *Technological Forecasting and Social Change*, 78, 785–796. Oxford. Elsevier.

UNEP. 2002. *Environmental Impact Assessment Training Resource Manual*, 2nd ed., UNEP, Geneva.

UNEP. 2007. *Life Cycle Management. A Business Guide to Sustainability*, United Nations Environment Programme, Paris.

UNEP. 2012. *Sustainable Public Procurement Implementation Guidelines*, United Nations Environment Programme, Paris.

United Nations Environmental Programme (UNEP)/International Chamber of Commerce (ICC)/International Federation of Consulting Engineers (FIDIC). (UNEP/ICC/FIDIC). 2001. *Environmental Management System Training Resource Kit*, 2nd ed., 334 p.

Waas, T., Hugé, J., Block, T., Wright, T., Benitez-Capistros, F. and Verbruggen, A. 2014. Sustainability Assessment and Indicators: Tools in a Decision-Making Strategy for Sustainable Development. *Sustainability*, 6, 5512–5534.

Weaver, P., Jansen, L., Van Grootveld, G., van Spiegel, E. and Vergragt, P. 2000. *Sustainable Technology Development*, Greenleaf Publishers, Sheffield.

WFEO. 2016. *World Federation of Engineering Organizations*, Code of Ethics. Available at: http://www.wfeo.org/ethics/ Accessed on 15 November 2019.

Worldwatch Institute. 2003. *Worldwatch Paper 166: Purchasing Power: Harnessing Institutional Procurement for People and the Planet*, July 2003. Available at: www.worldwatch.org.

Wrisberg, N., Udo de Haes, H.A., Triebswetter, U., Eder, P. and Clift, R. 2002. *Analytical Tools for Environmental Design and Management in a Systems Perspective*, Kluwer Academic Publishers, Dordrecht.

Xi, X. and Poh, K. 2013. Using System Dynamics for Sustainable Water Resources Management in Singapore. *Procedia Computer Science*, 16, 157–166.

Exercises

1. Describe the following sustainability assessment tools: Life Cycle Assessment (LCA), Environmental Risk Assessment (ERA), Material Flow Analysis (MFA), Environmental Impact Assessment (EIA), Cleaner Production Assessment (CPA), Environmental Management Systems (EMS), and Life Cycle Costing (LCC). Explain in which decision-making context each of these tools can be used.

2. The PLS Company (a fictitious name for a real company) prints rolls of thin plastic film with packaging labels and then glues the plastic film to a thin layer of metal foil. The final product is sold to food-processing firms for making food packaging. The production process produces a number of wastes, including solid scrap from start-up runs and full production runs, waste liquid ink, and solvent air emissions from printing ink and glues. PLS had the following ideas for managing the wastes:

   A. Chop up the scrap and sell it to another company for use as a stuffing material for bed mattresses

   B. Install an incinerator to burn the solid scrap

   C. Install a quality-control camera to catch printing errors earlier, thus reducing scrap generation

   D. Reuse waste ink by mixing it into the black ink

   E. Send the solid scrap to a landfill

   F. Re-use scrap film in printing start-up runs

   G. Change the employee compensation system to reward scrap reduction at each production unit

   H. Purchase a new distillation unit to recycle the waste solvents

   I. Use water-based glue instead of solvent-based glue, thus reducing solvent air emissions

Match the above options to the following environmental management hierarchy categories (You may choose to put an option into more than one of the four major categories. For any option you decide to consider cleaner production, place it into one or more of the sub-categories listed.)

Environmental Management Hierarchy Categories	Option
Cleaner Production, including the following:	
• Good housekeeping	
• Input substitution	
• Better process control	
• Equipment modification	
• Technology change	
• Product modification	
• Production of a useful by-product	
• On-site recycling/reuse	
Off-site Recycling/Reuse	
Pollution Control/Treatment	
Waste Disposal	

3. Conduct an Internet search to obtain a recent sustainability report for a major corporation following the GRI guidelines. Identify the performance indicators in the report.

4. Compact fluorescent light bulbs provide similar lighting characteristics to incandescent bulbs yet use just one-fourth of the energy. Assume that a 25-W fluorescent bulb provides the same illumination as a 100-W incandescent bulb. Calculate the mass of coal that would be required, over the 8,000-hour life of the fluorescent bulb, to generate the additional electricity required for an incandescent bulb. Assume transmission losses of 10%, 40% efficiency of electricity generation, and 30 MJ/kg for the heat of combustion of coal. Estimate the amount of carbon dioxide emissions avoided.

5. In evaluating the energy implications of the choice between reusable and single-use cups, the energy required to heat wash water is a key parameter. Consider a comparison of single-use polypropylene (PP) and reusable PP cups. The reusable cup has a mass roughly 14 times that of the single-use cup (45 g versus 3.2 g), which, in turn requires petroleum feedstocks.

   a. Calculate the number of times the reusable cup must be used in order to recoup the energy in the petroleum required to make the reusable cup.

   b. Assuming that the reusable cup is washed after each use in 0.27 L of water, and that the wash water is at 80°C (heated from 20°C), calculate the energy used in each wash if the water is heated in a gas water heater with an 80% efficiency. Calculate the number of times the reusable cup must be used in order to recoup both the energy required to make the reusable cup and the energy used to heat the wash water. Assume that 1.2 kg of petroleum is required to produce 1 kg of polypropylene, and that the energy of combustion of petroleum is 44 MJ/kg.

   c. Repeat Part b, assuming that an electric water heater is used (80% efficiency) and that electricity is generated from fuel at 33% efficiency.

C_p of water = 4.184 J/g K

6. Choose a common household appliance such as a refrigerator, TV set, an air conditioner, or a washing machine and describe its life stages, including who is primarily responsible for the environmental concerns at each stage.

7. Identify the specific sustainability assessment tools that have been used in this case study (Source: UN Global Compact Environmental Principles Training Package. www.unep.fr/shared/publications/other/dtix0601xpa/pdf/en/UNGC_DEL_ALL.pdf)

The "Re-Define" sofa and armchair, launched in Australia at the end of 2000, are the result of a demonstration project whose aim was to develop high-quality, "sustainable" furniture. The project, supported by a research grant from EcoRecycle Victoria in 1999, is a collaboration between three partners:

• Wharington International, an Australian company specializing in the manufacture of furniture frames and components;

• The Centre for Design at Royal Melbourne Institute of Technology (RMIT), a research group that specialises in business eco-efficiency as a source of innovation and responsible practice;

• MID Commercial Furniture, an Australian design practice headed by Danish architect Torben Wahl, which specialises in sustainable furniture design.

The aim of the Re-Define project was to develop a range of furniture whose environmental impacts were minimised across the entire life-cycle, including materials selection, manufacturing, distribution, use, re-use, recycling and disposal. Seven distinct areas (see below) were identified in which traditional furniture can cause damage to the environment. In the case of furniture, these impacts are almost exclusively related to the manufacture and disposal phases, rather than to the use of the product.

1. The release of toxic chemicals from glues, dyes, paints, and so on, during both the manufacture and the use of the product;
2. The production of timber waste during manufacture;
3. Greenhouse gas emissions arising from steel and aluminum production;
4. The consumption of rainforest timber and scarce hardwoods;
5. The use of timber from poorly-managed plantations, resulting in soil erosion, water pollution and habitat damage;
6. The creation of solid waste when the product is discarded;
7. The use of synthetic materials that constitute toxic or hazardous waste once discarded.

The design brief for Re-Define was drawn up by RMIT, including details of materials, manufacturing processes and resource recovery. The requirements set out in the brief included:

*   Minimise the quantity of material used;
*   Avoid toxic or hazardous substances;
*   Use metals with low "embodied energy";
*   Minimise the number of components and assemblies;
*   Replace glue and screws with simple "push, hook, and clip" assembly;
*   Avoid solvent-based adhesives;
*   Enable minor repairs to be carried out;
*   Avoid colors or designs that will go out of fashion quickly.

The final product incorporates plastic internal shells made from "Recopol" recycled resin, a material developed by Wharington International from the recycled casings of household appliances such as vacuum cleaners, telephones, computers, washing machines and refrigerators. Recopol replaces more traditional internal components that are manufactured from plywoods, hardwoods, plantation timber and virgin plastic. Each Recopol resin shell contains the equivalent of 45 recycled printer cases, or 39 kg of material that would otherwise have been incinerated or landfilled. At the end of its life, Wharington can even take back the shell for recycling.

Other features of the Re-Define range include:

*   Foam cushions are shaped so as to minimise scrap cuttings, with waste foam sent for use in carpet underlay material;
*   Stainless steel legs and mild steel bearers are 100% recyclable, and scrap from the manufacturing process is recycled;

- Metal components do not require any toxic coatings or finishings;
- Fabrics are made from recycled PET or natural wool;
- The nylon feet are designed to be recyclable;
- Fastenings are designed for easy removal and washing of the upholstery.

Forest products have been avoided, and Wharington claims that no toxic or hazardous materials are used at any stage of the manufacturing process. The product is designed to be easily maintained and repaired, in order to prolong its useful life. According to Wharington, manufacturing the Re-Define range carries no cost premium compared with conventional furniture. Only the design phase incurred extra expenditure. Although a Recopol shell is US$30-40 more expensive to produce than a plywood frame, it is substantially less labor intensive. In addition, says Wharington, a resin shell is cheaper to pad and upholster than a plywood frame.

According to the company, "Re-Define highlights that eco-design can produce sophisticated commercial furniture that meets the rigorous standards required in the corporate and government sectors."

8. Reading assignment: Engineering for Sustainable Development: Guiding Principles, Royal Academy of Engineering, Dodds, R., and Venables, R., Eds., 2005. Pages 12–24. (www.raeng.org.uk/publications/other/engineering-for-sustainable-development).

   Read the seven case studies. Choose one case study according to your field and identify which sustainability tools have been used. Identify the WFEO principles that guide engineering practice (WFEO, 2016) presented in Chapter 2, which have been used in this case study.

9. Taking transport as an example, explain "systems" in such a way that somebody unfamiliar with the concept would understand. How can system thinking and system dynamics modeling help in transportation?

10. The following exercise is adapted from the following case study: *Considering the Social Dimensions of Sustainability during the Construction Project Design* by Rodolfo Valdes Vasquez and Dr. Leidy Klotz.

(http://ecs.syr.edu/centers/sustainableengineering/modules/09-15_valdes-vasquez.pdf)

This is a group activity (three to four students) to consider social dimensions during the design of sustainable construction projects. The construction industry must support the sustainable development agenda by considering a holistic construction life cycle (acquisition of raw materials, planning, design, construction, operation, and deconstruction, disposal in landfills, reuse or recycling of materials). A central challenge of sustainable development is to integrate and manage various forms of disciplinary expertise in ways that lead to sustainable results. This comprehensive process aims to restore and maintain harmony between the natural and built environments, while creating communities that affirm human dignity and encourage economic equity. This integration improves not only the environment but also has an impact on the economic profitability and the relationships among stakeholder groups such as contractors, the production workforce, and operating managers. However, the construction industry frequently overlooks many of these social issues such as health, diversity, and workforce development. While these social dimensions require action during the construction phase, improved benefits are possible if they are also addressed during the planning and design phases. For instance, the design decision of large, open spaces for natural lighting, considered a sustainable feature, should also seek to minimize hazards to the workforce during the production (construction) and operating (maintenance) phases. Discuss the following dimensions of social sustainability and their impact on project stakeholders:

- Corporate Social Responsibility
- Community Involvement
- Social Design
- Safety Through Design

Annexes I to IV hereafter describes the above dimensions. Each student is required to explain one of the dimensions and will prepare to cover the following areas:

- Definition of the social dimension
- Impact of this dimension on project performance if it is not considered
- Its integration during the design phase of a construction project

Students are encouraged to research references on each topic and share them with the rest of the class.

Annex I: Corporate Social Responsibility

Sustainability reporting initiatives indicate the various views of social sustainability among companies. In 2008, the Global Reporting Initiative (GRI) presented a general overview of sustainability reporting in the Construction and Real Estate sector. This report, A Snapshot of Sustainability Reporting in the Construction and Real Estate Sector, was based on a review of sustainability reports from 16 global companies. The primary findings related to social aspects established that the aspects commonly reported were:

- Diversity by creating a more flexible working environment and increasing the diversity mix of the workforce.
- Equal employment opportunities.
- Health and safety education by minimizing risk of accidents through educational, programs in order for the employees and subcontractors to act responsibly.
- Community Involvement.

The study also reported that the main indicators reported for these 16 companies related to social aspects were fatal incident rate, accident rate, and percentage comparison of male and female employees. The Corporate Social Agenda is exemplified by Skanska. Its social agenda consists of four high-priority areas and their interactions with the workforce (employees and subcontractors), the marketplace, and the communities in which the company operates. The company has a diversity strategy designed to attract, recruit, develop, and retain the best possible female, ethnic, and non-engineer talent alongside the more traditional recruits. The Skanska safety and health agenda became its number one commitment to its work-force as seen in such statements as "[employees and sub-contractors] will return to their loved ones in the same condition that they left them." For example, each year Skanska organizes a Safety Week. In addition, 57% of Skanska Business Units operate in the International Health & Safety Management System OHSAS 18001.

The Skanska corporate community involvement agenda commits to acting as a responsible member of the global society and the communities in which it operates, including markets in Europe, the United States, and Latin America. Specifically, the company reports that it makes contributions to the local communities through charitable donations, sponsorships, volunteer work, and other initiatives focusing on community outreach activities in the three main areas of safety, education, and disaster relief. Its Code of Conduct, the umbrella for explicit expectations about ethical business behavior, was published in 2002 and was updated in 2008.

Reference

Skanska. 2008. Sustainable Development Proof of Progress. Available at: www.skanska.com/upload/About%20Skanska/Sustainability/Reporting/Sustainability_2008.pdf.

Annex II: Community Involvement

According to Solitare (2005), community involvement, also known as public participation or stakeholder engagement, occurs when "public constituencies attempt to influence governmental and private decisions." This social perspective of sustainability relates external stakeholder concerns to those decisions made by internal stakeholders such as owners and developers. Such methods for public participation in the U.S. include public hearings, reviews, and comment procedures in government decision-making. According to Innes and Booher (2004), one of the purposes of public participation is to promote equity and fairness, because individuals and groups excluded from the decision-making process are likely not to have their needs and preferences reflected in the outcome. In fact, those excluded may disproportionately rate the negative impacts of projects or policies, ignoring the positive. Thus, the key challenge for planning and designing sustainable projects is to facilitate a dialogue encouraging reflection on various impacts affecting this group of stakeholders in a framework where these can be openly discussed (Meppem and Bourke, 1999).

Since project stakeholders possess various values and, consequently, judge the characteristics of a project in different ways, these diverse expectations need to be identified and understood (Thomson et al., 2003). This expanded ownership is even more important in the context of sustainable built environments, where most of the gains for the end-users occur during the operating phase, requiring them to have significant buy-in to the solutions adopted in the design phase. Thus, the social choice of including end-users, the communities impacted by the project, and various public agencies has been argued as being crucial for implementing sustainable projects.

Numerous obstacles can impede meaningful participation. One of the biggest issues in participation is information, specifically who controls it and whether it is trustworthy (Hanna, 2000). While collaborative methods may seem costly because of the amount of time required to get the community involved, the cost of not using such methods can be even greater because the public can delay the plans. Over the last two decades, public agencies, such as EPA and the USDOT, have encouraged deliberative processes by providing resources and structures to assure inclusion of external stakeholders in the design of the project.

References

Hanna, K.S. 2000. The Paradox of Participation and the Hidden Role of Information. *Journal of the American Planning Association*, 66(4), 398–410.

Innes, J. and Booher, D. 2004. Reframing Public Participation: Strategies for the 21st Century. *Planning Theory and Practice*, 5(4), 419–436.

Meppem, T. and Bourke, S. 1999. Different Ways of Knowing: A Communicative Turn toward Sustainability. *Ecological Economics*, 30, 389–404.

Solitare, L. 2005. Prerequisite Conditions for Meaningful Participation in Brownfields Redevelopment. *Journal of Environmental Planning and Management*, 48(6), 917.

Thomson, D., Austin, S., Devine-Wright, H. and Mills, G.R. 2003. Managing Value and Quality in Design. *Building Research & Information*, 31(5), 334–345.

Annex III: Social Design

The social sustainability framework related to design includes on one side the assurance that not only are the designs energy-efficient and the infrastructures lasting but that it also includes to having a design that considers underrepresented groups, for example, accessibility for the elderly or people with a physical disability. The social design perspective also considers how designers approach sustainability and how their individual or collective ideas lead to decisions. For instance, McIsaac and Morey (1998) examined ideas about sustainable development, considering the values, assumptions, and culture of design professions such as engineering. The primary conclusions from the previous study encourage designers to develop an understanding and appreciation for goals, strategies, and values complementary to those in the engineering field to improve design decisions. The same hypotheses are true for other professionals such as architects and contractors. Design teams are challenged to create value during the entire delivery process, not just as an end-product. The life cycle of a construction project is larger than is usually portrayed. It involves the acquisition of raw materials to the disposal in landfills or reuse or recycling of materials. The decisions made by designers during these instances will influence social aspects as well. Another consideration for improving decisions during design is advancing building assessment practices by identifying four significant outcomes: integration, transparency, accessibility, and collaborative learning among stakeholders (Kaatz et al., 2005).

References

Kaatz, E., Root, D. and Bowen, P. 2005. Broadening Project Participation through a Modified Building Sustainability Assessment. *Building Research & Information*, 33(5), 441.

McIsaac, G. and Morey, N. 1998. Engineers' Role in Sustainable Development: Considering Cultural Dynamics. *Journal of Professional Issues in Engineering Education and Practice*, 124(4), 110–118.

Annex IV: Safety through Design

Designers (architects and engineers) can and should ensure worker safety by eliminating potential safety hazards from the work site during the design phase. These professionals can directly impact safety because they are involved in the selection of a procurement system, the preparation of contract documentation, the sequencing of the construction process, and the decisions regarding contract duration. Thus, waiting to consider safety until the construction phase, according to Toole and Gambatese (2008), is a lost opportunity to positively and effectively influence construction site safety. Previous research on the topic has found a link between project design and construction site injury and fatality incidents. For instance, Behm (2005) demonstrated that 42% of 224 fatality cases reviewed could be linked to the design for safety concept. This research was validated by a panel of experts with construction, design, and academic backgrounds (Gambatese et al., 2008). While design is one factor affecting safety, not all accidents can be prevented in this phase. Safety on construction sites is impacted by many factors including a holistic safety program. It is especially difficult to predict the outcomes when the project requires using new technologies and materials. Thus, there is a responsibility to have a rigorous monitoring system in place for obtaining feedback concerning the impact of a design to improve the future decisions made in the design phase.

At times, designers minimize their liability exposure by deliberately not addressing construction safety (Gambatese and Hinze, 1999). In addition, they believe they lack the skills and training to address construction worker safety. To address these issues, the following recommendations have been developed to improve the way safety is considered by designers:

(a) Design components to facilitate pre-fabrication in the shop or on the ground so that the pieces may be put in place as complete assemblies.

(b) Design parapets to be 42 inches (1.07 m) high.

(c) Design beam-to-column double-connections for continual support of the beams during the connection process by adding a beam seat, extra bolt holes, and/or other redundant connection points.

Procurement processes such as design-build are more conducive for implementing the concept. For instance, large design-builders such as Jacobs, Parsons, and the Washington Group have implemented programs that include designers in the consideration of construction worker safety (Gambatese et al., 2008).

References

Behm, M. 2005. Linking Construction Fatalities to the Design for Construction Safety Concept. *Safety Science*, 43(8), 589–611.

Gambatese, J. and Hinze, J. 1999. Addressing Construction Worker Safety in the Design Phase: Designing for Construction Worker Safety. *Automation in Construction*, 8(6), 643–649.

Gambatese, J.A., Behm, M. and Rajendran, S. 2008. Design's Role in Construction Accident Causality and Prevention: Perspectives from an Expert Panel. *Safety Science*, 46(4), 675–691.

Toole, T.M. and Gambatese, J. 2008. The Trajectories of Prevention through Design in Construction. *Journal of Safety Research*, 39(2), 225–230.

4

Fundamentals of Life Cycle Assessment

4.1 Introduction – Why and What Is LCA?

Environmentally sustainable solutions can only be found by taking a life cycle approach. The life cycle of a product consists of all the stages from raw material extraction to its disposal. Life cycle assessment (LCA) is an assessment tool for a product (in LCA the concept product refers not only to material products but also to services), which provides us a structured way to decide which alternative is more environmentally friendly. The International Organization for Standardization (ISO) defines LCA as "the compilation and evaluation of the inputs, outputs and the potential environmental impacts of a product system throughout its life cycle" (ISO, 2006a). Figure 4.1 presents the possible life cycle stages that can be considered in an LCA and the typical inputs/outputs measured. This holistic view makes LCA unique in the sustainability toolbox. LCA is mainly focused on the environmental impacts although it can also include both social and economic impacts for a sustainability assessment.

The strength of LCA is that it studies a whole product system and is an integrated information tool preventing problem shifting to other life cycle stages, to other substances, to other environmental problems, to other countries, and to the future. It helps to identify and include these trade-offs in the assessment. For example, compact fluorescent lamps reduce electricity consumption by about 75% but can cause mercury pollution. Bio-based fuels reduce greenhouse gas (GHG) emissions but contribute to air, water, and soil quality impacts in the agricultural phase. The other main characteristics of the LCA tool are as follows:

- It is about products or more specifically about product functions, such as driving a car or filling a beverage.
- It is based on science and is an engineering tool in the sense that technical systems are studied.
- It covers a broad range of environmental issues, typically around 15. In many cases, single attributes such as percentage of recycled

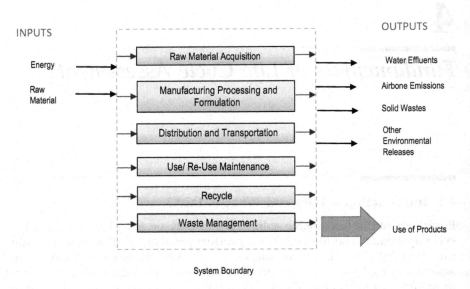

FIGURE 4.1
Life cycle stages in an LCA.

content or locally source materials are used to identify a product as environmentally profitable. However, these single attributes do not capture the total environmental picture – for example, a local product sourced in an inefficient and polluting factory might have a larger environmental impact than one produce far away in an efficient factory and transported to a site.

- Since a whole life cycle is studied, it is not site specific and environmental impacts cannot be modeled at a very detailed level. Environmental impact indicators express potential, not actual consequences.
- It is an assessment rather than an analysis to stress that value judgments are an intrinsic part of the procedure.
- It is a quantitative tool but only as much as possible, and nonquantifiable aspects can be incorporated as well.
- It is also a method to structure the large amount of complex data and to facilitate comparisons across product alternatives.

LCA is a relatively young discipline that started in the 1970s. For a good history of LCA, please refer to Bjørn et al. (2018) and Guinée et al. (2011). This chapter gives the fundamentals of LCA for the LCA beginner and is not a rule-describing manual. The reader can consult the standard for the rules and for the following books for a more advanced treatment of the topic:

Bauman, H and Tillman, A.M. (2004) *The Hitch Hiker's Guide to LCA: An Orientation in Life Cycle Assessment Methodology and Applications*, 1st ed., Studentlitteratur AB, Lund, Sweden.

Hauschild, M. Z, Rosenbaum, R.K, and S.I. Olsen (eds.) (2018) *Life Cycle Assessment – Theory and Practice*, Springer.

Curran, M. A. (ed.) (2015*) Life Cycle Assessment Student Handbook*, Scrivener Publishing, Wiley.

Other suggested reading materials are given at the end of the chapter. A case study of the comparison between an incandescent lamp and a compact fluorescent lamp (retrieved from www.lifecycleinitiative.org/resources/training/lca-life-cycle-assessment-training-kit-material) will be used as an example throughout this chapter to illustrate the fundamentals of the tool. Table 4.1 shows the complex information about the two products that LCA seeks to restructure to help in decision-making of which product is "more environmentally friendly."

4.2 The LCA Methodology

LCA also refers to a methodology on how life cycle studies are done and interpreted. The ISO has produced a series of standards and technical reports on LCA and which is referred to as the 14040 series and which specifies a structured framework and the rules, requirements, and considerations (see Figure 4.2). Specific data and calculation steps are not specified but much attention is given for transparency in reporting.

The structure of the LCA methodology, as Figure 4.3 shows, is organized into the following different phases:

1. Goal and scope definition – identifying the purpose and boundaries of the study, assumptions, and expected output.
2. Life cycle inventory (LCI) – construction of the life cycle model quantifying the energy use and raw material inputs and environmental emissions associated with each life cycle stage.

TABLE 4.1

Product Properties of Incandescent and CFL Lamps

Product property	Incandescent lamps	Compact fluorescent (CFLs)
Power consumption	60 W	18 W
Life span (average)	1,200 hours	8,000 hours
Mass	30 g	540 g
Mercury content	0 mg	2 mg

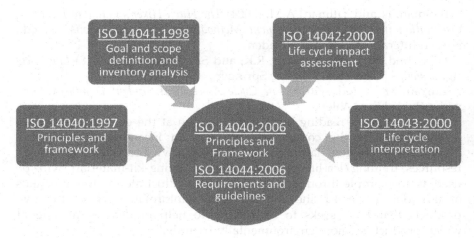

FIGURE 4.2
ISO standards combination.

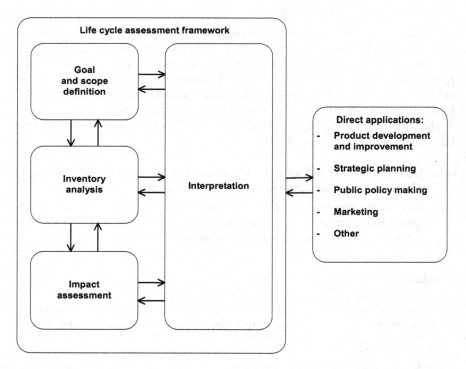

FIGURE 4.3
The ISO life cycle assessment framework.

Source: ISO 14040:2006

3. Life cycle impact assessment (LCIA) – assessing the impacts on human health and the environment associated with the LCI results.
4. Interpretation – analysis of the results of the inventory and impact modeling, and presentation of conclusions and findings in a transparent manner in relation to the goal of the study.

The bidirectional flows in Figure 4.3 emphasize that LCA is not a sequential process, starting with goal and scope definition and ending with interpretation. It is rather an iterative process, in which the goal and scope may be refined after inventory analysis, impact assessment, or interpretation, and similarly the inventory analysis may be refined after impact assessment or interpretation. LCA is also not a prescriptive process. It involves various assumptions and value-based judgments throughout the process and it is important to be explicit in the reporting phase about what assumptions and values were used. It is very important to maintain transparency in reporting an LCA study so that readers can see clearly what was done and decide if they agree with the approach. As illustrated in Figure 4.3, the ISO standard distinguishes the methodological framework of LCA from its different applications. The international standard lists the following applications: identification of improvement possibilities, decision-making (product or process redesign, support for regulatory measures, and policy instruments), choice of environmental performance indicators and market claims and communication (eco-labeling, environmental product declaration, and benchmarking). Another important application of LCA is that of learning more about the characteristics of a product system and the supply chain as well as to provide science-based defense to public concerns.

4.2.1 The LCA Goal and Scope Definition

ISO standards 14040 and 14044 (ISO, 2006a, 2006b) require that the goal of an LCA study be unambiguously stated and include

*   The intended application (what);
*   The reason for carrying out the study (why);
*   The intended audience (for whom);
*   Whether the results are intended for comparative assertions intended to be disclosed to the public (how).

A well-defined goal helps to define the scope and boundaries of the study. There must be a close communication between the LCA practitioner and the commissioner of the LCA. LCA standards do not define what the appropriate goals are but rather require that an LCA goal be clearly stated. A formal critical review of an LCA study can neither verify nor validate the goals but rather evaluates if the study conforms to the stated goals (ISO, 2006a). LCAs are often performed to do a "hot spot identification," that is, to identify the parts of a product system that contribute most to its environmental impact. The results can help identify opportunities to reduce impacts through re-design. Examples of specific evaluation LCAs include "What are the total potential life cycle environmental impacts of an office building?" or "What are the environmental impacts related to manufacturing PVC windows and what opportunities are there to reduce environmental impacts through changing manufacturing processes?". For example, a comprehensive goal statement for a PVC window could include the following points:

What – this LCA is a study of PVC window manufacturing.

Why – developed to understand the environmental impacts of the manufacturing process.

Who – for use by manufacturers.

How – to integrate into a more comprehensive LCA.

Benefit – to manufacturers looking to understand and reduce the impact of manufacturing.

Limits – the limitations of appropriate use of the LCA data (e.g., only for manufacturers' internal improvement or for publication for consumers to read).

LCA is also often used to compare two or more options and evaluate if one is preferable. These LCAs are designed to inform decisions and are called a comparative assertion. ISO standards require additional analysis and care when performing LCAs used to make comparative assertions to help avoid inaccurate or misleading claims. An example of a comparative LCA in the building industry is "Which structural system is preferable: steel, concrete, or wood?"

The goal definition sets the context of the LCA study and is the basis of the scope definition. The scope of an LCA requires the definition of what is included in and excluded from the analysis and defines the parameters of the study. ISO standard 14044 has specific requirements as to what must be included in the scope definition. The following is a summary of the key items that should be defined in an LCA scope as adapted from ISO (2006a):

(i) *The product to be studied*: Establishing function, performance, and unit of analysis (functional unit). A system is defined by the function it provides and can also be defined by the performance characteristics of a product. Defining the product to be studied is often not as simple as it might first appear, especially when comparing alternative options. An LCA typically identifies a unit of the product to analyze.

For a manufactured product such as a disposable cup, the unit might be one cup. However, one could not compare a disposable cup to reusable cup as they are not functionally equivalent. The term "functional unit" defines a unit of analysis that includes quantity, quality, and duration of the product or service provided. The choice of the functional unit and their reference flows directly steers the calculations. The reference flow is the amount of product that is needed to realize the functional unit.

The functional unit of a coffee cup might then be to deliver 200 ml (quantity) of 80°C liquid with insulation to be able to be held in hand (quality) up to twice in one day for ten years (duration). If an LCA study were based upon the assumption that 3,000 paper cups are used over ten years, the environmental impacts of one paper cup would be multiplied by 3,000 to compare with the equivalent single ceramic cup (provided one assumes that a ceramic cup can last that long). LCA results for two different products can thus be comparable if analyzing functionally equivalent systems. Light bulbs produce light, so in a comparison of two different light bulbs, the functional unit must be phrased in terms of light. To compare CFL to an incandescent lamp, an appropriate function would be performance based, that is, the amount of light needed to illuminate a standard room (such as 15 square meters with 1,000 lumen for 1 hour). Light bulb functional unit might be 1,000,000 lumen-hours of light.

Assuming a 60-W incandescent lamp with 900 lumens and a lifetime of 1,000 hours and a 15-W CFL lamp with 900 lumens and 8,500 lifetime hours and if the functional unit is 20 million lumen hours, then we will be comparing approximately 22 incandescent lamps with 3 CFL lamps.

As Hauschild (2018) pointed out, a functional unit involves answering the questions "'what,' 'how much,' 'for how long/how many times,' 'where,' and 'how well'?" For example, a comparison of outdoor paints may be based on the functional unit: complete coverage (what) of 1 m² (how much) primed outdoor wall (what) for 10 years (for how long) in Mauritius (where) in a uniform color at 99.9% opacity (how well?).

(ii) *The system boundary*: What is excluded from and included in the analysis (which unit processes are part of the system studied). LCA requires a clear definition of the system boundaries (which industrial and natural processes are to be included and excluded). While the purpose of an LCA study is theoretically a comprehensive evaluation, in reality, the preparation of an LCA requires that choices be made about which processes to include in the study, which inputs and outputs to measure or estimate, which environmental impacts to evaluate, and which data sources to use. The system boundary must be defined to support the stated goal and scope of the LCA. Some LCAs only analyze impacts from cradle-to-gate, while others include the whole life cycle. Often LCA studies will exclude the following processes:

- Manufacture of fixed equipment (the factory);
- Manufacture of mobile equipment (trucks);
- Hygiene-related water use (toilets and sinks for workers); and
- Employee labor (commuting).

When known LCA stages are excluded, the goal of the LCA must be carefully assessed to ensure that the limited scope is still adequate to meet the requirements of the goal. A system boundary diagram is a graphic representation of the scope of an LCA, and it should clearly indicate what is included in the LCA analysis and which inputs and outputs have been tracked.

(iii) *Methodological choices*: Including assumptions for multifunctionality and impact assessment and interpretation methods. The chosen impacts determine the parameters for which data will be collected during inventory analysis. The impact assessment categories should link the potential impacts to the entities that we aim to protect. The commonly accepted areas of protection (AoP) are natural resources, natural environment, and human health. It also involves deciding whether it is a consequential study or an attributional study.

(iv) *Level of detail in the study:* Sources of the data, data-quality requirements, and type of critical review (if any). As with the methodical choices, LCA standards permit a wide range of analysis methods, data sources, and assumptions. The standards require that these details be clearly defined within the LCA report to enable those reading and interpreting the data to understand the strengths and weaknesses of the study.

The goal and scope definition are taken into account when the results of the study are interpreted. They are constantly reviewed and refined during the process of carrying out an LCA, as additional data become available.

4.2.2 Life Cycle Inventory

The ISO standard for LCA defines LCI analysis as "the phase of life cycle assessment involving the compilation and quantification of inputs and outputs, for a given product system throughout its life cycle." This is often the most time-consuming part of an LCA. An LCI will include the following:

1. Construction of the flow model of the technical system on the basis of the unit process according to the system boundaries decided in the goal and scope definition. A unit process is the "smallest element considered in the life cycle inventory analysis for which input and output data are quantified" (Figure 4.4).

   Examples of unit processes include coal mining, cement production, use of a television set, transport by lorry, and so on. Each of these processes is described in quantitative terms as having inputs and outputs. The LCI is then the compilation of the different unit processes within the system under study. All unit processes of an LCI model belong to the technosphere (i.e., everything that is intentionally "man-made"). Environmentally relevant flows or elementary flows (use of scarce resources and emissions of substances considered harmful) go across the boundary between the technosphere and the ecosphere (or the "the environment" or "nature" i.e., which is not intentionally man-made). The ecosphere contains AoP or damage categories, which are qualities that LCA has been designed to protect, that is, ecosystems, human health, and resource availability.

2. Planning and data collection for all the activities in the product system. A number of sources are needed to collect data for the LCI. We can distinguish between primary data (on-site measurements, interviews, questionnaires or surveys, bookkeeping, and so on) and secondary data (databases, literature, statistics, industry data reports, specifications, and best engineering judgment). Data can also be classified by how they were created – site specific, modeled, calculated or estimated, non-site specific, that is, surrogate data or vendor data. In the collection of data, we also distinguish between unit processes belonging to the foreground and background system. The foreground system contains unit processes specific to the product

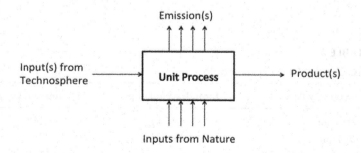

FIGURE 4.4
A unit process.

system being studied and it is largely modeled using primary data, that is, data collected first-hand by the LCA practitioner. Those processes that are not specific to the product system and are common in other product systems such as energy, transport, and waste management systems from part of the background system. The latter is typically modeled using LCI databases, which contain average industry data representing the process in specific nations or regions. In practice, an LCI contains a mixture of measured, estimated, and calculated data. Data quality information helps in the analysis of the uncertainty of the data. Weidema and Wesnaes (1996) and Cooper and Kahn (2012) have devised approaches for data quality analysis.

3. Calculation of the amount of resource use and pollutant emission of the system in relation to the functional unit. An LCI can be complicated by the fact that many technical processes produce more than one product, and the environmental load of such processes has to be allocated or partitioned between its different products. Handling of multifunctional processes is a methodological challenge in LCA. The assumption of linear technology is an important restriction of LCI, although it makes the calculations and data collection feasible (Curran, 2012).

The outcome of an LCI is a list of quantified physical elementary flows for the product system that is associated with the provision of the service or function described by the functional unit. ISO 14044 (2006a) recommends classifying the inventory data into four categories: (1) inputs (energy, materials, etc.), (2) products (including co-products and waste); (3) emissions (to air, water, and soil); and (4) other environmental aspects. This list can include hundreds of different emissions, which can be difficult for non-LCA experts to interpret. Table 4.2 shows that product data with LCA is less complex than without LCA. The Table also does no longer contain partially overlapping and compensating information as in Table 4.1.

TABLE 4.2

LCI for the Two Types of Bulbs

Elementary flow	Incandescent lamp	Fluorescent lamp
CO_2 to air	800,000 kg	50,000 kg
SO_2 to air	1,000 kg	80 kg
Copper to water	3 g	20 g
Crude oil from earth	37,000 kg	22,000 kg

4.2.3 Life Cycle Impact Assessment

Taking the LCI as a starting point, LCIA turns the inventory results into more environmentally relevant information (or impact category indicators) and at the same time aggregate the information from the LCI (which can contain hundreds of different items many of which require expert knowledge to understand their importance) in fewer parameters (typically about 15). It relates resource use and environmental emissions to estimations of natural resource, natural environment, and human health impacts using knowledge and models from environmental science. Compared to the three other LCA phases, LCIA is in practice largely automated by LCA software but the underlying principles must be understood by practitioners or users to properly interpret the results. The impact assessment consists of the following five elements (it should be noted that only the first three elements are mandatory according to the ISO 14040 standard).

1. *Selection* of impact categories as per the scope definition. For each impact category, a representative indicator is chosen together with an environmental model than can be used to quantify the impact of elementary flows on the indicator. In practice, this is typically done by choosing an already existing LCIA method.

2. *Classification* of elementary flows from the inventory by assigning them to impact categories according to their ability to contribute by impacting the chosen indicator. In practice, this is typically done automatically by LCI databases and LCA software.

3. *Characterization* using environmental models for the impact category to quantify the ability of each of the assigned elementary flows to impact the indicator of the category. The resulting characterized impact scores are expressed in a common metric for the impact category. This allows aggregation of all contributions into one score, representing the total impact that the product system has for that category. The collection of aggregated indicator scores for the different impact categories (each expressed in its own metric) constitutes the characterized impact profile of the product system. This is typically done automatically by LCA software.

4. *Normalization* is used to inform about the relative magnitude of each of the characterized scores for the different impact categories by expressing them relative to a common set of reference impact. It is a calculation of the magnitude of the category indicator results relative to a reference. Often the background impact from society is used as a reference. The result of the normalization is the normalized impact profile of the product system in which all category indicator scores are expressed in the same metric. This is an optional step in the ISO framework.

5. *Grouping and weighting* supports comparison across the impact categories. Grouping is the assignment of impact categories into one or more sets. *Weighting,* using weighting factors for each impact category, gives a quantitative expression of how severe it is relative to the other impact categories. The weighting allows aggregation of all the weighted impact scores into one overall environmental impact score for the product system. This can be useful in decision-making when LCA results are used together with other information like the economic costs of the alternatives. This is an optional step in the ISO framework.

While the unit process is the central element of the inventory analysis, the central element in impact assessment is the impact category. ISO defines it as "a class representing environmental issues of concern to which life cycle inventory analysis may be assigned," such as climate change or toxicity. A category indicator is a quantifiable representation of an impact category. The environmental mechanism is the system of physical, chemical, and biological processes for a given impact category, linking the LCI analysis results to category indicators. Impact pathway is the cause-effect chain of an environmental mechanism. The impact assessment categories should link the potential impacts and effects to the entities that we aim to protect. As indicated previously, the commonly accepted AoP are natural resources, natural environment, and human health.

A characterization model describes the relationship between the LCI results and the category indicators and is reflective of the environmental mechanism. The characterization model is used to derive a characterization factor, which is then applied to convert an assigned LCI analysis result to the common unit of the category indicator. An example of the terminology used for the environmental impact due to acidification would be as follows:

- Impact category: Acidification potential
- LCI result: 500 kg SO_2, 100 kg NO_x, 10 kg HNO_3 per functional unit
- Characterization model: TRACI 2.1 (TRACI is the Tool for the Reduction and Assessment of Chemical and Other Environmental Impacts)
- Category indicator: increase in acidity in the environment (moles H^+)
- Characterization factor: potential of each compound to cause acid deposition (such as acid rain) in relation to that of SO_2
- Category indicator result: kilograms of SO_2-equivalent
- Category endpoints: ecosystem effects such as acidic lakes, corrosion to buildings, and damage to plants

Figure 4.5 shows a graphical idea of an environmental impact pathway from the LCI results (top) to the endpoint (bottom) via a number of intermediate variables.

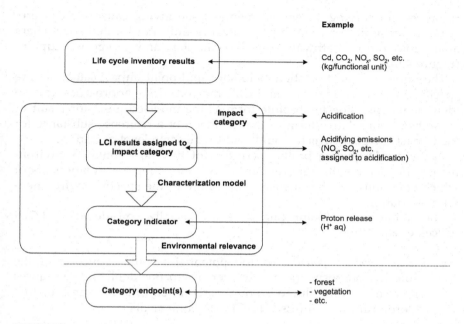

FIGURE 4.5
Example of an environmental impact pathway.

Source: www.lifecycleinitiative.org/wp-content/uploads/2013/06/Module-e-Impact-assessment.pdf

An elementary flow in the LCI flow has the ability to affect an indicator that can be selected along the cause-effect chain of the impact category. In general, the further down the cause-effect chain an indicator is chosen, the more environmental relevance it will have. However, uncertainty of the model and of parameters also increases further down the cause-effect chain, while measurability decreases. In the selection of an impact indicator a compromise must therefore be made between choosing an indicator of impact early in the environmental mechanism, giving a more measurable result but with less environmental relevance versus downstream in the environmental mechanism, giving more relevance but hardly verifiable information (ex-degraded ecosystems or affected human lifetime). This has led to two different types of impact categories, applying indicators on two different levels of the environmental mechanism: midpoint impact indicators and endpoint impact indicators. The midpoint approach has the advantage that it includes fewer debatable assumptions while the endpoint approach (or damaged oriented approach) has the advantage that it provides metrics that are better understood by policy makers. A category endpoint is an attribute or aspect of natural environment, human health, or resources, identifying an environmental issue giving cause for concern. A midpoint indicator is an impact category indicator located somewhere along the impact pathway between emission and category endpoint (Figure 4.6).

The units of characterization factors for mid-point impact categories are specific for each category and LCIA method. Two approaches can be identified: expression in absolute form as the modeled indicator result or expression in a relative form as that emission of a reference substance for the impact category, which would lead to the same level of impact.

Appendix D gives a brief description of the most common environmental impact results reported in LCAs, including information about which emissions and mechanisms contribute to the specific environmental degradation.

The following midpoint categories are usually considered in LCIA (Čuček et al., 2015):

- *Ozone depletion potential* is the potential for the reduction in the protective stratospheric ozone layer. The ozone-depleting substances are freons, chlorofluorocarbons, carbon tetrachloride, and methyl chloroform. It is expressed as CFC-11 equivalents.

- *Global warming potential* represents the potential change in climate attributable to increased concentrations of CO_2, CH_4, and other GHG emissions that trap heat. It leads to increased droughts, floods, losses of polar ice caps, rising sea level, soil moisture losses, forest losses, changes in wind and ocean patterns, and changes in agricultural

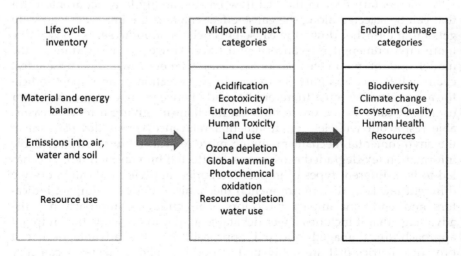

FIGURE 4.6
From LCI to endpoint damage categories.

production. The global warming potential is calculated in carbon dioxide equivalents meaning that the greenhouse potential of an emission is given in relation to CO_2. Since the residence time of gases in the atmosphere is incorporated into the calculation, a time range for the assessment is defined to be 100 years.

- *Acidification potential* is based on the potential of acidifying pollutants (SO_2, NO_x, HCl, NH_3, HF) to form H^+ ions. It leads to damage to plants, animals, and structures. It is expressed in SO_2 equivalents. The acidification of soils and waters occurs predominantly through the transformation of air pollutants into acids, which leads to a decrease in the pH value of rainwater and fog from 5.6 and below. Acidification potential is described as the ability of certain substances to build and release H^+ ions.

- *Eutrophication potential* leads to an increase in aquatic plant growth attributable of nutrients left by over-fertilization of water and soil, such as nitrogen and phosphorus. Nutrient enrichment may cause fish death, declining water quality, decreased biodiversity, and foul odors and tastes. It is expressed in PO_4^{3-} equivalents.

- *Photochemical ozone creation potential* is also known as ground-level smog, photochemical smog, or summer smog. It is formed within the troposphere from a variety of chemicals, including NO_x, CO, CH_4, and other volatile organic compounds (VOCs), in the presence of high temperatures and sunlight. It has negative impacts on human health and the environment and is expressed as C_2H_4 equivalents.

- *Ecotoxicity (freshwater, marine, terrestrial) potential* focuses on the emissions of toxic substances into the air, water, and soil. It includes the fates, exposures, and effects of toxic substances and is expressed as 2,4-dichlorophenoxyacetic acid equivalents.

- *Human toxicity potential* deals with the effects of toxic substances on human health. It enables relative comparisons between a larger number of chemicals that may contribute to cancer or other negative human effects for the infinite time horizon. It is expressed as 1,4-dichlorobenzene equivalents.

- *Abiotic depletion potential* is concerned with the protection of human welfare, human health, and ecosystems, and represents the depletion of nonrenewable resources [abiotic, nonliving (fossil fuels, metals, minerals)]. It is based on concentration reserves and the rate of de-accumulation and is expressed in kilogram antimony equivalents.

- Land use
- Water use

Endpoint impact indicators are chosen further down the cause-effect chain of the environmental mechanism closer to or at the very endpoint of the chains or the AoP. Typical endpoint indicators are as follows:

- Human Health (expressed as DALY (Disability-Adjusted Life Years)), which is a unit used by the WHO.

- Ecosystem quality or natural environment – expressed as $m^2/year$ or $m^3/year$ potentially disappeared fraction (PDF), which can be interpreted as the time and area (or volume) integrated increase in the disappeared fraction of species in an ecosystem per unit of midpoint impact indicator increase. It essentially quantifies the fraction of all species present in an ecosystem that potentially disappears over a certain area or volume and during a certain length of time.

- Natural resources and ecosystem services – there are different approaches used – some focus on the future costs for extraction of the resource as a consequence of current depletion and these divide into costs in the form of energy or exergy use for future extraction (measured in MJ) or monetary costs (measured in current currency such as USD or Euro).

To go from midpoint to endpoint indicator scores, additional midpoint to endpoint characterization factors are needed. Some LCIA methods only support endpoint characterization and the midpoint to endpoint characterization is combined in one characterization factor. Rosenbaum et al. (2018) recommend that instead of perceiving midpoint to endpoint characterization as two alternatives to choose from, it is recommended to conduct an LCIA on both midpoint and endpoint level using an LCIA method that provides both to support the interpretation of results obtained and which complement each other, respectively.

The following paragraph summarizes how one of the most common impact categories – climate change – is handled in LCIA. Climate change involves long sequence of causal mechanisms: emissions of GHGs lead to change in the composition of the atmosphere, which leads to a change in the temperature, which leads to a change in the climate, which leads to changes in ecosystems and human activities, and so on. The further we proceed in this causal chain, the more uncertain the knowledge becomes. To be able to quantitatively model the emissions of different GHGs into an impact indicator for climate change, we must do as follows:

- Choose a certain point in the causal mechanism. This can be upstream (change in the radiation balance), at the downstream end (change of biodiversity), or somewhere in between (change in temperature).

- A way must be found to convert the emission data into the chosen impact indicator. For example, the UN-based IPCC has developed quantitative models of the impact of GHG emissions – global warming potentials – which are quantitative measures of the strength of different GHGs.

A cause-effect chain for climate change is shown in Figure 4.7.

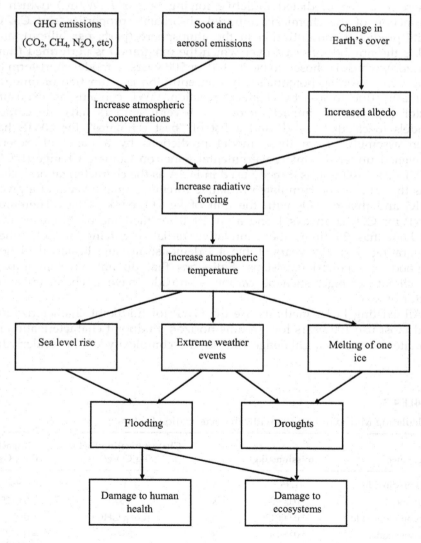

FIGURE 4.7
Cause-effect chain for climate change.

The unanimously used climate change indicator on midpoint level has been the global warming potential (GWP), an emission metric introduced by the IPCC. GWPs are calculated for each GHG according to:

$$GWP_i = \frac{\int_0^{TH} a_i c_i \, dt}{\int_0^{TH} a_{CO_2} c_{CO_2} \, dt}$$

where a_i is the predicted radiative forcing of gas i (W/m^2) (which is a function of the chemical's infrared absorbance properties and C_i), C_i is its predicted concentration in the atmosphere (parts per billion), and TH is the number of years over which the integration is performed (time horizon), which is chosen to be 20, 100, or 500 years to model short-term or long-term warming potentials. The concentration is a function of time (t), primarily due to loss by chemical reaction. GWP contains an exposure factor (C_i) and an impact factor (a_i). Several authors have developed models to calculate GWP and a list of "best estimates" for GWPs has been assembled from these model predictions by a panel of experts convened under the Intergovernmental Panel on Climate Change (IPCC, 2007). GWP 100 year is directly used in LCIA as the characterization factor. It is the ratio of the cumulated radiative forcing over 100 years of a given GHG and that of CO_2 with the unit of kg-CO_2-eq/kg GHG. Therefore, GWP for CO_2 is always 1 and a GWP100 for methane of 25 means that methane has 25 times the cumulative radiative forcing of CO_2 when integrating over 100 years. Table 4.3 demonstrates the basic calculation method of chemical emissions. Emissions that do not have an impact on climate change, such as carbon monoxide, have a characterization factor of zero.

All existing LCIA methods use the GWP for midpoint characterization and most use 100 years for the time horizon. Endpoint characterization of climate change is a challenge due to the complexity of the underlying

TABLE 4.3

Calculating Midpoint Impacts with Characterization Factors

Substance	Emissions (kg)		Characterization Factor (kg CO_2e/kg)	Impact (kg CO_2e)
Carbon dioxide	1.96	X	1	= 1.960
Methane	0.0005	X	25	= 0.0125
Carbon monoxide	0.00013	X	0 (not a GHG)	= 0
Nitrous oxides	0.0016	X	264	= 0.477

Total climate change estimated impact: 2.45 kg CO_2e

environmental mechanisms. Some endpoint methods have proposed end-point characterization factors (e.g., ReCiPe, LIME, Impact World+, and Ecoindicator 99). But they are associated with large uncertainties, which is why some endpoint methods (e.g., IMPACT 2002+) refrain from endpoint modeling for this impact category. Endpoint results for climate change must be taken with the greatest caution in the interpretation of results.

This type of characterization calculation is applied for each impact indicator category, which results in a list of scores or category indicator results. This complete list is called an LCIA profile or a characterization table. An LCIA method is a collection of individual characterization models (each addressing their separate impact category). Rosenbaum et al. (2018) give factors to be considered when selecting a particular LCIA method. Conducting LCIA has been greatly simplified by the increasing availability of software programs, which include these LCIA methods with their specific choice of impact categories. The most often used LCIA methods are as follows:

- IMPACT World+ (Midpoint and Endpoint)
- LIME (Midpoint and EndPoint)
- ReCiPe (Midpoint and Endpoint)
- TRACI (Midpoint)
- EPS 2000 (Endpoint)

All these methods comprise a recommended set of impact categories with a category indicator and a set of characterization factors. An example of an LCIA profile for the two types of lamps is shown in Table 4.4.

It shows how the use of florescent lamps instead of incandescent lamps may reduce climate change and acidification impacts but increases potential for ecotoxicity. Although the characterization clearly helps to understand the inventory results, it is insufficient to tell the decision-maker what to do. The impact category results are still difficult to understand due to

TABLE 4.4

LCIA Profile for the Two Types of Lamps

Impact Category	Incandescent Lamp (60 W)	Fluorescent Lamp (18 W)
Climate change	120,000 kg CO_2-eq	40,000 kg CO_2-eq
Ecotoxicity	320 kg DCB-eq	440 kg DCB-eq
Acidification	45 kg SO_2-eq	21 kg SO_2-eq
Depletion of resources	0.8 kg antinomy-eq	0.3 kg antinomy-eq

difference in units, difference in scale, and so on. The normalization step relates the results to a reference value, for example, to the total world impacts in 2002 and it will result in the normalized environmental profile shown in Table 4.5.

It must be realized that the ratios between the score for the two products is the same as in the characterization table. Normalization only rescales each indicator result, so that they are more comparable. They have the same unit and a different order of magnitude implies a different share in the reference total. It should be noted that when the normalized score for ecotoxicity is 10^{-10} and for acidification it is 10^{-11}, it does not mean that the acidification aspect is less serious. It only means that this product has a lower share in the acidification than in ecotoxicity. Although a score of 10^{-10} or 10^{-11} suggest a negligible value, it must be remembered that the functional unit is about one lamp and all the millions of lamps used create the total problem.

Although normalization removes the arbitrary units that are present in the characterization results, its result is still not sufficient to make a clear decision. *Weighting* using weighting factors for each impact category gives a quantitative expression of how severe it is relative to the other impact categories. Quantitative weighting allows aggregation of all the weighted impact scores into one overall environmental impact score for the product system (Table 4.6).

It looks simple, but interpretation is needed to place this single-number information in the right perspective. A number of techniques have been suggested for use in weighting. One principle widely used to derive weighting factors is the social science-based perspective, which does not

TABLE 4.5

Normalized LCIA Profile for the Two Types of Lamps

Impact Category	Incandescent Lamp	Fluorescent Lamp
Climate change	1.2×10^{-11} yr	4×10^{-12} yr
Ecotoxicity	1.6×10^{-10} yr	2.2×10^{-10} yr
Acidification	9×10^{-11} yr	4.2×10^{-11} yr
Depletion of resources	24×10^{-12} yr	9×10^{-13} yr

TABLE 4.6

Weighted Index for the Two Types of Lamps

Weighed index	Incandescent lamp	Fluorescent lamp
Weighted index	8.5×10^{-10} yr	1.4×10^{-10} yr

represent the choices of a specific individual, but regrouping typical combinations of ethical values and preferences in society into a few internally consistent profiles. It applies three cultural perspectives, the Hierarchist, the Individualist, and the Egalitarian. For each cultural perspective, weighting factors are developed. It must be pointed out that due to a number of problems at both philosophical and practical levels associated with using these techniques, there is no consensus at present on how to aggregate the environmental impacts into a single environmental impact function. It is also argued that valuation should not be carried out at all since considering the impact in a disaggregated form enhances the transparency of decision-making based on LCA results.

4.2.4 Interpretation and Presentation of Results

ISO (2006a) defines interpretation as the "phase of life cycle assessment in which the findings of either the inventory analysis or impact assessment, or both, are evaluated in relation to the defined goal and scope in order to reach conclusions and recommendations." This LCA phase includes the following steps:

- identification of major burdens and impacts;
- identification of "hotspots" in the life cycle;
- sensitivity analysis; and
- evaluation of LCA findings and final recommendations.

Critical thinking and careful evaluation of the strengths and weaknesses of the LCA are essential during the interpretation phase in order for the conclusions, limitations, and recommendations resulting from an LCA to be meaningful. ISO (2006a) requires a minimum of three evaluation methods to be summarized and adapted, as shown below:

(i) *Completeness check*: Evaluating the LCA information and data to ensure that it is complete: that there are no missing emissions to environment. If items are found to be missing, new data and analysis are required.

(ii) *Sensitivity check*: Evaluating how sensitive the analysis is to changing assumptions or choices to assess the reliability of the final results and conclusions. This might lead to the need to obtain better data in order to understand key aspects of the LCA.

(iii) *Consistency check*: Evaluating the study to ensure the analysis is internally consistent and matches the established goal and scope.

The interpretation must be done with the goal and scope definition in mind and respect the restrictions that the scoping choices impose on a meaningful interpretation of the results, for example, due to geographical, temporal, or technological assumptions. By comparing the contribution of the life cycle stages or groups of processes to the total result, and examining them for relevance, contribution analysis determines which inventory data or impact indicator has the biggest influence. An example of a contribution analysis to climate change for the two types of lamps is given in Table 4.7. The results show that the main contribution of emissions occurs during electricity production for the use stage of the lamp.

Uncertainty and variability are inherent properties of the models, data, assumptions, and choices that are required when performing an LCA. Uncertainty and its reduction is the very reason for the iterative nature of LCA. Sensitivity analysis and uncertainty analysis are applied as part of the interpretation to guide the development of conclusions from the results, to appraise the robustness of the conclusions, and to identify the focus points for further work in order to further strengthen the conclusions. Huijbregts (1998) gives a useful classification of the types of uncertainties in LCA. One method commonly used for analyzing uncertainty is the semi-quantitative pedigree matrix approach used, for example, by the database eco-invent for the quantification of variability and uncertainty of LCI data. Sensitivity analysis can be used to indicate which parameters are most important to the analysis. Monte Carlo simulation is a widely used method to perform uncertainty and sensitivity analysis. It uses statistical sampling techniques to approximate the probability of certain outcomes by running multiple trail runs, called simulations, using random variables. LCA software performs these calculations easily and the results approximate the full range of possible outcomes, and the likelihood of each. A simulation can typically involve more than 10,000 evaluations of the model.

4.2.5 The Iterative Nature of LCA

In Figure 4.3, the arrows show that rather than being a sequential process, LCA is rather an iterative process. Each phase of the methodology

TABLE 4.7

Contribution Analysis to Climate Change for the Two Types of Lamps

Process	Incandescent lamp	Fluorescent lamp
Electricity production	88%	60%
Copper production	5%	15%
Waste disposal	2%	10%
Other	5%	15%
Total climate change	120,000 kg CO_2-eq	40,000 kg CO_2-eq

provides feedback to the previous phases of the study and helps target the next iteration of the LCA. Sensitivity and uncertainty analyses are performed in the interpretation phase but also throughout the study as part of both inventory analysis and impact assessment in order to identify the key figures or key assumptions of the study and the data that are associated with the largest uncertainties. The first iteration is often a screening that covers the full life cycle, but in terms of inventory data largely based on easily accessible data from available databases. The impact assessment results allow the identification of the parts of the product system that contribute most to the impacts and the scoping and system boundaries may have to be refined. The screening LCA also allows identifying those inventory data or assumptions made in the inventory analysis that have the largest influence on overall results or for which the uncertainties are so large and these data should be the target of the next iteration, where effort should be focused on testing and refining these assumptions or data and get more representative or recent data. Based on the revised inventory a new impact assessment is performed, and the sensitivity analysis is performed once more to see which are now the key figures and key assumptions. The uncertainty of the LCA results is reduced through the repeated iterations, and these are carried out until the remaining uncertainty of the results is sufficiently small to meet the goal of the study.

4.2.6 Methodological Choices

Although the ISO provides a general framework for conducting an assessment, it leaves much to interpretation by the practitioner. LCA standards permit a wide range of methods to be employed provided that the analysis clearly documents the methods used for the study. There are a number of issues that the engineer should bear in mind in order to carry out an LCA or to interpret the information correctly and to be critical. One should always remember that an LCA results refer to a goal definition with a specific purpose and a methodology is chosen to answer the question posed. The engineer reading an LCA report should be aware of the implications of the issues discussed below and should appreciate that there are many different applications of LCA and that different applications put different requirements on the methodology. Methodological choices should be guided by the purpose of the study. The following are critical choices of methodology that determine the outcomes of an LCA study:

(1) *Functional unit*: Critically checking the functional unit is important. Is it a relevant one? Does it allow fair comparison?

(2) *Multifunctionality*: In the real world there is hardly any product system that exists in isolation. In order to solve multifunctionality

issues, the ISO 14044 standard presents the following hierarchy of solutions.

(i) Subdivision of unit process – The first choice is to try to solve the problem through increasing the resolution of the modeling by dividing the multifunctional unit process into minor units so as to separate the production of the product from the production of the co-product.

(ii) System expansion – In a comparison of two processes, this means expanding the second process with the most likely alternative way of providing the secondary function of the first process.

(iii) Allocation – When the above methods are not feasible, the ISO 14044 standard recommends dividing the inputs and outputs of the multifunctional process or system between the different products or functions. This is called allocation. The environmental impacts should be partitioned between the system's "different products or functions in a way which reflects the underlying physical relationships between them" (ISO 14041, 1998). If this approach is not possible, allocation should be on a basis "which reflects other relationships" between the products. The "other relationships" are most commonly the economic values of the co-products as they leave the process in question.

(3) *Type of data*: Data representing how the system reacts to changes are relevant for change-oriented studies, whereas for other types of studies data representing the average behavior of the system are more relevant. There are also questions concerning choice of data, for example, between site-specific data and data representing an average of a population of processes. In order to interpret and apply the results of an LCA, the data must be assessed for quality and applicability.

(4) *Impact assessment method*: An important part of impact assessment methods is characterization methods, that is, the way environmental problems are modeled and measured. Characterization methods are much better developed for some impact categories than for others. The engineer should be aware of the limitations of these models when interpreting results. Also, some environmental impacts are not quantitative but they should be included in a qualitative assessment through life cycle thinking.

(5) *Cut-off rules*: Cut-off is a solution to the problem that the system is theoretically infinitely large. Cut-off rules are typically expressed in terms of mass, for example "the study will account for at least 95% of the total mass of inputs and no input shall be excluded that

individually contributes 1% or more of the mass. However, one should be careful that a material with a small mass contribution may have significant impacts on energy or environmental impacts. Another cut-off decision involves capital goods and infrastructure, that is, the buildings and equipment used to manufacture the product or the vehicles used to transport products. Usually, it has been assumed that the contributions is small when allocated over the total amount of output over the life time of the infrastructure – however, recent research has shown that their contributions may be significant in certain industry sectors, for example, nonfossil electricity generation (Frischknecht et al., 2007).

(6) *Attributional and consequential modeling*: Consequential and attributional LCAs answer different questions. Attributional LCA answers the question "What are the total emissions from the processes and material flows used during the life cycle (production, consumption and disposal) of a product, at the current level of output?" In contrast, consequential LCA seeks to answer the question "What is the change (either positive or negative) in total emissions which results from a marginal change in the level of output (and consumption and disposal) of a product?". Attributional LCA is useful for comparing the emissions from the processes used to produce (and use and dispose of) different products. It is also valuable for identifying opportunities for reducing emissions within the life cycle or supply chain, through improvements in processing efficiency or new technologies. However, it is not suitable for quantifying the total change in emissions that result from changes to the output (and other life cycle stages) of a product. This is because there may be indirect impacts that are outside its scope. The difference in the application of the two types of LCAs was illustrated by Searchinger et al. (2008), who found that on the basis of a conventional attributional LCA, US corn-based ethanol gave a 20% emissions' saving compared to gasoline. However, on the basis of a consequential LCA of the increase in output demanded by the US Energy Independence and Security Act, they predicted a 47% increase in emissions compared to gasoline. The expected increase in GHG emissions was attributed to land-use changes induced by higher prices of corn, soybeans, and other grains, predicted as a consequence of the additional demand for corn starch for ethanol production. The process of system expansion is an inherent part of consequential LCA studies. Choosing between an attributional and a consequential LCA is decided by the defined goal of the study.

(7) *Use of input–output and hybrid analysis in LCA*: In economic sciences, an input–output (IO) model represents the interdependencies between

different branches of a national economy or of different regional economies. For example, the steel sector in a country is buying X million dollars of goods or services from the transportation sector in that same country, and Y million dollars from other countries. We can couple this economic information with statistical impact data, for example, on environmental emissions or social characteristics. For example, the production of X million dollars of steel is associated with Y tonnes of CO_2 emissions and employment of low-skilled workers. "Input-Output databases" or "IO-databases" for short are databases for LCA that are based on national economic and environmental statistics and a number of IO-databases are now available in LCA software. The main advantage of these IO-databases is that one does not have to make cut-offs, that is, to exclude parts of the product system. The disadvantage of IO-databases for LCA is that processes are relatively aggregated, that is, at the level of product groups rather than individual products. However, combining process-based LCA and IOA in what has become known as "hybrid analysis" can yield a result that has the advantages of both methods (i.e., both detail and completeness).

4.2.7 LCI Databases and LCA Softwares

Since early 1990s, LCI databases have proliferated in response to the growing demand for life cycle information and these data sources come mainly from North America, Western Europe, and northeast Asia. Table 4.8 lists some national inventory databases. It is also possible to extract meaningful information from public databases.

TABLE 4.8

Available National Life Cycle Inventory Databases (Source: Curran, 2012)

Name	Website	Availability	Geographic origin
Ecoinvent	www.ecoinvent.ch	License fee	Global/Europe/Switzerland
German Network on Life Cycle Inventory Data	www.lci-network.de	Free	Germany
Australian Life Cycle Inventory data project	www.auslci.com.au/	Free	Australia
US LCI database	www.lcacommons.gov/nrel/search	Free	USA
EDIP	www.lca-center.dk	License fee	Denmark
Thailand National LCI database	www.thailcidatabase.net/	Free	Thailand

With even moderately large systems, data handling and calculation can be complex and a software will help in reducing the time needed for assessments, prevent errors, assist conversion of data to functional unit basis, increase capabilities (e.g., Monte Carlo simulations, sensitivity analyses), and organize systems and data. However, some software/database packages can be fairly expensive and they also require a learning curve to be used effectively.

There are several commercially available LCA software packages that provide databases, LCIA methods as well as facilitate the conduct of an LCA, which typically requires information about hundreds of processes and their input and output flows. In recent years, Simapro from PreConsultants (www.pre.nl/simapro) and Gabi from PE international (www.Gabi-software. com) have the highest market share on a worldwide level. Prices are in the range of several thousand Euros with academic licenses offered at a much lower price. Some tools are still freely available such as CMCLA (www. cmlca.eu) and Gemis (www.iinas.org/gemis-download-en.html) though not offering all features of commercial LCA systems. Open LCA is an internationally used, open source software program (http://openlca.org).

4.2.8 Strengths and Limitations of LCA

Several characteristics make LCA appealing to engineers. It is based on science, deals with environmental issues in a highly structured manner, and it is a quantitative tool focused on technical systems. It takes a life cycle perspective and as such it allows identifying and preventing the burden shifting between life cycle stages. It covers a broad range of environmental issues, typically, around 15 so as to avoid burden shifting that can happen when efforts to lower one type of environmental impact unintentionally increase other types of environmental impacts. However, one should be aware of the following limitations:

- The comprehensiveness of LCA is also a limitation, as it requires simplifications in the modeling of the product system and the environmental impacts. LCA does not calculate *actual* environmental impacts but it calculates *potential* impacts. It can answer the question "should we use PVC or aluminum for a window?" but not the question "Do current emissions from a specific factory lead to pollutant concentrations above the standards in a nearby river?"

- LCA models are based on the average performance of the processes and do not support the consideration of risks of extreme events such as industrial accidents.

- While LCA can help us understand which product system is more environmentally friendly, it cannot tell us if it is "good enough." If an LCA shows that a product has a lower environmental impact compared to another product, we cannot conclude that the product is environmentally sustainable. Therefore, the results of an LCA study should be used together with other information in the decision-making process when assessing the trade-offs with cost and performance.

- Performing an LCA can be resource and time intensive. Gathering the data can be problematic, and the availability of data can greatly impact the accuracy of the final results. Therefore, it is important to weigh the availability of data, the time necessary to conduct the study, and the financial resources required against the projected benefits of the LCA.

- LCI can rarely, if ever, include every single process and capture every single input and output due to system boundaries, data gaps, cut-off criteria, and so on.

- The LCI data collected contain uncertainty and characterization models, which are far from perfect.

4.3 Environmental Life Cycle Costing, Social Life Cycle Assessment, and Life Cycle Sustainability Assessment

4.3.1 Introduction

The question "what is a sustainable product?" should be based on criteria relative to the three dimensions of sustainability. A product with an eco-label, for example, cannot be called "sustainable" if it is produced in social and economic conditions outside international norms of behavior. The debate on biofuels clearly demonstrates the shortcomings in the current way of applying LCA for sustainability evaluations. There is a need to deepen the scope of mechanisms and to broaden the scope of indicators and/or the object of analysis. Life cycle sustainability assessment (LCSA) is a proposed approach that integrates the information developed by LCA, life cycle costing (LCC), and social life cycle assessment (S-LCA) to include economic, environment, and social considerations in decision-making (Kloeppfer, 2008).

4.3.2 Environmental Life Cycle Costing

LCC is a technique that assesses costs over the life cycle of a product or system. It is also referred to as whole-life costing or total cost of ownership. We can distinguish LCC into three types:

(i) *Conventional LCC* – also known as financial LCC – is the original method. External costs are usually not included. It is mainly applied as a decision-making tool to support acquisition of capital equipment and long-lasting products with high investment costs (Hunkeler et al., 2008). It is done from the perspective of a single actor, often the user of the solution.

(ii) *Environmental LCC* – is aligned with LCA in terms of system boundaries, functional unit, and methodological steps. It takes the perspective of a functional unit and considers the whole life cycle, including all actors in the value chain.

(iii) *Societal LCC* – includes monetization of externalities, including both environmental and social impacts and is still at a young stage of its development. Its aim is to support decision-making on a societal level including governments and public authorities.

Environmental LCC is aligned with the ISO standards 14040 and 14044 on LCA and it takes the perspective of a functional unit and considers the whole life cycle, including all actors in the value chain. It supports LCA by covering the economic dimension and helps identify hotspots in terms of both cost and environmental impacts. Besides the internal costs borne by actors in the life cycle, environmental LCC may also include external costs that are expected to be internalized in the near future (Rebitzer and Hunkeler, 2003). The overall approach is very similar to the standardized LCA. Rödger et al. (2018) give a step-by-step application of environmental LCC.

4.3.3 Social Life Cycle Assessment

S-LCA, as part of an overacting LCSA, is an impact assessment technique that deals with the social and socioeconomic aspects of products, specifically their potential positive and negative social impacts throughout their life cycle, from "cradle to grave." The methodology is still in its infancy. The different aspects assessed in an S-LCA are those that may directly affect stakeholders during the life cycle of a product. Five main stakeholder categories have been identified: workers/employees, local community, society, consumers, and value-chain actors. Additional categories of stakeholders or subgroups may be added. A comprehensive set of sub-categories is identified under such stakeholder:

- For the stakeholder "Worker" – for example, "child labor" and "fair salary"
- For the stakeholder "Consumer" – for example, "health and safety"
- For the stakeholder "Local Community" – for example, "cultural heritage" and "local employment"
- For the stakeholder "Society" – for example, "consumption"
- For the stakeholder "Value-chain Actors" – for example, "supplier relationships"

S-LCA methodology is still in its infancy and the existing S-LCA literature thus presents a broad variety of approaches. To date, the most important step toward the standardization of SLCA has been the development of the "Guidelines to S-LCA" under the UNEP-SETAC Life Cycle Initiative (Benoît and Mazijn, 2009).

4.3.4 Life Cycle Sustainability Assessment

LCSA is a sustainability framework that integrates the information developed by LCA, LCC, and SLCA (Kloeppfer, 2008). The three methods, which have a different degree of development (LCA is the most mature life cycle–based method), are applied at product level, and independently from another. An important requirement of LCSA is that the three pillars of sustainability must be assessed using the same system boundaries.

4.4 LCA Applications in Engineering

This section introduces the relevance of LCA for decision-making in various fields of engineering. Further reading material on case studies is given, which explains the key methodological issues, challenges, limitations, and good practice of applying LCA within the field.

4.4.1 Environmental Product Declarations and Product Category Rules

An environmental product declaration (EPD) presents quantified environmental data for products or systems based on information from an LCA that was conducted using a standard approach defined by ISO. The format is governed by ISO 14025, Environmental Labels and Declarations – Type III Environmental Declarations – Principles and Procedures. In short, an EPD is the equivalent of a nutrition label for products and materials and is issued by independent third-party organizations that ensure uniformity and transparency in the process. EPDs allow a comparison between products being used for the same purpose. An EPD is not a certification, a green claim or a promise – it simply shows product information in a consistent way, certified to a public standard and verified by a credible third party. EPDs are product- and company-specific and thus owned by the company that develops them. An EPD is thus a document primarily based on the LCA methodology and is developed by following a "Product Category Rule." PCRs are a set of specific rules, requirements, and guidelines for developing Type III EPDs for one or more product categories, or a group of products that can fulfill an equivalent function, thus enabling fair product comparisons. PCRs make calculation rules as clear as possible so that impacts are quantified consistently across multiple studies.

4.4.2 Carbon and Water Footprinting

Footprints are life cycle based, narrow scoped, and environmental metrics focusing on an area of concern. They are widely used and easily applicable, as well as understood by nonenvironmental experts and easy to communicate. However, they have a narrow scope and limited representativeness for a comprehensive set of environmental indicators.

4.4.2.1 Carbon Footprinting

In order to quantify and manage GHG emissions to prevent global warming, carbon footprint as an indicator widely used. Carbon footprint analysis is also known as GHG accounting or GHG inventory. Because of its ease of conveying information about the GHG intensity of a variety of products and activities among the general public, carbon footprinting also offers a simple mode of communication to different stakeholders. Carbon footprint represents the net emissions of CO_2 and other GHGs over the full life cycle of a product, process, service, or organization (Wiedmann and Minx, 2007). The life cycle concept of carbon footprint means all direct (on-site, internal) and indirect (off-site, external) emission sources, sinks, and storage, within the specified spatial and temporal system boundary need to be taken into account. It is expressed as a CO_2 equivalent (usually in kilograms or tonnes per unit of analysis) and as such it is equivalent to the LCA impact category of GWP. To increase transparency and consistency in measuring and reporting carbon footprints, several guidelines and standards have been developed.

Everything has a carbon footprint. So carbon footprint can be assigned to a product, a production plant, and an organization or a business. Standards and guidance are available for GHG accounting depending on the scale of analysis. At the product level, the widely used standards are Publicly Available Specifications (PAS) 2050 (BSI, 2011), GHG Protocol – Product Life Cycle Accounting and Reporting Standard (WRI/WBCSD, 2011), and ISO 14067 (ISO 14067, 2018). These standards are all based on the LCA approach. The GHG Protocol is the most commonly used international tool for business leaders and governments to comprehend, quantify, and control GHG emissions. The accounting framework looks at both direct (Scope 1) and indirect emissions (Scopes 2 and 3), which are explained further below:

- Scope 1 – Direct GHG emissions – these occur from sources that are owned or controlled by the company, for example, emissions from combustion in owned or controlled boilers, furnaces, and vehicles, or emissions from chemical production in owned or controlled process equipment.
- Scope 2 – Electricity and heat indirect GHG emissions – this account for GHG emissions from the generation of purchased electricity and

heat consumed by the company. Purchased electricity is defined as electricity that is purchased or otherwise brought into the organizational boundary of the company. Scope 2 emissions physically occur at the facility where the electricity is generated.

- Scope 3 – Other indirect GHG emissions – this is a reporting category that allows for the treatment of all other indirect emissions. Scope 3 emissions are a consequence of the activities of the company, but occur from sources not owned or controlled by the company. Some examples of Scope 3 activities are the extraction and production of purchased materials, the transportation of purchased fuels, and the use of sold goods and services.

4.4.2.2 Water Footprinting

In recent years, water use has become an area of increased interest in life cycle inventories and assessments. Water footprinting requires a different approach than carbon and it can be more complicated to calculate. Water is not an emission, but a resource, so we must consider both supply and regionality. The first step in calculating a water footprint is the inventory stage where we add up the total volume of water used to produce a good in the supply chain. This provides valuable insight on its own, as it is important to understand how much water a product, process, or company consumes. The next step is impact assessment. What does using this much water mean? And what are the implications and impact of this? Water quality and water availability are the main considerations that need to be taken into account using impact assessment. Water is unequally distributed, meaning that using water in one area might be worse than using water in another area. In addition, water scarcity is driven by competition – the more water you use, the less there is left for others. Opinions differ on how best to take these considerations into account. Therefore, while there is basically one accepted method for calculating a carbon footprint, there are several methods for calculating a water footprint. For guidance on how to calculate a water footprint, ISO published ISO 14046: *Water footprint: Principles, requirements, and Guidelines* in August 2014. This is an international standard that defines the principles, requirements, and guidelines for conducting an LCA-based water footprint for products, processes, or organizations.

4.4.3 Energy Systems

Energy systems embody a wide range of systems and technologies and can be regarded as a supporting sector, that is, a sector that feeds all other applications sectors, for example, transportation, building sectors, and industrial sectors. The application of LCA to energy systems is a vibrant research pursuit (see Box 4.1 for example) that is likely to continue as the world seeks ways to meet growing electricity demand with reduced environmental and human health impacts. Energy systems can be considered relevant to nearly all LCA studies and can be categorized into two major groups (1) electricity and heat production systems and (2) fuels for transportation. LCA studies on electricity and heat systems can roughly be divided into three main categories:

(i) Studies assessing a specific energy technology/source/system at a power plant level or sub-power plant level (e.g., specific component of the system). The goals typically include eco-design opportunities, reporting/documentation of environmental performances of a newly developed technology, and benchmarking. There is now a good amount of research reporting the life cycle environmental and economic aspects of specific renewable energy technologies.

(ii) Studies providing accurate and transparent LCA data for electricity supply in a region, thereby increasing the robustness of LCA results for a multitude of products producing or consuming electricity throughout the life cycle.

(iii) "Whole systems" appraisal of national energy sectors and transition pathways with goals oriented toward policy making. For example these studies look at energy scenarios and penetration of renewables into electricity grid mixes or assesses changes in impacts due to policy initiatives such as increased public transport and increased CHP production.

Private transportation is increasingly responsible for a significant share of GHG emissions. In this context, biofuels and electric vehicles (EVs) are considered to be key technologies to reduce the environmental impact caused by the mobility sector. We distinguish between first-generation biofuels (derived from crops, such as corn, soybeans, sugarcane, sugar beets, wheat, barley, cassava) and second-generation biofuels (derived from agricultural lignocellulosic biomass, which are either nonedible residues of food crop production or nonedible whole plant biomass). The liquid biofuels that are largely produced and consumed are biodiesel and ethanol, which substitute fossil diesel and gasoline, respectively. The expanding biofuels global market has raised concerns about its sustainability due to the potential impact of biofuels on food security and agricultural commodity prices as well as social and environmental impacts such as deforestation, monoculture, water resource depletion, and labor conditions. These concerns pose challenges to the development of the biofuel market, and recently, there has been an explosion of LCA studies of biofuels to support biofuel policy making. LCA is normally used to measure the environmental impact of EVs and to identify potential problem shifting. EVs do offer an opportunity to decrease significantly the production of GHGs in the tailpipe. However, the production of the battery system is responsible for around 50% of the total CO_2 -eq. emissions of the vehicle's manufacturing stage. In addition, there is the risk of shifting the problem to other life cycle stages or areas of impact through the increased requirements for metals such as copper and aluminum for the battery system as well as rare earth metals for the production of electric motors. Moreover, the energy source to power an EV has an ultimate influence on the environmental impact caused during the vehicle's use stage.

BOX 4.1 GREEN ENERGY CHOICES: THE BENEFITS, RISKS, AND TRADE-OFFS OF LOW-CARBON TECHNOLOGIES FOR ELECTRICITY PRODUCTION

Global demand for energy is expected to double by 2050, requiring an estimated investment of 2.5 trillion USD a year over the next 20 years in new energy installations and energy conservation initiatives. Production of electricity, a major energy carrier, is currently responsible for 25% of anthropogenic greenhouse gas (GHG) emissions, as well as other negative impacts on the environment and on human health. Fortunately, technologies with lower carbon emissions have become available and started to penetrate the market. There is a risk that the massive deployment of low-carbon energy technologies, while effective in reducing GHG emissions, could lead to new environmental and social impacts, such as toxic metal pollution, habitat destruction, or resource depletion. Given the scale of the investments and infrastructure development required, it will be important to strategically plan the appropriate energy mix of each country or region, in order to avoid "lock-ins" of resource-intensive technologies and infrastructure that will be difficult to change. Building an optimal energy system equipped with low-carbon technologies should consider maximizing energy output, mitigating GHG emissions and at the same time addressing a range of environment and resource problems. The IRP's report Green Energy Choices: The Benefits, Risks and Trade-Offs of Low-Carbon Technologies for Electricity Production represents the first in-depth international comparative assessment of mainstream renewable and nonrenewable power-generation technologies, not only analyzing their GHG mitigation potential, but also other benefits, risks, and trade-offs, which include:

- environmental impacts: ecotoxicity, eutrophication, and acidification.
- human health impacts: particulates, toxicity, smog.

- resource use implications: iron, copper, aluminum, cement, energy, water, and land.

The report provides a comprehensive comparison of nine electricity-generation technologies including coal and gas with and without CO_2 capture and storage (CCS), photovoltaic solar power, concentrated solar power, hydropower, geothermal, and wind power. It takes a whole life cycle perspective, from the production of equipment and facilities to the extraction and use of fuel, from the operation of power plants to their dismantling. The assessment analyses and benchmarks the environmental impacts of the technologies per unit of power produced. It also assesses global environmental, health and resource implications of implementing the IEA's Blue Map (or 2°C) mitigation scenario in comparison to a business as usual scenario. The scenario envisions replacing fossil fuels for power generation with renewables on a large-enough scale to keep global warming to 2°C. From the life cycle perspective, the GHG emissions of electricity produced from renewable sources are less than 6% of those generated by coal or 10% by natural gas. Human health impacts from renewable energy electricity production are only 10–30% of those from the state-of-the-art fossil fuel power. Damage to the environment from renewable energy technologies is 3 to 10 times lower than from fossil fuel-based power systems. Natural-gas combined cycle plants, wind power, and roof-mounted solar power systems have low land-use requirements, while coal-fired power plants and ground-mounted solar power require larger areas of land. Site-specific environmental impacts, such as the ecological impacts of coalmines, hydropower dams and wind turbine installations, vary greatly, depending on the significance of the species and habitats affected and may be mitigated or offset by proper site selection and planning. n Coal- or gas-fired systems with CCS are a promising way to reduce GHG emissions, but will exacerbate various environmental pollution impacts by 5–80%.

Under the business as usual scenario for energy production, GHG emissions would double; so would the human health impacts; the impacts on ecosystems and land use would increase substantially. Under the 2°C scenario, GHG emissions from power generation decline by a factor of 5 while electricity generation doubles; human health impacts stabilize or decline moderately; environmental pollution, ecosystem impacts, and land use stabilize or decline moderately. Building the worldwide low-carbon power systems, as envisaged in the 2°C scenario will require an increased use of steel, cement, and copper in comparison with the BAU scenario. However, the amount of cement required is small relative to current levels of world production. The iron and copper required until 2050 represent 1–2 years' worth of current global production, which is significant but manageable. Renewable energy technologies rely on several functionally important metals, such as silver, indium, tellurium, and rare earth elements. The demand for these materials will be significant, especially given competing uses, associated with the 2°C scenario. Potential possible solutions to resource supply constraints include materials efficiency, material recycling, material substitution and alternative technologies, as well as new sources. The literature reviewed does not agree on the severity of potential supply constraints of critical materials. The use of variable renewables, such as wind and solar, poses challenges to balancing generation and demand in an electricity grid. It often requires fossil fuels to offset the fluctuations in supply, causing additional emissions. Variability can be addressed through larger, more versatile grids, demand-side management and energy storage.

Methane emissions from hydropower facilities are an important concern. Yet, the emissions are unevenly distributed, with few power plants responsible for a large share of emissions. Avoiding large reservoirs that produce relatively little energy and reducing the influx of biomass and nutrients can largely address these concerns. Some wind power plants have been in the focus because of collisions of birds and bats with rotating blades. Such collisions can be avoided, in part through avoiding habitats for such birds and bats and in part through the slowdown of wind turbines when birds are detected. 80% of the pollution associated with photovoltaics and wind power is caused by the production of the conversion devices. Increasing power production through good siting and avoiding downtimes reduces the impact per unit of energy delivered. Producing materials with clean energy reduces production related emissions. Coal mining, transport, and storage is responsible for 70% or more of the freshwater pollution resulting from coal power and a substantial share of associated emission of particulate matter. Pollution varies widely depending on geological factors and operational measures. Sourcing fuels only from low-polluting sites and putting in place appropriate mitigation measures can substantially reduce the environmental impacts of fossil fuel power. The key to sound energy decisions lies in selecting the right mix of technologies according to local or regional circumstances and putting in place safeguard procedures to mitigate and monitor potential impacts. This demands careful assessment of various impacts of different alternatives, so as to avoid the unintended negative consequences, and to achieve the most desirable mix of environmental, social, and economic benefits. Life cycle assessment is of central importance to making the right energy choices. Sound criteria are essential to distinguish between different actions and technology choices in terms of their ultimate sustainability. These criteria will help ensure that overall sustainability goals are met and that the actions taken are in line with global targets, such as the 2°C target under the UN Framework Convention on Climate Change and species protection targets under the Convention on Biological Diversity. Green energy choices lay the foundation for developing such sustainability criteria, and so making good decisions about the energy sources that will influence the human future and that of life on Earth, in an era of growing pressure over scarce resources.

Source: International Resource Panel. www.resourcepanel.org/reports/ green-energy-choices-benefits-risks-and-trade-offs-low-carbon-technologies-electricity

4.4.4 Buildings and the Built Environment

LCA is used in the evaluation of the environmental impacts of building materials, building products, whole buildings, and the built environment. The built environment is understood as a collection of autonomous buildings along with the infrastructure and human activity between those buildings (e.g., mobility, leisure, etc.). There are four main applications of LCA in the building industry: (1) evaluating manufacturing processes; (2) developing environmental product declarations and product labels; (3) comparing materials and methods; and (4) analyzing whole buildings.

A manufacturer or group of manufacturers can commission on LCA of their manufacturing process. This may be a cradle-to-gate impacts or only focus on gate-to-gate impacts that a specific manufacturer can control. Often these studies report the inventory of emissions for the process under the study (e.g., for reinforcing steel) and reported in database that can be integrated into an LCA of a whole building. Many LCA studies have been performed to compare a material (e.g., a steel or concrete structure) or a method (precast vs. cast in place concrete). Often funded by industry trade organizations, these studies look to identify an environmentally preferable method. Based on established LCA standards, LCA data should only be used to make decisions between options (termed comparative assertions) unless very specific analysis requirements are met. A whole building LCA can assess the impacts of manufacturing and construction in context with the operations and maintenance impacts. These whole building LCAs can be performed to evaluate methods as discussed above, to develop an understanding of the largest drivers of environmental impacts and identify opportunities for improvement. Government agencies and green building rating systems are beginning to test the potential of using LCA to evaluate whole building performance. In order to analyze a building, assumptions must be made about its life. The establishment of building life, including its use, maintenance, refurbishment, and end-of-life treatment is all uncertain and can be made based on conjecture, statistical data for past buildings or testing variable alternatives. Modeling assumptions should match current practice or evaluate the impacts of different scenarios.

We can differentiate four fields of the built environment: buildings, open spaces (roads, green spaces), networks (water, telecommunications, sewage, heating distribution, electricity distribution), and mobility. There are numerous LCA studies on each on these four fields, but there is a growing interest for the city or neighborhood scale in the field of urban sustainability assessment. It is a typical operational scale for urban eco-design addressing key issues such as shared equipment (e.g., district heating), urban density, or mobility issues. For instance, decisions made at the settlement level (orientation, compactness, and urban density) largely affect the energy consumption of an urban area. Environmental and energy concerns that nowadays focus mainly on buildings are being transferred to neighborhood planning and LCA.

4.4.5 Drinking Water Supply and Wastewater Treatment

Water supply systems are becoming more complex with increasingly more intensive treatment methods. LCA has been used to assess the potentials and reveal hotspots among the possible technologies and scenarios for water supplies for the future. Studies have been performed along the water cycle of abstraction, production, transport, and distribution as well as on entire urban water systems. LCA is used as follows:

1. To identify hotspots in water supply: Several studies have shown that water production has the highest contribution to the impacts of the entire water supply system and desalination in particular. However, in some cases, the distribution subsystem can have the highest impacts due to pumping electricity consumption.

2. LCA of water reuse: LCA has been used to show the reduced environmental burden from water systems turning toward water reuse instead of expanding surface water treatment or turning to desalination.

3. LCA as a tool in water supply management: LCA is also used in cases where there is a lack of water and the need for strategic choices in the planning of future water supplies.

LCA in the field of wastewater treatment has been applied more on tertiary treatment (removing nitrogen and phosphorus) and processes for removing micro-pollutants including ozonation and activated carbon treatment. Wastewater is also becoming more and more considered as a resource of, for example, energy (biogas from anerobic digestion of sludge); nutrients (especially phosphorus); and polymers (sludge) and LCA is used for identifying trade-offs and comparing the relative sustainability of alternative treatment systems.

4.4.6 Solid Waste Management

Following the waste hierarchy will generally lead to the most resource-efficient and environmentally sound choice. However, in some cases, refining decisions within the hierarchy or departing from it can lead to better environmental outcomes. Waste management is an area where local conditions often influence the choice of policy options. LCA can be used to decide on policy options in a specific local or regional situation such as whether it is better to recycle waste or to recover energy from it or are the GHG emissions created when collecting waste justified by the expected benefits. LCA has been used to determine the impact of municipal solid waste management under different scenarios such as landfilling with and without biogas collection, incineration, composting, anerobic digestion, recycling, and different combinations of recycling, landfill, composting, anerobic digestion, and incineration.

4.4.7 Chemicals and Chemical Production

The welfare of modern society largely builds on extensively mining minerals and fossil fuels including coal, petroleum, and natural gas to produce large quantities of synthetic chemicals. Consequently, the chemical industry faces significant challenges for the management of environmental and human health impacts related to chemicals production and use. LCAs are used extensively by the chemical industry for internal support and strategic decision-making, as well as for communication with stakeholders. LCA has been applied to evaluate the environmental performance of chemicals as well as of products and processes where chemicals play a key role. The life cycle stages of chemical products are differentiated into extraction of abiotic and biotic raw materials, chemical synthesis and processing, material processing, product manufacturing, professional or consumer product use, and finally end-of-life. A large number of LCA studies focus on contrasting different feedstocks or chemical synthesis processes, thereby often conducting a cradle to (factory) gate assessment. While typically a large share of potential environmental impacts occurs during the early product life cycle stages, potential impacts related to chemicals that are found as ingredients or residues directly in products can be dominated by the product-use stage. In many studies, LCA is discussed in relation to other chemicals' management frameworks and concepts including risk assessment, green and sustainable chemistry, and chemical alternatives assessment

4.4.8 Automotive Industry

Reducing consumption of energy and material resources is crucial for achieving sustainability. The automotive industry is considered one of the most strategic contributors to reach this objective. On the one hand, the automotive sector plays an important role for the economic stability and social welfare of several countries; on the other hand, with its huge supply chain, it is often a cause of negative environmental as well as social impacts. The environmental impacts of vehicles and their components show more significant contributions during the use-phase compared to the manufacturing. Automotive manufacturers are investing more and more resources in vehicles light-weighting, which brings important environmental advantages in terms of fuel consumptions and emissions reductions during the use-phase of cars. However, the environmental effects of light materials involve not only the use-phase, but also production and end-of-life treatment, so that a correct evaluation of environmental advantages – in terms of fuel consumptions and emissions reductions – must be extended to the whole life cycle of the vehicle.

4.4.9 Food and Agriculture

Food production systems and consumption patterns are among the leading drivers of impacts on the environment and LCA has been used extensively to assess and improve food-related supply chains as much as possible. The life cycle of a food product is split into six stages: production and transportation of inputs to the farm, cultivation, processing, distribution, consumption, and waste management. A large number of LCA studies focus on the two first stages in cradle-to-farm gate studies, as they are the stages where most impacts typically occur, due to animal husbandry and manure handling, production and use of fertilizers and the consumption of fuel to operate farm machinery. In the processing step, the raw agricultural product leaving the farm gate is converted to a food item that can be consumed by the user. Distribution includes transportation of the food product before and after processing. In the consumption stage, environmental impacts arise due to storage, preparation, and waste of the food. In the waste management stage, food waste can be handled using a number of technologies, such as landfilling, incineration, composting, or digestion. LCA is an excellent tool for generating the evidence to help the selection of interventions toward more sustainable agri-food.

4.4.10 Electronics and Information and Communication Technologies (ICT)

Electronics products have undergone a tremendous change in the recent decades with rapid technology development. Many LCAs have been conducted to determine the environmental implications of this development for consumer electronics products. Multiple LCA studies have been performed on the environmental implications of consumer electronics and LCA has been used successfully to develop eco-design strategies in the electronics industry. When there is a possibility for making a choice between metal types, availability of life cycle impact data for the metals can constitute important decision support in green electronics.

Compared to many products and services sold in the world today, ICT can have both positive and negative environmental impacts. On the one hand, ICTs have an environmental impact at each stage of their life cycle, for example, from energy and natural resource consumption to e-waste. On the other hand, ICTs can enable vast efficiencies in lifestyle and in all sectors of the economy by the provision of digital solutions that can improve energy efficiency, inventory management; and business efficiency by reducing travel and transportation, for example, tele-working and video conferencing; and by substituting physical products for digital information, for example, e-commerce. LCA is suitable to identify hotspots, that is, to identify parts of the life cycle of ICT equipment. The benefits of ICT can be quantified by the comparison of LCA results between the ICT goods, networks and services product system and the reference product system performing the same function.

4.5 Summary

This chapter gives the fundamentals of the LCA methodology according to the ISO 14040 standard. The main elements of each of the four phases are presented. Although the ISO provides a general framework for conducting an LCA, it leaves much to interpretation by the practitioner. LCA standards permit a wide range of methods to be employed provided that the analysis clearly documents the methods used for the study. There are a number of issues that the engineer should bear in mind in order to carry out an LCA or to interpret the information correctly and to be critical. Examples of the applications of LCA in various engineering disciplines are given. Lifecycle environmental performance is becoming today as predominant as safety and quality are today in the design and development of products, technologies and services. Engineers should understand the need to view solutions in a life cycle perspective. Those who design product and technical systems should be able to critically read and evaluate life cycle information about the alternatives they are considering and those who specialize in sustainability should be able to perform LCA studies.

SUSTAINABLE ENGINEERING IN FOCUS: LCA FOR PLANNING OF DISPOSAL OF PET BOTTLES IN MAURITIUS

by Dr R.K Foolmaun and Toolseeram Ramjeawon, University of Mauritius

Polyethylene terephthalate (PET) has become the most favorable packaging material worldwide in the beverage industries. PET is increasingly being used for soft drinks, mineral water, energy drinks, ice teas as well as for more sensitive beverages such as beer, wine, and juices (Welle, 2011). The reason for this development is the excellent properties of the PET material, especially its unbreakability and the very low weight of the bottles compared to glass bottles of the same filling volume. The global PET production in 2014 amounted to some 41.56 million metric tonnes, and it is forecasted that by 2020 its production will be approximately 73.39 million metric tonnes (www.statista.com/statistics/650191/global-polyethylene-terephthalate-production-outlook/). Likewise, in Mauritius PET bottles' production has increased from 97,213,115 units (or 2,946 tonnes) in 2014 to 128,301,572 units (or 3,888 tonnes) in 2018. Disposing the increasing amount of post-consumer PET bottles is becoming a nuisance in the environmental landscape, as a significant amount of these bottles are discarded irresponsibly by some consumers in watercourses, remote areas, and barelands. Such undesirable practice, over and above being eyesores, results into numerous environmental and health-related problems. Moreover, in the absence of relevant studies on disposal of post-consumer PET bottles in Mauritius, it has been difficult to define a suitable disposal route, which in addition to being cost-effective, has the least environmental and social impacts. It is in this endeavor that a Life Cycle Sustainability Assessment (LCSA) was conducted on the following four disposal options of post-consumer PET bottles:

- *Scenario 1*: 100% landfilling (all post-consumer PET bottles are sent to the landfill);
- *Scenario 2*: 75% incineration with energy recovery and 25% landfilling (75% of post-consumer PET bottles are sent to the incinerator for energy recovery and the remaining 25% landfilled);
- *Scenario 3:* 40% flake production and 60% landfilling (this is the current situation where 40% of post-consumer PET bottles are diverted to the flake production industry, while the rest are sent to the landfill); and

- *Scenario 4*: 75% flake production and 25% landfilling (similar to scenario 3, the collection rate and hence flake production rate is increased to 75% and the remaining 25% is to the landfill).

The three assessment tools LCA, LCC, and SLCA were applied to the four defined scenarios. The four phases of life cycle, i.e., goal and scope definition, inventory analysis, life cycle impact assessment, and interpretation were followed for each assessment tool. Table 4.9 provides a summary of the analysis undertaken. For detailed assessment, readers are advised to read the publications by Foolmaun and Ramjeawon (2012a, 2012b).

TABLE 4.9

Summary of the Three Assessment Tools Applied on the Four Scenarios

Tool	LCA	LCC	SLCA
Method used	ISO 14040:2006	Code of Practice on LCC	UNEP/SETAC Guidelines for Social Life Cycle Assessment of Products;
Goal and scope definition			
− Goal of study	To carry out LCA of four defined disposal scenarios of post-consumer PET bottles, with a view to determine the scenario with the least environmental impact	To carry out the environmental life cycle costing of four disposal scenarios of post-consumer PET bottles and to determine the least costly method for disposing post-consumer PET bottles in Mauritius	To carry out the social life cycle assessment of the four disposal scenarios of post-consumer PET bottles and to determine the disposal option that is socially more attractive/beneficial
− Functional unit	Disposal of 1 tonne of post-consumer PET bottles	Same as LCA	Same as LCA
− System boundary	From the point consumers dispose their post-consumer PET bottles up to the moment they lose,	System boundary defined in LCA was extended so as to include the period of setting up of the	Same as LCA

(Continued)

TABLE 4.9 (Cont.)

Tool	LCA	LCC	SLCA
Method used	ISO 14040:2006	Code of Practice on LCC	UNEP/SETAC Guidelines for Social Life Cycle Assessment of Products;
	totally, their value, that is, either through landfilling or incinerating the post-consumer PET bottles (Figure 4.8)	respective disposal facilities, that is, landfill, incinerator and flake production industry	
Stakeholders and subcategories selected	Not applicable	Not applicable	Three stakeholders (workers, society, and local community) and eight subcategories (child labor, fair salary, forced labor, health and safety, social benefits/ social security, discrimination, community engagement, and contribution to economic development) were selected for the four disposal scenarios
Life cycle inventory			
– Data collection	Data were collected mainly during site visits to industries; from technical reports and Central Statistics Office reports	Economic data such as operating costs (e.g., labor cost, maintenance cost, fuel cost, etc.), investment costs (e.g., cost of infrastructure, cost of land, cost of mobile equipment, etc.) were collected	Data were collected in the form of opinions and attitudes of the main stakeholders involved in the four defined scenarios, through survey questionnaire. The population of workers involved in

(Continued)

TABLE 4.9 (Cont.)

Tool	LCA	LCC	SLCA
Method used	ISO 14040:2006	Code of Practice on LCC	UNEP/SETAC Guidelines for Social Life Cycle Assessment of Products;
		during site visits, technical reports, and from literature	waste management was estimated to be 4,650. The representative sampling size was found out to be 140. 140 on-site interviews were conducted with workers. Interviews were also held with the respective management in order to verify the veracity of the information provided by workers while ensuring confidentiality of the information of both parties
– Data analysis	Software Sima Pro 7.1 was used	Collected economic data were entered into an excel spreadsheet model constructed to calculate the various cost categories including operating costs, net present values (NPVs), damage costs, and breakeven year or payback period for each scenario	Collected data were entered into an excel spreadsheet

(Continued)

TABLE 4.9 (Cont.)

Tool	LCA	LCC	SLCA
Method used	ISO 14040:2006	Code of Practice on LCC	UNEP/SETAC Guidelines for Social Life Cycle Assessment of Products;
Life Cycle Impact Assessment			
– LCIA method selected	Eco-indicator 99	There is no comparable impact assessment phase	A new methodology based on scoring was proposed.
– Impact categories used in the model	Carcinogens, respiratory organics, respiratory inorganics, climate change, ecotoxicity, ozone layer, acidification/ eutrophication, mineral and fossil fuel	Not applicable	Not applicable
Interpretation			
Results	Single score results is presented in Figure 4.9	Comparison of net present values and breakeven year is shown in Table 4.10	Final scores for the disposal scenarios are shown in Table 4.11
Observation	Scenarios 2, 3, and 4 showed negative values (−102, −76.1, and −154 pt, respectively). Negative values imply net environmental benefits owing to avoided emissions	Scenario 4 had the highest NPV and lowest payback period	Scenario 4 had the highest score (25.37)
Conclusion	Scenario 4 was the option with the least environmental impact	Scenario 4 was the most cost-effective disposal route	Scenario 4 was the option with the least social impact

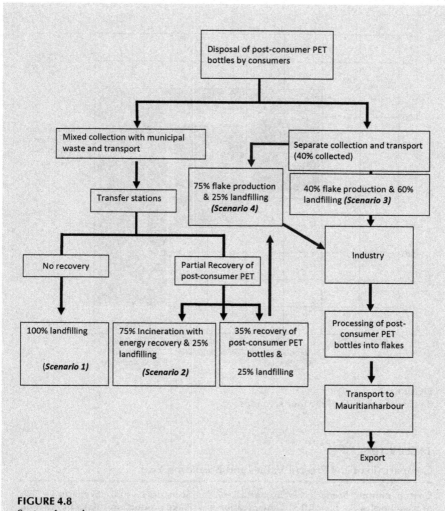

FIGURE 4.8
System boundary.

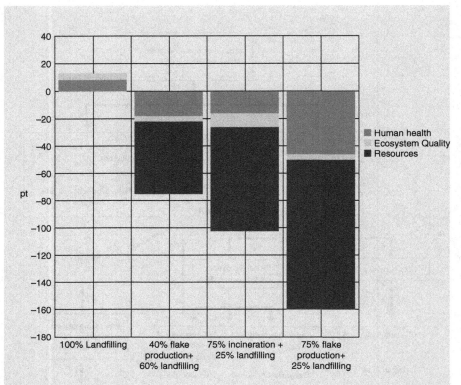

FIGURE 4.9
Single-score results of the four scenarios.

TABLE 4.10

Comparison of Net Present Values and Breakeven Year

Costs per tonne of used bottles disposed ($)	Scenario 1 – Landfill (100%)	Scenario 2 – 75% incineration + 25% landfilling	Scenario 3 – 40% Flake production + 60% landfilling	Scenario 4 – 75% flake production + 25% landfilling
Present value	17.47	29.59	55.78	96.24
Net present value	1.35	2.34	25.47	56.83
Breakeven year	18.96	16.13	9.93	7.64

TABLE 4.11

SLCA Final Scores for the Disposal Scenarios

Disposal scenarios	Score	Hierarchy
Scenario 1: 100% landfilling	20.5	4
Scenario 2: 40% flake production and 60% landfilling	23.30	3
Scenario 3: 75% incineration and 25% landfilling	23.13	2
Scenario 4: 75% flake production and 25% landfilling	25.37	1

The three assessment tools were combined using multi-criteria analysis (MCA). MCA provides a systematic approach for ranking options against a range of decision criteria. The various criteria can be weighted to reflect the relative importance of each criterion. The weighted sum of the different chosen criteria is used to rank the options. Within the MCA, a powerful and flexible decision-making process to help users set priorities and make the best decision when both qualitative and quantitative aspects of a decision need to be considered is the analytic hierarchy process (AHP). AHP is a theory of measurement through pairwise comparisons and relies on the judgments of experts to derive priority scales. After combination of the three domains (environment, economic, and social) using AHP, the final ranking of the four studied scenarios are shown in Table 4.12. This table shows that scenario 4 is the best disposal option from environmental, economic, and social points of view, that is, scenario 4 was found out to be the sustainable disposal route of post-consumer PET bottles in Mauritius.

TABLE 4.12

Hierarchy of Disposal Scenarios for Post-consumer PET Bottles in Mauritius

Hierarchy of disposal scenario	Assessment tools				Overall rank
	E-LCA	LCC	S-LCA	LCSA	
Most preferred	Scenario 4	Scenario 4	Scenario 4	Scenario 4	1
↓	Scenario 2	Scenario 3	Scenario 2	Scenario 2	2
	Scenario 3	Scenario 2	Scenario 3	Scenario 3	3
Least preferred	Scenario 1	Scenario 1	Scenario 1	Scenario 1	4

The three tools: LCA, LCC, and S-LCA and AHP results showed that scenario 4 was the best scenario among the four disposal routes studied. The study showcased that partial recycling scenario is not only beneficial from the environmental and economic viewpoint but is also equally socially valuable. Scenario 4, however, is one of the two potential future scenarios. It is therefore recommended that policy measures be taken to increase the flake production rate from 40% to 75%.

References

Baumann, H. and Tillman, A.M. 2004. *The Hitch Hikers's Guide to LCA: An Orientation in Life Cycle Assessment Methodology and Application*, Studentlitteratur, Lund, Sweden, ISBN 978-91-44-02364-9.
Benoît, C.A. and Mazijn, B. 2009. *Guidelines for Social Life Cycle Assessment of Products – Social and Socio-Economic LCA Guidelines Complementing Environmental LCA and Life Cycle Costing, Contributing to the Full Assessment of Goods and Services within the Context of Sustainable Development*, United Nations Environment Programme, Paris, France, ISBN 9789280730210.
Bjørn, A., Owsianak, M., Molin, C. and Hauschild, M.Z. 2018. LCA History. In M. Hauschild, R. Rosenbaum and S. Olsen (Eds.), *Life Cycle Assessment – Theory and Practice*, Springer, Cham, Switzerland, 17–30.
BSI. 2011. *The Guide to PAS 2050:2011, How to Carbon Footprint Your Products, Identify Hotspots and Reduce Emissions in Your Supply Chain*, British Standard Institute, London.
Cooper, J. and Kahn, E. 2012. Commentary on Issues in Data Quality Analysis in Life Cycle Assessment. *The International Journal Life Cycle Assessment*, 17, 499–503.
Čuček, L., Klemeš, J.J. and Zdravko, K. 2015. Overview of Environmental Footprints. In *Assessing and Measuring Environmental Impact and Sustainability*, Elsevier BV, Amsterdam, Netherlands, 131–193.
Curran, M.A. (Ed.). 2012. *The Life Cycle Assessment Handbook: A Guide for Environmentally Sustainable Products*, Scrivener-Wiley, Salem, MA.
Environment Agency. 2011. *Life Cycle Assessment of Supermarket Carrier Bags: A Review of the Bags Available in 2006*, UK. Available at: https://assets.publishing.service.gov.uk/government/uploads/system/uploads/attachment_data/file/291023/scho0711buan-e-e.pdf (Accessed 21st August 2019). www.gov.uk/government/publications/life-cycle-assessment-of-supermarket-carrierbags-a-review-of-the-bags-available-in-2006.
Foolmaun, R.K. and Ramjeawon, T. 2012a. Comparative Life Cycle Assessment and Life Cycle Costing of Four Disposal Scenarios for Used Polyethylene Terephthalate Bottles in Mauritius. *Environmental Technology*, 33(17), 2007–2018.
Foolmaun, R.K. and Ramjeawon, T. 2012b. Comparative Life Cycle Assessment and Social Life Cycle Assessment of Used Polyethylene Terephthalate (PET) Bottles in Mauritius. *The International Journal of Life Cycle Assessment*, 18, 155–171.
Frischknecht, R., Althaus, H.-J., Dones, R., Hischier, R., Jungbluth, N., Nemecek, T., Primas, A. and Wernet, G. 2007. Renewable Energy Assessment within the Cumulative Energy Demand Concept: Challenges and Solutions. In *Proceedings*

from: SETAC Europe 14th LCA Case Study Symposium: Energy in LCA – LCA of Energy, 3–4 December 2007, Gothenburg, Sweden.

Guinée, J.B., Heijungs, R., Huppes, G., Zamagni, A., Masoni, P., Buonamici, R., Ekvall, T. and Rydberg, T. 2011. Life Cycle Assessment: Past, Present, and Future. *Environmental Science & Technology*, 45, 90–96.

Hauschild, M.Z. 2018. Introduction to LCA Methodology. In M. Hauschild, R. Rosenbaum and S. Olsen (Eds.), *Life Cycle Assessment – Theory and Practice*, Springer, Cham, Switzerland, 59–66.

Huijbregts, M.A.J. 1998. Application of Uncertainty and Variability in LCA. A General Framework for the Application of Uncertainty and Variability in Life-Cycle Assessment. *The International Journal of Life Cycle Assessment*, 3(5), 273–280.

Hunkeler, D, Lichtenvort, K and Rebitzer, G (Eds.). 2008. *Environmental Life Cycle Costing*, SETAC, Pensacola, FL (US) in collaboration with CRC Press, Boca Raton, FL.

IPCC. 2007. Climate Change 2007: The Physical Science Basis. In S. Solomon, D. Qin, M. Manning, Z. Chen, M. Marquis, K.B. Averyt, M. Tignor and H.L. Miller (Eds.), *Contribution of Working Group I to the Fourth Assessment Report of the Intergovernmental Panel on Climate Change*, Cambridge University Press, Cambridge, UK and New York. Available from: www.ipcc.ch.

ISO. 2006a. *ISO 14040:2006 – Environmental Management – Life Cycle Assessment – Principles and Framework*, International Standards Organization, Geneva.

ISO. 2006b. *ISO 14044:2006 -Environmental Management – Life Cycle Assessment – Requirements and Guidelines*, International Standards Organization, Geneva.

ISO 14041. 1998. *Environmental Management – Life Cycle Assessment – Goal and Scope Definition and Inventory Analysis*, International Standards Organization, Geneva.

ISO 14067. 2018. *Greenhouse Gases – Carbon Footprint of Products – Requirements and Guidelines for Quantification*, International Standards Organization, Geneva.

Kloeppfer, W. 2008. Life Cycle Sustainability Assessment of Products: (With Comments by Helias A. Udo de Haes, p. 95). *The International Journal of Life Cycle Assessment*, 13, 89–95.

Rebitzer, G. and Hunkeler, D. 2003. Life Cycle Costing in LCM: Ambitions, Opportunities and Limitations. *The International Journal of Life Cycle Assessment*, 8, 253–256.

Rödger, J.-M., Kjær, L.L. and Pagoropoulos, A. 2018. Life Cycle Costing: An Introduction. In M.Z. Hauschild, R.K. Rosenbaum and S. Irving Olsen (Eds.), *Life Cycle Assessment: Theory and Practice*, Springer, Cham, Switzerland, 373–399.

Rosenbaum, R.K., Hauschild, M.Z., Boulay, A.-M., Fantke, P., Laurent, A., Núñez, M. and Vieira, M. 2018. Life Cycle Impact Assessment. In *Life Cycle Assessment: Theory and Practice*, Springer, Cham, Switzerland, 167–270.

Searchinger, T., Heimlich, R., Houghton, R., Dong, F., Elobeid, A., Fabiosa, J., Tokgoz, S., Hayes, D. and Tun-Hsiang, Y. 2008. Use of U.S. Croplands for Biofuels Increases Greenhouse Gases Through Emissions from Land-Use Change. *Science*, 319, 1238–1240.

WBCSD/WRI. 2011. *Greenhouse Gas Protocol: Product Life Cycle Accounting and Reporting Standard*. World Business Council for Sustainable Development and World Resources Institute, Geneva, Switzerland and Washington, DC.

Weidema, B. and Wesnaes, M. 1996. Data Quality Management for Life Cycle Inventories-An Example for Using Data Quality Indicators. *Journal of Cleaner Production*, 4(3–4), 167–174.

Welle. 2011. Twenty Years of PET Bottle to Bottle Recycling—An Overview. *Resources Conservation and Recycling*, 55(11), 865–875.

Wiedmann, T. and Minx, J. 2007. *ISAUK Research Report 07-01*, Centre for Integrated Sustainability Analysis, ISAUK Research & Consulting, Durhau.

Exercises

1. Name the four main phases of a life cycle assessment according to ISO 14040. Note the content of each phase.

2. LCA is an important tool for identifying burden shifting. Through a literature review, give examples of potential trade-offs that may occur from one media to another, or from one life cycle stage to another, or from one environmental impact to another or from one geographical area to another.

3. What is a carbon footprint? What are the six GHG emissions it accounts for and what are their sources? What is the unit of measurement? What is the difference between a carbon footprint and an LCA? What is the GHG protocol and distinguish between scopes 1, 2, and 3.

4. "Carbon footprinting are easier to conduct and interpret than more complete LCAs because they greatly simplify the process". Discuss the benefits and disadvantages of carbon footprinting and other single-issue approaches.

5. What is the life cycle tree for a plastic water bottle? (includes the label and cap; exclude the water). You may wish to discuss with students in other engineering disciplines.

6. Define appropriate functional units to compare the following systems:

    (i)   Soft drink packaging
    (ii) Waste management options
    (iii) People transportation
    (iv) Goods transportation
    (v)   Remote control for your TV with either one-way or rechargeable batteries
    (vi) Online or conventional dictionary (book)

7. If you want to compare electronic billing vs. paper bills, in order for you to choose whether you should or should not opt for electronic billing from your electricity bills, what functional unit would you use to assess the life cycle of the bills?

8. Determine what functional unit should be used for an LCA comparing gasoline and ethanol production (hint: think energy). Explain your choice.

9. Consider the use "stage of a life cycle assessment on an incandescent light bulb". Assume that the only flow within the system during that stage is the electricity needed to operate the bulb. The bulb consumes 1 kWh of electricity to produce 16,000 lumen hours of light. Each kWh of electricity has the following simplified inputs and outputs to and from nature:

    Inputs: 0.356 kg coal

    Outputs: 1.01 kg CO_2, 1.60 × 10^{-3} kg NO_x, 1.22 × 10^{-2} kg SO_2, and 9.26 × 10^{-6} kg PM_{10}

    Considering the functional unit is 20,000,000 lumen hours, convert the LCI data into the quantities of inputs and outputs based on the functional unit.

10. Draw a simple system diagram for generation of electricity in a coal-fired power plant and in a solar PV plant. Look up online what some of the major inputs, outputs, and processes are (detailed inputs and outputs are available from the US LCI website for example). Be sure to include a system boundary and place a process for use of the electricity outside of this boundary.

11. System A is a multi-output process, which produces 10 kg of product A and 5 kg of product B per minute. 1 kg of product A is worth 20 USD, 1 kg of B is worth 5 USD. The process emits 120 kg CO_2 per hour. Product B can also be produced using another process, which produces 0.05 kg CO_2 per kg of product B. Calculate the CO_2 emissions for two different allocation procedures for product A.

12. What is the system expansion approach for use in co-product allocation?

13. Through a literature search, explain, with examples, the difference between closed-loop and open-loop recycling. How does each affect modeling choices?

14. Explain the difference between modeling to the midpoint and modeling to the endpoint. What are the advantages and disadvantages of each approach?

15. Why do you think normalization, grouping, and weighting are optional steps in the ISO standards framework?

16. Distinguish between uncertainty analysis, sensitivity analysis and contribution analysis in the Interpretation phase of an LCA.

17. Mobility is one of the measures of quality of life that citizens of many developed nations value highly, ranked behind only food and shelter as necessities for life. Mobility is also a key factor in sustainability

because of the cumulative effects of providing mobility on the environment, on resource depletion, and on the economy. Compare two modes of transportation, for example, car and bus, in terms of energy consumption and carbon dioxide emissions. Assume 2 persons per car and 30 persons per bus. Assume from the literature typical gasoline and diesel energy content and consumption for the automobile and bus, respectively. You have to define an appropriate functional unit for a comparison of the bus and car transportation table for personal mobility and then calculate the energy consumption and CO_2 emissions per functional unit.

18. Each country has a unique profile of electricity generation types, and this characteristic is also true for cities within these countries. Using the table of electricity generation sources in your country available from statistical agencies, calculate in a table the global warming index for your country's electricity based on 1 kWh generated. Typical emissions of gaseous pollutants based on the production of 1 kWh of each electricity type are available in the literature LCI databases and software. These inventory data represent a compilation over the entire life cycle from extraction of raw materials from nature to the output of electricity from the power plant: a "cradle-to-gate" inventory. The inventory values for each pollutant must be multiplied by the appropriate GWP characterization factor, which is also available from the literature or databases

19. Discuss the following statement: *LCA cannot determine if a product is "sustainable" or "environmentally friendly"*

20. Why is the following LCSA equation described as "symbolic"?

$$LCSA = LCA + LCC + SLCA$$

21. Find a paper or an article on an application of LCA in a journal or online and examine its goal statement, scope, assumptions, inventory results, LCIA method, and interpretation.

22. Life cycle assessments of carrier bags have been carried out in several countries and regions to aid the local debate on carrier bag use. In addition to the environmental impacts, several of the studies have also included various social impacts, such as potential impacts on littering, industry, recycling, and consumers. The following case study was conducted in 2004 by Ecobilan for the French supermarket chain Carrefour (Source: Environment Agency, 2011).The aim of the study was to quantify the environmental impacts associated with Carrefour carrier bags in the different countries where Carrefour is represented (mainly France, Belgium, Spain, and Italy). The results were intended for use in Carrefour policy development. The study was carried out in

conformance with ISO 14040 and included a critical review conducted by the French Environment Agency (ADEME). The carrier bags involved in the study are shown in Table 4.13, which includes bag weight and usable volume.

The functional unit for the study was the carrier bags required for the packing of 9,000 L of purchases. This was based on the typical annual volume of purchased goods per customer. The study is representative of France and primary data were collected from suppliers of bags to Carrefour and supplemented using secondary data from the Ecobilan database. The modeling of end-of-life activities was based on French household waste processing, which was split between landfill (51%) and incineration (49%). A recycling rate of 45% was included for waste paper. The Impact categories/indicators considered were the consumption of nonrenewable energy resources, water, the emission of greenhouse gases, atmospheric acidification, and the formation of photochemical oxidants, eutrophication, residual solid waste, and littering. A sensitivity analysis was also carried out on the following parameters: (i) reuse of 32.5% and 65%, respectively, of HDPE bags as bin liners; (ii) reuse of paper bags once; (iii) 100% landfill of used bags; (iv) 100% incineration of used bags, with an without energy recovery; and (v) recycling of 30% of used LDPE bags.

The relative performances of the different carrier bags against the environmental indicators assessed are shown in Table 4.14.

TABLE 4.13

The Carrier Bags Included in the Study

	HDPE bag	LDPE bag	Paper bag	Biodegradable bag
Materials	HDPE, virgin LLDPE TiO_2 ink adhesive	LDPE, virgin TiO_2 ink	Paper, recycled ink adhesive	50% starch 50% PCL ink
Weight	6.04 g	44 g	52 g	17 g
Useable volume	14 l	37 l	20.48 l	25 l
Country of manufacture	Malaysia, France, Spain	France	Italy	Italy
Reuse scenarios	No Yes, in sensitivity analysis (reused as bin liner)	Yes (reuse rates of 1, 2, 3, 4, and 20 investigate)	Generally no. Yes, in sensitivity analysis	No

TABLE 4.14

The Results of the Study over 8 Indicators (>1 equals worse than HDPE bag, <1 equals superior to HDPE bag)

Impact categories	HDPE	LDPE Used 2x	Used 4x	Used 20x	Paper bag	Bio-degradable bag
Consumption of nonrenewable energy sources	1	1.4	0.7	0.1	1.1	0.9
Consumption of water	1	1.3	0.6	0.1	4	1
Emission of greenhouse gases	1	1.3	0.6	0.1	3.3	1.5
Acidification	1	1.5	0.7	0.1	1.9	1.8
Formation of photochemical oxidants	1	0.7	0.3	0.1	1.3	0.5
Eutrophication of water	1	1.4	0.7	0.1	14	12
Production of solid waste	1	1.4	0.7	0.1	2.7	1.1
Risk of littering	High	Average to low			Low	Average to low

Analyze the above data and recommend the best option(s).

Recommended reading and websites

- Books
  - The Life Cycle Assessment Handbook: A Guide for Environmentally Sustainable Products (2012). MA Curran (ed.) Scrivener-Wiley; ISBN 9781118099728; 625 pp.
  - Life Cycle Assessment: Principles, Practice and Prospects (2009). Ralph Horne, Tim Grant and Karli Verghese (eds) CSIRO Publishing; 192 pp.
  - Why Take a Life Cycle Approach? (2004). Prepared for the UNEP/SETAC Life Cycle Initiative, United Nations Publications ISBN 92-807-24500-9; 24 pp.
- Scientific journals with case studies. The most important sources for this are:
  - International Journal of Life Cycle Assessment
  - Journal Cleaner Production
  - Journal of Industrial Ecology

- There are a number of organizations and networks that aim to facilitate and disseminate LCA. The most important ones are listed below:

  - The UNEP/SETAC Life Cycle Initiative (http://lcinitiative.unep.fr/): the United Nations Environment Program, UNEP and the Society for Environmental Toxicology and Chemistry, SETAC launched an International Life Cycle Partnership, known as the Life Cycle Initiative, to enable users around the world to put life cycle thinking into effective practice.

  - The SETAC (www.setac.org/): the Society of Environmental Toxicology and Chemistry (SETAC) is a nonprofit, worldwide professional society comprising individuals and institutions engaged in the study, analysis, and solution of environmental problems, the management and regulation of natural resources, environmental education, and research and development. LCA is one of their themes.

  - The European Platform on Life Cycle Assessment (http://lca.jrc.ec.europa.eu/): The "European Platform on Life Cycle Assessment" is a project of the European Commission. The purpose is to improve credibility, acceptance, and practice of LCA in business and public authorities, by providing reference data and recommended methods for LCA studies.

- LCA information portals providing links to documents, data, software, conferences, and so on.

  - American Center for Life Cycle Assessment (www.lcacenter.org/)
  - Australian LCA Network (http://auslcanet.rmit.edu.au/)
  - EPA-NRMRL – Life-Cycle Assessment (www.epa.gov/ORD/NRMRL/lcaccess/)
  - European Platform on Life Cycle Assessment (http://lca.jrc.ec.europa.eu/)
  - LCA hotlist – Gabor Doka (www.doka.ch/lca.htm)
  - UNEP-SETAC Life Cycle Initiative (http://lcinitiative.unep.fr/)

5

Introduction to Environmental Economics

5.1 Introduction – What Is Environmental Economics?

To produce the goods and services that humans want, resources such as land, labor, and raw materials are necessary. However, these resources exist in limited supply and scarcity arises when the demand exceeds supply. At any point in time, there is only a finite amount of resources available. Economics is the study of the allocation of scarce resources, of how markets function, and of how incentives can affect the behavior of people, businesses, and institutions. It had been the practice in the past to consider the resources and services provided by the environment to be limitless, having no cost, and not scarce. However, this is not the case. Environmental economics deals specifically with the application of the principles of economics to the study of how environmental and natural resources are developed and managed. It focuses on the incentives rational humans have when deciding to use our natural environment's scarce resources. Although economics is often blamed for many of the problems we are currently facing, it helps us to explain the perverse incentives present in society that increase unsustainable behavior and can help us change these incentives that support the desired, more sustainable behavior. In 2002, for example, Ireland introduced a tax on the consumer for plastic bags. The results were significant: one year after the introduction of the fee, the consumption of plastic bags decreased by more than 90% (Convery et al., 2007). By putting a tax on plastic bags, the Irish government created economic incentives for environmentally friendly behavior.

The understanding of the interaction between economies and their environments is a basic aspect of environmental economics. Fundamental to an understanding of sustainable development is the fact that the economy is not separate from the environment in which we live (Pearce et al., 2013). There is an interdependence because the way we manage the economy, have impacts on the performance of the economy. Economists had traditionally focused only on economic activities. However, as Figure 5.1 shows, the natural environment serves three main functions.

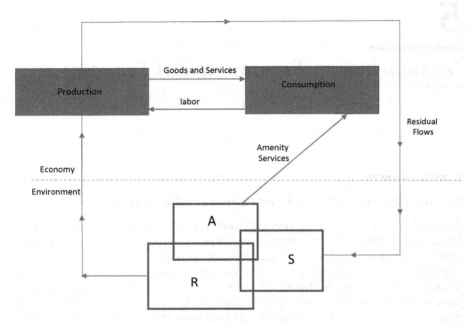

FIGURE 5.1
The linkage between environment and economic activities.

Economic activities need resources (R) in the form of materials and energy from the environment, and the latter also act as a sink (S) for the emissions occurring from production and consumption. Amenity (A) services such as living space, natural beauty, and recreational space flow from the environment directly to consumption. Economic activities and the environment are thus not mutually exclusive, as the debate on climate change is currently revealing. This two-way interaction is absolutely fundamental to sustainable development thinking. Economy affects the environment and the environment in turn affects the economy. It must be noted that in environmental economics we distinguish between two types of costs: private costs and social costs, depending on who is evaluating costs – a private individual or the society. Social costs (or economic costs) represent the real and full value of resources used. For example, driving a car involves private costs (which consist of fuel, oil, maintenance, etc.), while social costs include, besides the above private costs, the costs of congestion and pollution. In environmental economics, we always evaluate projects from the perspective of society.

The physical effect of an emission into the environment can be biological, chemical, or auditory (noise). This will lead to a human reaction showing up as an expression of distaste, unpleasantness, concern, anxiety, and is called a loss of welfare. For example, consider an upstream industry, which discharges wastes to a river, causing a deterioration of the water quality, which in turn causes a decrease of the fish population, bringing about financial and/or recreational losses to fishermen downstream. If the fishermen are not compensated for their loss of welfare, the upstream industry will continue its activities as if the damage done downstream was irrelevant to it. The industry is said to create an external cost. An external cost is also known as a negative externality. Thus, as Pearce and Turner (1989) put it, an external cost exists when the following two conditions prevail:

1. An activity by one agent causes a loss of welfare to another agent.
2. The loss of welfare is uncompensated.

A negative externality is therefore an uncompensated, human-caused harm to others in society. If the loss of welfare is compensated by the agent causing the externality, the effect is said to be internalized. Pollution is the prime example of a negative externality. Internalization of externalities should be done to incorporate these costs and benefits in the decision-making process. This will help reflect the true cost of environmental resources, which are usually treated as "free goods" and hence lead to optimal social allocation of resources. The "Polluter Pays Principle" (PPP), which is a guiding rule for environmental policy in many countries, intends to deter the free use of resources, making polluters "internalize the externalities" caused by their activities. The objective is to integrate all environmental degradation and resource depletion in the price of goods and services. There are two possible forms of PPP:

(i) the extended one, in which the polluter compensates those affected by the pollution that s/he generates, and

(ii) the standard one, in which the polluter bears the expenses of carrying out the abatement measures decided by a public authority.

Market failure is defined as the inability of markets to reflect the full social costs or benefits of a good or service. This "failure" needs to be addressed by correcting prices so they take into account the "external" costs. It is important for an engineer to understand the key concepts of environmental economics for the following reasons:

- The context in which organizations now take decisions is rapidly changing and environmental economics is part of that context.
- Nature provides many freely available ecosystem services such as erosion control and climate regulation. It is important to understand how these ecosystem services can be valued and how environmental externalities can be quantified.

- Understand regulatory and market-based instruments devised by policy makers
- Have a better understanding of the full cost of business and society's decisions.

5.2 Valuing the Environment

Assessing the economic value of the environment is a major topic in environmental economics and it raises technical questions reflecting the nature of environmental goods. Because environmental goods are typically not priced, estimates of the value of environmental changes – improvements or degradation – must be measured by some sort of indirect means. Some of the benefits cannot readily be quantified at all, let alone in monetary terms. Furthermore, there are different conceptual types of "value", all of which should be taken into account. Figure 5.2 gives the components of total economic value. The first component is use value in use of environmental goods, such as the value in use of water for irrigation, or the value of wetlands in providing natural filtration and treatment of water. Use value refers to attributes of the environment that provide some service now. In addition, there is the

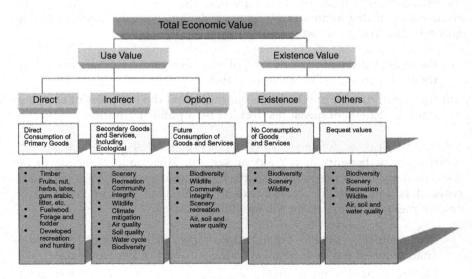

FIGURE 5.2
Taxonomy of total economic value.

(Source: Pearce and Moran, 1994)

concept of option value – there may be a resource that may have value in the future but is not currently used. For example, within the natural forest areas there may be plants or insects which potentially have a value for medicine or other products, but which are not currently used as such. Thirdly is the concept of existence value – the notion that people want something conserved or restored simply so that it exists, regardless of whether they ever actually use it. Associated with these latter two is the notion of altruistic or bequest value – environmental quality is valued for the next generation. There are difficulties in estimating all these values, although perhaps existence value is the most difficult.

Consider a plan for the construction of a dam and the opposition of people due to the deforestation it will entail. We can try to put a monetary value on the cost of losing the forest. While people derive "happiness" or "quality of life" from the forest, certain aspects of it do not have a market price as they are not sold in the market. We do not pay for the beautiful view of the forest unless it is part of a park where we have to pay an entrance fee. The absence of a market price is synonymous as if it had no value. Environmental economics tries to measure the value of these aspects and give a monetary value to the environment so that the value can be taken into consideration. We can give a monetary value to the water-retaining function of the forest by estimating the cost of construction of a dam, which has the same amount of water-retaining function as the forest. Similarly, we can give a monetary value to the recreation function of the forest by looking at how much people are willing to pay as the travel cost to go to the forest. The satisfaction of going to the forest is evaluated as higher than the travel cost. The monetary value of clean air can be estimated by considering the differential prices of houses with the same characteristics but in areas with different degrees of air pollution. For specific functions of the forest that are not used by humans, such as providing habitat to an endangered animal, we can give monetary value to the importance by conducting a survey to ask people how much they are willing to pay for the forest preservation for the animal. Through such methods, environmental economics tries to give a monetary value to the environment so that the values can be taken into consideration in the decision-making.

The main method used for valuation of the environment in a project is cost-benefit analysis (CBA). This analysis is basically compiling the costs of a project as well as the benefits, then translating them into monetary terms and discounting them over time (discounting is the process of determining the present value of future benefits and costs). Ideally, only projects with benefits greater than costs would be acceptable. The following methodologies are used to estimate the economic value associated with ecosystems ("Dollar-Based Ecosystem Valuation Methods," n.d.):

1. *Market price method*: This method estimates economic values for eco-
    system products or services that are bought and sold in commercial
    markets. For example, a cultural site could be valued based on the
    entrance fees collected.

2. *Productivity* method: This method estimates economic values for
    ecosystem products or services that contribute to the production of
    commercially marketed goods. For example, the benefits of different
    levels of water quality improvement would be compared to the costs
    of reductions in polluting runoff.

3. *Hedonic pricing method*: This method estimates economic values for
    ecosystem or environmental services that directly affect market prices
    of some other good. Most commonly applied to variations in housing
    prices that reflect the value of local environmental attributes.

4. *Travel cost method*: This method estimates economic values associated
    with ecosystems or sites that are used for recreation. Assumes that
    the value of a site is reflected in how much people are willing to pay
    to travel to visit the site. For example, adding up the costs people
    would expend to travel and recreate at a particular area.

5. *Damage cost avoided, replacement cost, and substitute cost Method*:
    This method estimates economic values based on costs of avoided
    damages resulting from lost ecosystem services, costs of replacing
    ecosystem services, or costs of providing substitute services, for
    example, the costs avoided by providing flood protection.

6. *Contingent valuation method*: This method estimates economic values
    for virtually any ecosystem or environmental service. It is the most
    widely used method for estimating non-use, or "passive use" values.
    It asks people to directly state their willingness to pay for specific
    environmental services, based on a hypothetical scenario. For exam-
    ple, people would state how much they would pay to protect a
    particular area.

7. *Contingent choice method*: This method estimates economic values for
    virtually any ecosystem or environmental service. It is based on
    asking people to make trade-offs among sets of ecosystem or envir-
    onmental services or characteristics. It does not directly ask for will-
    ingness to pay – this is inferred from trade-offs that include cost as an
    attribute. For example, a person would state their preference between
    various locations for siting a landfill.

8. *Benefit transfer method*: This method estimates economic values by
    transferring existing benefit estimates from studies already com-
    pleted for another location or issue. For example, an estimate of the
    benefit obtained by tourists viewing wildlife in one park might be
    used to estimate the benefit obtained from viewing wildlife in a
    different park.

5.3 Market-based Incentives (or Economic Instruments) for Sustainability

The importance of the internalization of external costs in sustainable development and the critical role of economic instruments in bringing it about was duly recognized by the United Nations Conference on Environment and Development in Rio de Janeiro, June 1992. Principle 16 of the Rio Declaration states:

> National authorities should endeavor to promote the internalization of environmental costs and the use of economic instruments, taking into account the approach that the polluter should, in principle, bear the cost of pollution with due regard to public interest and without distorting international trade and investment.
>
> (Retrieved from: https://sustainabledevelopment.un.org/ index.php?page=view&type=111&nr=1709&menu=35)

As a result of market failures, producers and consumers of products and services do not receive the correct signals about the true scarcity of resources they deplete or the cost of environmental damage they cause. This leads to overproduction and overconsumption of commodities that are resource-depleting and environment-polluting, and underproduction and underconsumption of commodities that are resource-saving and environment-friendly. Thus, the emerging pattern of economic growth and the structure of the economy is one that undermines its own resource base, and is ultimately unsustainable (Panayotou, 1998).

Market-based incentives (MBIs) or economic instruments can be defined as instruments that affect costs and benefits of alternative actions open to economic agents, influencing behavior in such a way that it is favorable to the environment. The purpose of economic instruments is to ensure that all economic agents incorporate all costs and benefits deriving from the actions in their decision-making process. In many circumstances, economic instruments are applied via prices, in order to guarantee the proper pricing of environmental resources and to encourage their more efficient use and allocation. In case of pollution, a price must be set on emissions resulting from economic activities, so that generators of pollution also bear the costs of environmental protection. The main purpose of economic instruments is to create incentives in order to change behavioral patterns, but these can also be a source of financial resources that may be used for environmental protection. There are five main types of economic instruments:

(i) **Taxes or Charges**
    These are utilized to discourage polluting activities and/or to provide financial assistance for pollution reduction. Taxes are a direct method

of putting prices on the use of the environment. The most important charges are emission or effluent charges, user charges, and product charges.

Emission charges are charges on the discharge of pollutants into air, water or on the soil, and on the generation of noise. They are calculated according to the quantity and toxicity of the pollutant. A carbon tax on fossil fuels according to their carbon content is an example of an emission charge. Effluent or emission charges are suitable when there are stationary point-source discharges of pollutants to the air, water or the soil, as well as when there is a generation of noise. They are also appropriate when monitoring is viable at a reasonable cost and when pollution abatement is feasible.

User charges have been defined as payments for the costs of collective or public treatment of effluent (including waste). User charges (including charges for recovery of administrative costs) have a revenue-raising purpose, by definition, and only those who are connected to the public service are charged. They can be directly related to the amount of pollution discharge. The intention is that payments for such services will reflect the costs of providing the service. This includes the costs of complying with the environmental requirements imposed on the plants. User fees are typically calculated and collected by the units that own and operate the plants in question. Administrative taxes are payments for authority services that are associated with the administration of related environmental regulation. Administrative taxes can, in a sense, be considered as a user fee.

Product charges are levied on products that are harmful to the environment when used in production processes, consumed or disposed of. They can be applied to products that cause environmental problems either because of their volume or as a consequence of their toxicity of certain harmful contents such as heavy metals, polyvinyl chloride (PVC), chlorofluorocarbons (CFCs), halogenated hydrocarbons, nitrogen, and phosphorus. Product charges can act as substitute for effluent charges when direct charging for emissions is not possible. They may be applied to raw materials, intermediate or final (consumer) products. These charges are appropriate for products used in large quantities and are used mainly for mobile sources. Differentiated taxes on vehicle fuels thus are widely applied to stimulate, for example, the use of low-sulfur diesel or the use of unleaded petrol.

(ii) **Marketable or Tradable Permits**
Within the economic approach to environmental protection, it is possible to use marketable permits as alternatives to taxes. In this case, the environmental agency will set limits to waste discharge corresponding to target levels. Permits will then be issued allowing

for only agreed upon quantity of pollution (or quotas) to be generated. The permits can then be marketed, and a market clearing price will result which will indicate the opportunity cost of waste emissions. Their primary advantage is that they can reduce the cost of compliance. Marketable permits allow companies to pollute at a level that is marginally cost-effective. It allows them to buy additional permits as needed if they fail to meet their targets internally, and to sell excess permits if they exceed their internal pollution reduction targets (Stavins, 2003). Marketable permits are potentially applicable in all media and economic sectors. Examples or existing systems include:

- "Bubbles," in which two existing stationary sources of air pollution are permitted to readjust their assigned emission limits (with one source's limit being raised while the source's limit is lowered) as long as the resulting emission limits yield equal or better environmental results.
- "Offset" programs under which a firm can enter or expand a polluting activity in a geographical area in which an increase in pollution levels is prohibited. Under the "Offset" program, the firm wishing to increase its emissions can buy allowances from other firms already located in the area, which must then abate their emissions by the amount that is necessary to at least maintain ambient environmental level.
- Production quotas, in which assigned levels of production, say CFC, can be traded among CFC producers. Trades can be external (between different plants, products etc. of the same firm). In the latter case, no financial trading transactions take place.

(iii) **Deposit–Refund systems**
These are mainly used to encourage reuse and/or more "environmentally friendly" disposal. In deposit–refund systems, on purchase of a product, a deposit is paid on potentially polluting products. When pollution is avoided by returning the products or their residuals, a refund follows. This instrument has the attractive element of rewarding good environmental behavior. Deposit-refund systems have operated since long in the field of beverage containers. Their origin is purely economical, as returnable bottles are normally cheaper than nonreturnable bottles. Since substantial parts of household waste consist of packaging, deposit–refund systems can considerably reduce waste volumes. Furthermore, such systems can contribute to prevention of the release of toxic substances into the environment, for instance, from disposal of batteries, incineration of plastics or residuals from pesticide containers. Deposit–refund system may be a desirable part of integrated life cycle management

for certain products for proper handling. This system is appropriate for products for which recycling and reuse are technically feasible.

(iv) **Subsidies**

Subsidies can take the form of grants, soft loans, and tax allowances, which may be used to encourage more environmentally conscious behavior. For example, a subsidy could be offered for the purchase of clean technology in order to achieve a reduction in overall pollution levels. Subsidies provide the opportunity for financial assistance to motivate individuals or enterprises to act more environmental-friendly per se. Subsidies may also be used to reduce compliance costs in relation to specific environmental regulations. In many cases, the removal of misdirected subsidies can prove to be an effective environmental protection measures – for example, subsidies on fossil fuels and fertilizers.

(v) **Financial enforcement incentives**

These instruments are penalties designed to induce polluters to comply with environmental standards and regulation. They include instruments such as noncompliance fees (i.e., fines) and performance bonds (payments made to regulatory authorities before a potentially polluting activity is undertaken, and then returned when the environmental performance is proven to be acceptable), which provide an additional inducement to comply with existing environmental regulations. This category of economic instruments is sometimes considered to be a legal, rather than an economic instrument.

Table 5.1 and Box 5.1 give examples of the typical application of economic instruments

TABLE 5.1

Typical Application of Economic Instruments

Market-based Instrument	Source of Pollution
Charge system	Industrial air and water pollution from large units
Deposit refund	☐ Waste management households (glass and plastic, car batteries)
Taxes	☐ Air pollution mostly from large units ☐ Fuel use ☐ Traffic congestion ☐ Halting deforestation via a "forestry tax"
Subsidies	☐ Air pollution from both large and small units ☐ Used to incentivize reforestation and adoption of cleaner technologies
Tradable permits	☐ Air pollution (SO_2) from large units ☐ Water use by large units ☐ Car use/congestion in megacities

BOX 5.1 THE FRENCH ECOLOGICAL "BONUS–MALUS" SYSTEM FOR THE PURCHASE OF PRIVATE CARS

The bonus–malus system is a financial instrument that provides a financial reward (bonus) for purchasers of fuel-efficient new cars and a financial penalty (malus) for those buying cars emitting high levels of CO_2. The amount of the bonus or malus depends on the amount of CO_2/km emitted by the vehicle. For instance, a bonus of €200 to €1,000 is provided for vehicles emitting a maximum of 130 g CO_2/km and €5,000 for those emitting no more than 60 g CO_2/km. Similarly, a malus of €200 to €2,600 is applied to those vehicles emitting over 160 g CO_2/km (consuming around 6.9 liters of fuel per 100 km) and even more for the least fuel efficient vehicles. A super bonus scrapping fee of up to €300 is paid to people delivering a car that is at least 15 years old to the breakers yard and buying a vehicle eligible for the bonus. The overall goal of the bonus–malus system is to reduce the average level of CO_2 emissions from new vehicles in France to 130 g/km of CO_2. The more detailed objectives of the system include the acceleration of the removal of obsolete and polluting vehicles from French roads, while encouraging manufacturers, retailers, and car owners to produce, sell, and purchase greener vehicles (ETC/SCP and EEA, 2010a). The bonus–malus mechanism will be progressively tightened by lowering the thresholds of eligibility for the bonus and imposition of the malus by 5 g of CO_2/km every two years, thus allowing manufacturers to adapt their production (ETC/SCP and EEA, 2010b). According to an assessment of the results of the bonus–malus scheme done in March 2008, just three months after the implementation of the scheme, the sales of the most polluting cars had decreased by 70% and sales of less polluting cars had increased by 38%. An evaluation undertaken in February 2009 showed that in just one year of implementation the bonus–malus led to a decrease in the average CO_2 emissions of new sold cars by 7 g, which allowed France to reach the EU objective of 140 g/km. By December 2009, average emissions went down to 132.8 g/km (ETC/SCP and EEA, 2010b). The combination of the bonus–malus scheme and the scrapping premium has had a significant effect in changing the structure of private car sales in France since 2008. Consumers and manufacturers have responded beyond expectations to the price signal.

(Source: UNEP (2012))

5.4 Command-and-Control versus Economic Instruments

Environmental regulations in most countries have traditionally relied on command-and-control policies where standards or limits are set by governments and which are applied uniformly to a broad category of sources. We distinguish between the following three types of command-and-control mechanisms: ambient, emissions, or technology standards. An ambient standard sets the amount of a pollutant that can be present within a specific environment. Emission standards seek to limit the amount of emissions released by a firm, industry, or area. Finally, a technology-based standard can be implemented, which will force polluters to use a particular pollution control technology that they deem reasonably cost-effective, such as installing scrubbers on smokestacks.

The main advantage of the command-and-control approach is that the final outcome is defined. Compliance monitoring is simple since regulators only have to make sure that the standard has been met or else the polluters pay a fine. There are three main drawbacks of command-and-control mechanisms. First, it can be very costly for regulators to gather all the necessary information about polluters. Second, polluters have very little choice about how to meet the standard and there is no incentive for innovation to further reduce emissions. However, in the case of emission standards, sources are often able to decide how they can best meet the standard. Finally, since command-and-control mechanisms are uniformly applied across broad categories of sources, it is unlikely that it can be the most cost-effective way to decrease pollution levels or emissions. As the marginal costs for reducing pollution will vary among the sources, it means that equity will not be achieved.

Regulations, especially in developing countries, are generally based on command-and-control instrument which are often not effective due to the fact that monitoring and enforcement are inadequate. Even if monitoring may be technically feasible, this is impeded by economic and cultural factors, such as limited budget, manpower and administrative constraints, centralization, and, of course, corruption. The potential to reduce costs by applying economic instruments instead of regulations is an important consideration for polluters due to the fact that there is more flexibility for selection of optimal levels of pollution. Economic instruments provide continuing motivation to reduce pollution below the levels set by regulations. They also encourage new pollution control technology, production processes, and new nonpolluting products by providing incentives for research and development. While command-and-control regulation is still common, more and more legislation is beginning to use market mechanisms, or a combination of command-and-control along with market mechanisms.

5.5 A Simple Model of Pollution Control

Abatement costs are costs incurred to reduce pollution to a lower level so that there are fewer damages. They include labor, capital, energy needed to lessen emissions as well as opportunity costs from reducing levels of production or consumption. Marginal abatement costs are the additional cost caused by an additional unit of abatement. For example, if total abatement costs increase from \$10,000 to \$20,000 when abatement increases from 5 tonnes per week to 10 tonnes per week, the marginal abatement cost is \$2,000 per tonne. The marginal abatement cost function represents the costs of reducing pollution by one more unit. In Figure 5.3, E_1 represents the level of pollution that would be generated in absence of any government intervention. As pollution is reduced below E_1, the marginal abatement cost increases.

Marginal abatement costs rise as cheaper options for reducing pollution are exhausted and more expensive steps must be taken. It slopes upward to the left meaning the higher the emission reduction, the greater the marginal abatement cost. The total abatement costs of achieving a reduction in emissions are measured by the area below the curve. Different firms may have different marginal abatement cost (MAC) functions because of different technology of abatement available to them. For

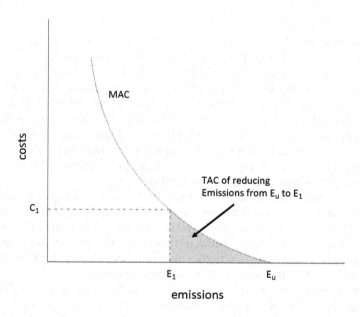

FIGURE 5.3
Marginal abatement cost function.

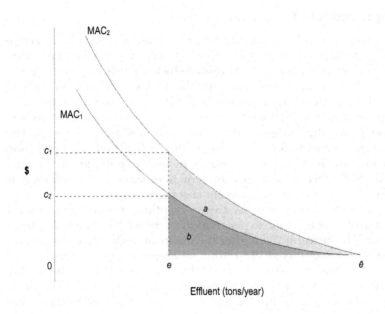

FIGURE 5.4
Marginal abatement cost curve.

the same source, the MAC function may shift with time periods because of improvement in technology of pollution control (Figure 5.4).

Marginal damage (MD) is the additional damage caused by an additional unit of emission. For example, if total damages increase from $20,000 to $30,000 when emissions increase from 8 tonnes per week to 10 tonnes per week, the MD is $5,000 per tonne. The MD function (Figure 5.5) is a relationship between quantity of emissions and the damage caused by emissions. It is an upward-sloping MD curve, assuming that MD increases with increasing emissions. There is a threshold below which MD is zero. The area below the function measures total damage.

The MD function may shift with time because of changes in natural environment. It is also a population-specific function; it may shift with an increase in the number of people exposed to the pollutants. Figure 5.6 shows two marginal emissions damage functions. For purposes of simplicity, the graph refers to a strictly noncumulative pollutant with all damages occurring in the same period as the emissions.

Consider a MD and a MAC function, which have been evaluated from the perspective of society. MD can be interpreted as the social marginal benefit of reducing emissions (damages saved), and MAC is the social marginal cost of reducing emissions. The efficient level of

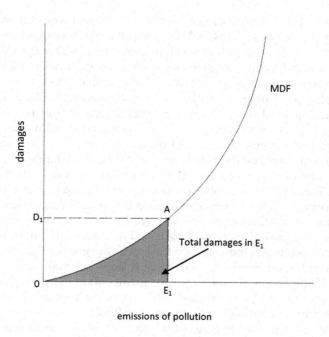

FIGURE 5.5
Marginal damage function.

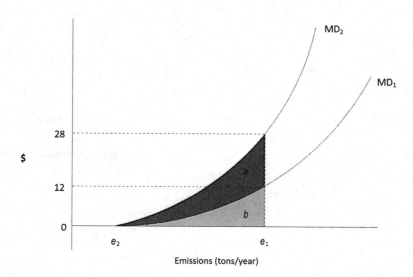

FIGURE 5.6
Typical marginal damage function.

emissions (the level that maximizes social net benefits) would then be where MD and MAC are equal (MAC = MD), the point of intersection of MAC and MD (Figure 5.7). Alternatively, we can interpret both MD and MAC as costs. No matter which level of emissions a society chooses, it has to incur costs on bringing down emissions to that level, and at that level, there are some damages still remaining. A society incurs abatement costs plus costs of damages at any level of emissions and that society would be efficient if it chooses that level of emissions which is associated with the minimum possible total of abatement costs and damages. The area below the MAC is total abatement costs and the area below the MD is total damages. It is only at the intersection of the two curves (MAC and MD), the sum of the two costs is minimum. For a particular pollutant being released at a particular place and time, the socially efficient level of emissions is therefore found where the MD function and the MAC function are equated. At this level of emissions, the net social costs – the total of abatement costs and damages – are minimized. If the level of emissions is less than E_0, then the MACs are greater than the MDs that the unit of pollution would have caused. It does not make sense to reduce pollution. If the levels of emissions are greater than E_0, then the MDs are greater than the MACs associated with reducing pollution by one unit. Society is better off eliminating that unit of pollution.

The MAC–MD model is conceptual and allows us to examine a wide variety of cases. In the real world, every pollution problem is different. Many different factors lie behind the MD and MAC functions, and when any of these underlying factors change, the functions themselves will shift

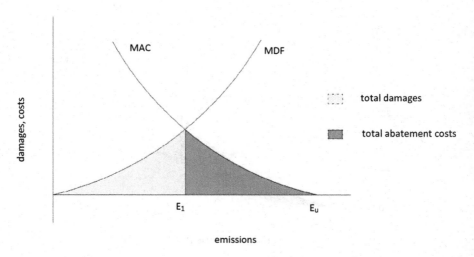

FIGURE 5.7
Optimal or efficient level of emissions.

and E_0 will change. E_0, the level that balances abatement costs and damage costs, is presented as a desirable target for public policy. It is unlikely that in the economy the level of emissions will be at E_0 without some form of government intervention. Unless persuaded to take into account the damages they inflict on society, polluters will have no incentive to incur any abatement costs. They will simply produce at the maximum pollution level. Section 5.3 presented briefly some instruments that induce polluters to reduce emissions to move toward a socially efficient equilibrium.

The equimarginal principle is an important one in pollution control and says the following: *To minimize the total cost of abating pollution by a given level we must allocate abatement among multiple sources such that marginal costs of abatement are equalized.* To illustrate this principle, consider Table 5.2, which shows the MAC for two firms emitting a particular pollutant into the environment. If there is no pollution control, they will each emit 12 tonnes/week. If Source A were cut 50% to 6 tonnes/week, its MAC at this level would be $6,000/week, whereas at this level of emissions, the MAC of Source B would be $20,000/week. Total abatement costs (TAC) of the 12-tonne total are $21,000/week for Source A and $56,000/week for Source B, or a grand total of $77,000/week. A and B must have different emission rates but together emit no more than 12 tonnes of effluent and have the same MACs. This condition is satisfied if Source A emits 4 tonnes and Source B emits

TABLE 5.2

The Equimarginal Principle

Emissions (tonnes/week)	Marginal abatement costs ($1,000/week)	
	Source A	Source B
12	0	0
11	1	2
10	2	4
9	3	6
8	4	10
7	5	14
6	6	20
5	8	25
4	10	31
3	14	38
2	24	58
1	38	94
0	70	160

(Source: Data from Field (1994))

8 tonnes. These rates add up to 12 tonnes total and give each source an MAC of \$10,000/week. The TAC = \$39,000/week for Source A; TAC = \$22,000/week for Source B and TAC = \$61,000/week grand total.

The model described in this section is very general and risks giving an overly simplistic impression of pollution problems in the real world. There are few actual instances where the MD and marginal abatement to functions are known with certainty. However, the simple model is useful for thinking about the basic problem of pollution control and to distinguish between the various approaches to environmental policy.

5.6 Summary

The objective of this chapter was to explore some basic linkages between the economy and the environment, which are interdependent. Environmental economics is the subset of economics that is concerned with the efficient allocation of environmental resources. The key to the environmental economics approach is that there is value from the environment and value from the economic activity and the goal is to balance the economic activity with environmental degradation by taking all costs and benefits into account. Because environmental goods are typically not priced, estimates of the value of environmental changes – improvements or degradation – must be measured by some sort of indirect means. Ecosystem services can be valued and environmental externalities can be quantified. The study of environmental economics supplies us with economic instruments that affect costs and benefits of alternative actions open to economic agents, influencing their behavior in such a way that it is favorable to sustainable development.

SUSTAINABLE ENGINEERING IN FOCUS: EXECUTIVE SUMMARY OF THE REPORT OF THE HIGH-LEVEL COMMISSION ON CARBON PRICES

The purpose of this Commission is to explore explicit carbon-pricing options and levels that would induce the change in behaviors – particularly in those driving the investments in infrastructure, technology, and equipment – needed to deliver on the temperature objective of the Paris Agreement, in a way that fosters economic growth and development, as expressed in the Sustainable Development Goals (SDGs). This report does not focus on the estimation and evaluation of the climate change impacts that would be avoided by reducing carbon emissions. While the Commission also covers other policies relevant and important to carbon-pricing design and delivery on the Paris agreement, its primary focus is on pricing. This report has been prepared based on the Commission's assessment of the available evidence and literature as well as on its members' judgment, developed through their extensive international policy experience. While the commissioners are in broad agreement on the overall thrust of the arguments presented in the report, they may not necessarily support every single assertion and conclusion.

1. **Tackling climate change is an urgent and fundamental challenge**. At COP21 in Paris, in December 2015, nearly 200 countries agreed to hold "the increase in the global average temperature to well below 2°C above pre-industrial levels and to pursue efforts to limit the temperature increase to 1.5°C." The goal of stabilizing the temperature increase well under 2°C is largely motivated by concerns over the immense potential scale of economic, social, and ecological damages that could result from the failure to manage climate change effectively. These temperature targets require a large-scale transformation in the structure of economic activity – including a major change in energy systems (especially power generation); industrial processes; space heating and cooling systems; transport and public transportation systems; urban forms; land use (including forests, grasslands, and agricultural land); and the behaviors of households. However, climate policies, if well designed and implemented, are consistent with growth, development, and poverty reduction. The transition to a low-carbon economy is potentially a powerful, attractive, and sustainable growth story, marked by higher resilience, more innovation, more livable cities, robust agriculture, and stronger ecosystems. To succeed, that is, to deliver efficiently and fully realize the potential benefits of climate policies, careful policy design is essential.

2. **A well-designed carbon price is an indispensable part of a strategy for reducing emissions in an efficient way**. Carbon prices are intended to incentivize the changes needed in investment, production, and consumption patterns, and to induce the kind of technological progress that can bring down future abatement costs. There are different ways to introduce a carbon price. Greenhouse gas Report of the High-Level Commission on Carbon Prices (GHG) emissions can be priced explicitly through a carbon tax or a cap-and-trade system. Carbon pricing can also be implemented by embedding notional prices in, among other things, financial instruments and incentives that foster low-carbon programs and projects. For instance, specific project-based credits, building upon the experience of the Clean Development Mechanism (CDM) of the Kyoto Protocol and on the mechanism

established under Article 6 of the Paris Agreement, can provide similar incentives by applying a price to a unit of GHG emissions. Explicit carbon pricing can be usefully complemented by shadow pricing in public sector activities and internal pricing in firms. Reducing fossil fuel subsidies is another essential step toward carbon pricing – in effect, these subsidies are similar to a negative emissions price. Governments can enhance the effectiveness of carbon pricing by establishing an enabling environment, building technical and institutional capacity, and establishing an appropriate regulatory framework. As carbon-pricing mechanisms take time to develop, countries should begin doing so immediately.

3. **Achieving the Paris objectives will require all countries to implement climate policy packages.** These packages can include policies that complement carbon pricing and tackle market failures other than the GHG externality. These failures are related to knowledge spillovers, learning and R&D, information, capital markets, networks, and unpriced co-benefits of climate action (including reducing pollution and protecting ecosystems). Some countries may conclude that the carbon-pricing trajectories required, if carbon pricing were the sole or dominant instrument, could entail excessive distributional or adjustment costs. Others may conclude that, given the uncertainties, requirements for learning, and scale and urgency of the transformation, rapid and more equitable change could be achieved more efficiently and effectively in other ways. The design of these policies will thus vary and always have to take into account national and local circumstances.

   International cooperation – including international support and financial transfers, carbon-price-based agreements, and public guarantees for low-carbon investments – to promote consistency of action across countries can help lower costs, prevent distortions in trade and capital flows, and facilitate the efficient reduction of emissions (as well as the achievement of other Paris Agreement objectives, such as those related to the "financial flows consistent with a pathway towards low greenhouse gas emissions and climate-resilient development").

4. **The Commission explored multiple lines of evidence on the level of carbon pricing that would be consistent with achieving the temperature objective of the Paris Agreement, including technological roadmaps, analyses of national mitigation and development pathways, and global integrated assessment models, taking into account the strengths and limitations of**

these various information sources. Efficient carbon-price trajectories begin with a strong price signal in the present and a credible commitment to maintain prices high enough in the future to deliver the required changes. Relatively high prices today may be more effective in driving the needed changes and may not require large future increases, but they may also impose higher, short-term adjustment costs. In the medium to long term, explicit price trajectories may need to be adjusted based on the experience with technology development and the responsiveness to policy. The policy dynamics should be designed to both induce learning and elicit a response to new knowledge and lessons learned. Price adjustment processes should be transparent to reduce the degree of policy uncertainty.

5. **Explicit carbon-pricing instruments can raise revenue efficiently because they help overcome a key market failure: the climate externality.** The revenue can be used to foster growth in an equitable way, by returning the revenue as household rebates, supporting poorer sections of the population, managing transitional changes, investing in low-carbon infrastructure, and fostering technological change. Ensuring revenue neutrality via transfers and reductions in other taxes could be a policy option. Policy decisions will need to duly take into account the country's objectives and specific circumstances, while keeping in mind the development objectives and commitments agreed in relation to the Paris Agreement objectives.

6. **Carbon pricing by itself may not be sufficient to induce change at the pace and on the scale required for the Paris target to be met, and may need to be complemented by other well-designed policies tackling various market and government failures, as well as other imperfections.** A combination of policies is likely to be more dynamically efficient and attractive than a single policy. These policies could include investing in public transportation infrastructure and urban planning; laying the groundwork for renewable-based power generation; introducing or raising efficiency standards, adapting city design, and land and forest management; investing in relevant R&D initiatives; and developing financial devices to reduce the risk-weighted capital costs of low-carbon technologies and projects. Adopting other cost-effective policies can mean that a given emission reduction may be induced with lower carbon prices than if those policies were absent.

Conclusion: Countries may choose different instruments to implement their climate policies, depending on national and local circumstances and on the support they receive. Based on industry and policy experience, and the literature reviewed, duly considering the respective strengths and limitations of these information sources, this Commission concludes that the explicit carbon-price level consistent with achieving the Paris temperature target is at least US$40–80/tCO$_2$ by 2020 and US$50–100/tCO$_2$ by 2030, provided a supportive policy environment is in place. The implementation of carbon pricing would need to take into account the nonclimate benefits of carbon pricing (such as the use of revenues derived from it), the local context, and the political economy (including the policy environment, adjustment costs, distributional impacts, and political and social acceptability of the carbon price). Depending on other particular policies implemented, a carbon price could have powerful co-benefits that go beyond climate, for instance, potential improvements in air pollution and congestion, the health of ecosystems, access to modern energy, and so on. Further, in a realistic context where domestic and international compensatory transfers are limited, imperfect, and costly, it is impossible to disregard distributional and ethical considerations when designing climate policies. In view of this, the appropriate carbon-price levels will vary across countries. In lower-income countries, they may actually be lower than the ranges proposed here, partly because complementary actions may be less costly and the distributional and ethical issues may be more complex. It is of vital importance to the effectiveness of climate policy, particularly carbon pricing, that future paths and policies be clear and credible. New data will emerge continually and new knowledge be generated, and these facts and lessons learned should be taken into account – indeed, carbon pricing should foster learning and technological progress. It will be important to monitor and regularly review the evolution of emissions, technological costs, and the pace of technological change and diffusion so that carbon prices can be adjusted, particularly upward, if actual prices fail to trigger the required changes. Policy adjustments should be made based on criteria that are transparent and sound: policies should be "predictably flexible." It is desirable that the carbon-price range across countries narrow over the long term, in a time frame that depends on several factors, including the extent of international support and financial transfers, and the degree of convergence in living standards across countries. The temperature objective of the Paris Agreement is also achievable with lower near-term carbon prices than indicated above if needed to facilitate transitions; doing so would require stronger action through other policies and instruments and/or higher carbon prices later, and may increase the aggregate cost of the transition. The carbon pricing and complementarity measures indicated here are substantially stronger than those in place at present (85% of global emissions are currently not priced, and about three quarters of the emissions that are covered by a carbon price are priced below USD10/tCO$_2$). This statement is consistent with the observation that the Nationally Determined Contributions (NDCs) for 2030 associated with the Paris Agreement represent emission reductions that are substantially smaller than those necessary for achieving the Paris target of "well below 2°C."

Source: High-Level Commission on Carbon Prices. 2017. Report of the High-Level Commission on Carbon Prices. Washington, DC: World Bank. License: Creative Commons Attribution CC BY 3.0 IGO

References

Convery, F., McDonnell, S. and Ferreira, S. 2007. The Most Popular Tax in Europe? Lessons from the Irish Plastic Bags Levy. *Environmental Resource Economy*, 38, 1–11.

"Dollar-Based Ecosystem Valuation Methods". n.d.. www.ecosystemvaluation.org/dollar_based.htm.

ETC/SCP and EEA. 2010a. European Topic Centre/SCP and European Environment Agency. SCP Policy Factsheets for Selected EEA Member Countries. http://scp.eionet.europa.eu/facts/factsheets_scp.

ETC/SCP and EEA. 2010b. European Topic Centre/SCP and European Environment Agency. Examples of National Policies to Promote Sustainable Household Consumption The Case of the France. http://scp.eionet.europa.eu/facts/factsheets_waste/fs_scp/pdf/policy_france.

Field, B.C. 1994. *Environmental Economics: An Introduction*, McGraw Hill International Edition, New York.

Panayotou, T. 1998. *Instruments of Change: Motivating and Financing Sustainable Development*, Earthscan Publications Inc, London.

Pearce, D., Markandya, A. and Barbier, E. 2013. *Blueprint 1: For a Green Economy*, Routledge, Abingdon-on-Thames, UK.

Pearce, D. and Turner, R. 1989. *Economics of Natural Resources and the Environment*, The John Hopkins University Press, Baltimore, MD.

Pearce, D.D.W. and Moran, D. 1994. *The Economic Value of Biodiversity*, Earthscan, London.

Stavins, R. 2003. Experience with Market-Based Environmental Policy Instruments. *Handbook of Environmental Economics*, 1, 355–435.

UNEP. 2012. *Global Outlook on Sustainable Consumption and Production Policies: Taking Action Together*, United Nations Environment Programme, Paris. https://sustainabledevelopment.un.org/content/documents/559Global%20Outlook%20on%20SCP%20Policies_full_final.pdf.

Introduction to Sustainability for Engineers

Exercises

1. Through a literature search explain what is the environmental Kuznets curve? Do you think developing countries should follow such an economic growth pattern?

2. Give and describe briefly five examples where economic instruments are used for environmental protection in your country to correct market failures.

3. Discuss how the establishment of a new paper mill closer to a river bank will create an externality. What type of externality will this be? With a diagram explain why this will not lead to a socially optimum allocation of resources. What can be done to ensure the paper mill will take into account the externality?

4. Discuss the Hedonic pricing method and the Travel cost method of environmental valuation. Include a case study for both methods to illustrate your answer.

5. Use a case study to outline the contingent valuation approach to environmental valuation, including details both of the survey and of the data analysis. Suppose you want to determine the aggregate willingness to pay among students of your university for reducing litter on the university grounds. How might you do this?

6. Environmental protection programs are sometimes designed to require all polluters to cut back emissions by a certain percentage. What are the perverse incentives built into this type of program? Use the marginal abatement cost (MAC) and marginal damage (MD) functions to explain your answer.

7. Effluent taxes may be beneficial in driving pollution prevention activities. Subsidies or tax credits can have a similar impact. In your opinion which system is preferable? Why?

8. When pollution regulations are imposed, governments incur enforcement costs that are part of social costs. Assume that enforcement costs are a constant amount, independent of the amount of pollution reduced. How would this change the location of the socially efficient equilibrium? Show graphically and explain.

9. Consider a firm which has two plants A and B. The table below shows MACs of reducing pollution in the two plants (Source: Data from Field (1994)).
   If the total targeted reduction of pollution is 10 units, what should be the allocation of abatement between the two plants?

10. Consider two sources of a particular type of emission (Source A and Source B) and assume that these emissions after leaving their

Quantity of Emission Reduction, units	MAC of Plant A $ per unit	MAC of Plant B $ per unit
1	0.60	0.70
2	0.65	0.75
3	0.70	0.80
4	0.80	1.00
5	0.90	1.15
6	1.00	1.35

respective sources are uniformly mixed together, so that the emissions of the two plants are equally damaging in the downwind impact area. The MACs for the two sources are given in the table. Compare the efficiency of a policy for the introduction of an effluent tax of USD30/tonne on each source to that of a policy that each source should reduce their emissions equally by 50%.

Emission Level (tonnes/month)	MAC for Source A	MAC for Source B
20	0.0	0.0
19	1.0	1.9
18	2.2	4.5
17	3.3	9.2
16	4.5	15.1
15	6.2	32.6
14	7.5	54.8
13	9.5	83.0
12	11.6	116.4
11	13.8	157.2
10	16.4	205.0
8	22.5	333.6
7	25.8	407.6
6	28.8	486.9
5	33.0	580.0
4	36.5	680.5
3	41.2	789.8
2	45.6	906.9
1	50.3	1034.9
0	55.3	1191.8

Recommended Reading and Websites

- Field B.C. 1994. Environmental Economics an Introduction. McGraw Hill International Edition. Chapter 5: The Economics of Environmental Quality is an essential read for an engineer.

- Hanley, N; J. F. Shogren and B. White. 2001. Introduction to Environmental Economics. New York: Oxford University Press.

- Pearce, D.W. and R.K. Turner. 1990. Economics of Natural Resources and the Environment. Prentice Hall.

- The World Business Council for Sustainable Development (WBCSD) and IUCN have developed a free role-playing game called "Buy, Trade, Sell" which shows how ecosystem markets work (www.wbcsd.org).

- The Australian government has a useful resource describing market-based incentives (www.marketbasedinstruments.gov.au).

- The Green Economy Initiative of UNEP (www.unep.org/green economy).

- Stern Review on the Economics of Climate Change (www.hm-treasury.gov.uk/stenreview_index.htm).

- Professors Dennis King from the University of Maryland and Marisa Mazzotta from the University of Rhode Island provide the following website for noneconomists with clear, nontechnical explanations of ecosystem valuation concepts, methods, and applications. www.ecosystemvaluation.org/index.html.

- A variety of reports from EPA's National Center for Environmental Economics examine the interest and use of economic incentive mechanisms for environmental management over the past 20 years in both the United States and abroad. (yosemite.epa.gov/ee/epa/eed.nsf/webpages/EconomicIncentivesPollutionControl.html).

6

Integrating Sustainability in Engineering Design

6.1 Problem Solving in Engineering

Problem solving is the foundation of engineering activities and it is the process of determining the best possible action to take in a given situation. We can distinguish between two types of engineering problems:

1. Analytic problems (or exercise solving) are solved by analysis and logical deductions. They are typically closed-ended problems that determine the properties of a given device or system and generally have only a single correct solution. Convergent thinking or a problem narrowing approach is applied when undertaking modeling and analysis to solve such problems, usually occurring at the detailed design stage.

2. Open-ended problems do not have a unique solution and do not lend themselves to the analytic approach. There is no single "correct" solution and the engineer looks for the best solution. Divergent thinking or a problem-widening approach is applied when confronted with such problems.

Engineers usually deal with ill-defined problems that are vaguely or ambiguously formulated by the community or clients. Sometimes, engineering problems are also stated in overly specific terms so that a solution is stated unintentionally in the formulation. This is also an ill-defined problem. Problems faced by the engineer are therefore usually initially ill-defined and open-ended, and the "best solution" has to be worked out among alternative, potentially acceptable solutions. Also, at least in the initial stages, there is a significant lack of information. A methodology for solving such complex problems is as follows (Dandy et al., 2017):

- Formulate the problem clearly and in general terms and ensure it is not over-specific.
- Develop a wide range of promising approaches through creativity techniques and gather information to understand the state of the art.
- Choose criteria for ranking the alternative approaches.
- Discard the least promising approaches using simplified evaluations.
- Discard progressively with more detailed evaluations until a short list remains.
- Choose the best approach from the short list using detailed evaluations.
- Develop the best approach into a detailed solution and communicate the results through either oral/written communication or engineering drawings.
- Implement the solution.

Planning and design, together with management, construction and operation or production, are key activities of an engineering project from conceptual stage to successful implementation. They are activities in which the details of an engineering project or a process are worked out to the extent necessary for implementation. Planning involves working out the steps needed to change the present state into the new state. Design, the essence of engineering, is the process of determining the details of any required new system. Both planning and design are essentially problem-solving processes, and the above methodology is used to carry out the engineering design process for new things or to improve things. The quality of the planning and design work will have a decisive effect on the success of the project. A common characteristic of all engineering works is that they should be undertaken in minimum time using limited resources such as people, materials, and money. Engineers now have also to be creative and responsive in facing the challenge of sustainability. One of the first steps in engineering work is to identify the real problem to be solved as distinct from community perceptions and views and statements of individuals. Hence, engineers must also be problem framers in addition to being problem solvers (Dandy et al., 2017). And sustainability must be part of the problem framing.

6.2 The Conventional Engineering Design Process

Although design is one among many other tasks conferred to engineers, such as project management, operation and maintenance management, site supervision, quality control, research and development, etc., it remains a central part of engineering practice. It can be defined as the process of devising a system, component, or process to meet desired needs, converting resources optimally and considering the limitations imposed by practicality, regulation, safety, cost, and sustainability. Design problems are usually ill-defined and open-ended. The problem statement often gives no indication of what a solution must be, and it is this uncertainty that makes designing a challenging activity. According to Dieter and Schmidt (2009), one way to summarize the challenges presented by the designing environment is to think of the following *four Cs of design*:

- Creativity – requires creation of something that has not existed before
- Complexity – requires decisions on many variables and parameters
- Choice – requires making choices between many solutions at all levels, from basic concept to the smallest detail
- Compromise – requires balancing multiple and sometimes conflicting requirements

A design process is a systematic problem-solving strategy which takes into account the constraints while developing possible solutions to solve or satisfy human needs or wants and deciding on a final choice. The process follows the problem-solving steps: problem definition, search for alternatives, analysis, selection of the "best" alternative and implementation. Design projects will include all these phases and in addition will include conceptual design, embodiment design and detailed design. The conventional design process is summarized in Table 6.1, while Figure 6.1 illustrates the iterative nature of the design process as well as important points and data. Besides the iterative nature of the design process, the importance of communication with all partners involved should be emphasized.

6.3 From Conventional to Sustainable Engineering Design Process

Design cost varies little in terms of the overall product cost, but its decision has a major impact on the overall cost. As design proceeds toward the detailed phase, design decisions become more and more specific and have less impact on overall costs. Figure 6.2 illustrates why it is important to start as early as possible in the design process to enhance sustainability. The early design phases are the ones which have the greatest influence on the project as a whole due to the fact that project planning is more flexible at this stage. As the project evolves, flexibility is reduced and the chance of making changes is smaller, and making changes involves higher costs. During

TABLE 6.1

Generic Conventional Design Process Tasks

Design Phase	Tasks
I – Planning and problem definition	1. Team formation 2. Definition of the problem, objectives, and context 3. Identification of constraints and other preliminary data 4. Planning of subsequent phases
II – Conceptual design	1. Identification of the functions the systems must serve 2. Generation of alternative concepts 3. Definition of design specifications, based on the functions or mandatory requirements such as regulations
III – Preliminary design	1. Elaboration of the alternative concepts 2. Evaluation of the concepts, including technical performance, cost estimation, risk analysis, etc. 3. Selection and communication of the best concept
IV – Detailed design	1. Specific data gathering and detailed elaboration of the chosen alternative 2. Further evaluation and optimization 3. Identification of requirements for the manufacturing, construction, operation, and maintenance phases 4. Documentation and communication of the final design

the early phases, there is more potential for studying different alternatives, reducing costs, implementing changes, and improving performance. The design phase can thus be considered as one of the key phases in achieving sustainability.

The most advanced green designs can only be accomplished through integrative design in which project teams work together to identify synergies among strategies.

To move toward a more sustainable practice of engineering, the design process must be modified to enable engineers to tackle sustainability issues in a structured manner. There have been many attempts to incorporate sustainability principles into engineering design. The Sandestin Sustainable Engineering Principles (Abraham and Nguyen, 2003) and the 12 Principles of Green Engineering (Anastas and Zimmerman, 2003) capture similar, but also complementary, elements of sustainability and engineering design, which are summarized in Table 6.2. These guiding principles must be considered by engineers in all design phases of a project through well-integrated components to the conventional approach.

Sustainable engineering design requires a systems approach whereby sustainability is systematically integrated in the design rather than considered as an "add-on." Based on the literature (Azapagic and Pedan,

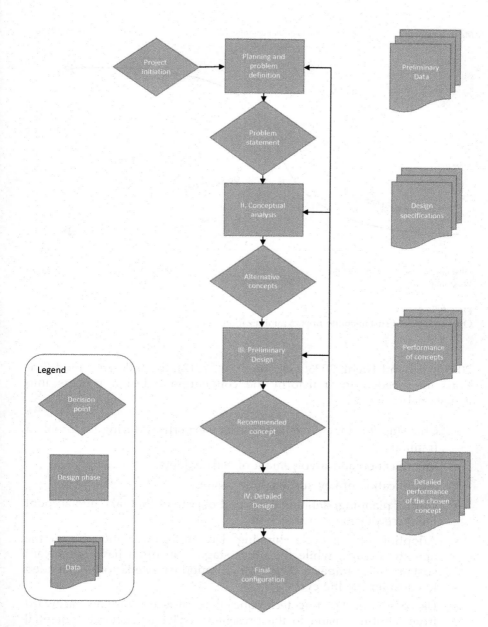

FIGURE 6.1
The conventional engineering design process.

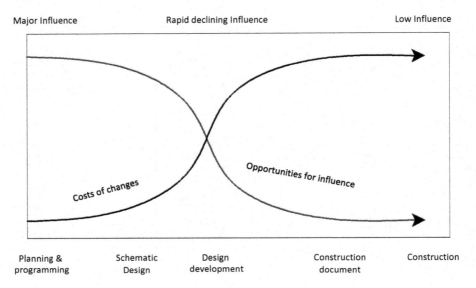

FIGURE 6.2
Opportunities of sustainability impact minimization along the process life cycle.

2005; Crul and Diehl, 2006; Gagnon et al., 2012), the following modifications are needed to transform the conventional design process into a sustainable one:

- Choosing team members with various expertises during the problem framing.
- Identification and involvement of stakeholders.
- Identification of key sustainability issues.
- During planning, defining goals and objectives that take into account the three pillars of sustainability.
- Adoption of life cycle thinking. Design is often focused on the operation stage, while the other stages can often have significant sustainability impacts. The system boundary needs to be extended to consider the life cycle.
- Use of the backcasting technique. It consists of working backward from a future vision to the present in order to elaborate potential solution pathways and then evaluating the different options.
- Identifying criteria and indicators for measuring sustainability.
- Use of a multi-criteria approach to handle compromise between various environmental, economic, social, or technical issues.

TABLE 6.2

Principles for Sustainable Engineering Design

Sandestin Sustainable Engineering Principles	The 12 Principles of Green Engineering
Principle 1: Engineer processes and products holistically, use system analysis, and integrate environmental impact assessment tools.	Principle 1: Designers need to strive to ensure that all material and energy inputs and outputs are as inherently nonhazardous as possible.
Principle 2: Conserve and improve natural ecosystems while protecting human health and well-being.	Principle 2: It is better to prevent waste than to treat or clean up waste after it is formed.
Principle 3: Use life-cycle thinking in all engineering activities.	Principle 3: Separation and purification operations should be designed to minimize energy consumption and materials use.
Principle 4: Ensure that all material and energy inputs and outputs are as inherently safe and benign as possible.	Principle 4: Products, processes, and systems should be designed to maximize mass, energy, space, and time efficiency.
Principle 5: Minimize depletion of natural resources.	Principle 5: Products, processes, and systems should be "output pulled" rather than "input pushed" through the use of energy and materials.
Principle 6: Strive to prevent waste.	Principle 6: Embedded entropy and complexity must be viewed as an investment when making design choices on recycle, reuse, or beneficial disposition.
Principle 7: Develop and apply engineering solutions, while being cognizant of local geography, aspirations, and cultures.	Principle 7: Targeted durability, not immortality, should be a design goal.
Principle 8: Create engineering solutions beyond current or dominant technologies; improve, innovate, and invent technologies to achieve sustainability.	Principle 8: Design for unnecessary capacity or capability (e.g., "one size fits all") solutions should be considered a design flaw.
Principle 9: Actively engage communities and stakeholders in development of engineering solutions.	Principle 9: Material diversity in multicomponent products should be minimized to promote disassembly and value retention.
	Principle 10: Design of products, processes, and systems must include integration and interconnectivity with available energy and materials flows.
	Principle 11: Products, processes, and systems should be designed for performance in a commercial "afterlife."
	Principle 12: Material and energy inputs should be renewable rather than depleting.

- Selecting strategies and creativity techniques to guide idea generation during the conceptual phase.
- During preliminary design, a preliminary sustainability assessment of potential solutions is carried out and during the final design a full sustainability assessment is carried out.
- Sustainability of the product is communicated at its launch
- Post-project monitoring is carried out with relevant criteria and indicators.

In the sustainable engineering design process, it is important to use appropriate tools for the analysis of potential solutions and the synthesis of the data gathered. Incremental innovation is unlikely to solve sustainability challenges, and radical innovation needs creativity. Key creativity tools in the field of sustainability (such as brainstorming, mind mapping, backcasting, industrial ecology, eco-design strategies and Factor X engineering) must be given the same attention as tools for analysis (such as EIA, SEA, LCA, CBA, LCC, SLCA, etc.) and synthesis (such as MCA). Stakeholder engagement is a cross-cutting tool to be employed in the creativity, analysis, and synthesis activities. Sustainable projects need to be as technically sound as conventional projects, and these tools must thus be used in conjunction with existing approaches. Table 6.3 presents the critical tasks in each design phase deemed necessary to move from the conventional design toward sustainable design.

The large number of criteria that need to be considered when integrating sustainability issues with conventional technical issues can be problematic. Azapagic and Pedan (2005) indicate that most MCA studies recommend that a maximum of 10 criteria be considered in order to simplify assessment and decision making. In contrast, a sustainability assessment framework can prescribe more than 10 criteria or indicators in addition to technical specifications. To address this situation, Gagnon et al. (2012) recommend the consideration of criteria in a sequential manner, reducing the number of them to be tackled simultaneously. Four categories are proposed for the classification of criteria (see Table 6.4) and to help in the decision-making process.

6.4 Design for Life Guidelines and Strategies

It is crucial to the successful delivery of sustainable development to realize that the problem definition stage is where rigorous consideration of sustainable development issues and in particular life-cycle thinking approach will generate the greatest benefits. By incorporating the principles in the

TABLE 6.3

Critical Tasks in Each Design Phase for a Sustainable Engineering Design Process

Design Phase	Critical Tasks
I – Planning and problem definition	1. Form a multidisciplinary design team 2. Define sustainability principles 3. Identify sustainability issues associated with the defined problem 4. Analyze stakeholders and plan stakeholder's involvement
II – Conceptual design	5. Define sustainability criteria in line with the sustainability issues previously identified, in parallel with technical functions 6. Develop a vision for the future in which functions are fulfilled respecting the sustainability principles 7. Generate at least one alternative concept radically different from conventional ones using sustainability creativity tools 8. Define sustainability indicators derived from the issues or criteria, in parallel with technical specifications derived from functions 9. Identify the analysis tools with which data will be generated for each of indicators 10. Chose a multi-criteria decisions aid method
III – Preliminary design	11. Assess the performance of alternative concepts according to the sustainability criteria or indicators, including one "benchmark alternative" representative of current practice 12. Validate the multi-criteria decision aid method chosen and use it to recommend a preferred concept
IV – Detailed design	13. Refine the assessment of the preferred concept and optimize its performance along sustainability criteria or indicators. Do a full assessment of the sustainability of the chosen design 14. Communicate recommendations for the manufacturing, construction, use and end-of-life phases 15. Generate the set of sustainability indicators for monitoring

Adapted from Gagnon et al. (2012).

problem statement and design brief, its importance will trickle down throughout various aspects of the design. Table 6.5 gives the guidelines for design from a life-cycle perspective.

The D4S strategy wheel in Figure 6.3 illustrates the seven general D4S strategies which parallel the stages of the product life cycle and give improvement directions. The seven strategies have been extended with sub-strategies in Table 6.6. These basic suggestions are useful during the brainstorming of sustainable design options and can be used as a checklist or as a source of inspiration.

TABLE 6.4

Categories of Criteria in Sustainability Assessment (SA)

Category	Purpose	Description
Mandatory, non-discriminating	Screening of concepts (conventional design)	These criteria are associated with regulations, codes, standards, etc. and relate to various dimensions of sustainability (environmental standards, health and safety requirements, purchase of regional labor and materials, etc.) or technical issues. The respect of such criteria is mandatory in conventional as well as in sustainable design, so they cannot serve to discriminate the former from the latter.
Relative, non-discriminating	Improvement of concepts	This category of criteria guides the improvement of a given concept and serves to assess, in a qualitative or quantitative manner, the relative improvement realized compared to an initial configuration. These criteria thus help making each alternative concept more sustainable, but do not allow comparison between the concepts.
Threshold, discriminating	Screening of concepts	Threshold criteria set objectives (more demanding than mandatory requirements) that concepts need to respect in order to be considered acceptable (go/no-go testing). Concepts which meet the criteria can pass on to subsequent design phases (go) and are discriminated from those who cannot (no-go).
Weighted, discriminating	Identification of an optimal concept	These criteria are considered in a single decision step, with the help of one of the many weighted MCA methods available. An optimal concept outperforming others is identified, the outcome depending on the criteria considered and the decision rules particular to the chosen MCA method.

Source: Gagnon et al. (2012).

In practice, the design may incorporate several ideas, and several concepts may be developed at the same time. The "Morphological Box" techniques are useful for combining several ideas in one product concept in a systematic way.

6.5 Measuring Sustainability

Efforts should be made to be as quantitative as possible when evaluating alternative designs. Indicators and metrics that can be used to measure and quantify environmental sustainability need to be developed to provide

TABLE 6.5

Design for Life Guidelines

Raw materials	Design for resource conservation Design for low impact materials
Manufacturing	Design for cleaner production
Use	Design for energy efficiency Design for water conservation Design for minimal consumption Design for low impact use Design for service and repair Design for durability
Distribution	Design for efficient distribution
End of life	Design for reuse Design for remanufacture Design for disassembly Design for recycling Design for safe disposal

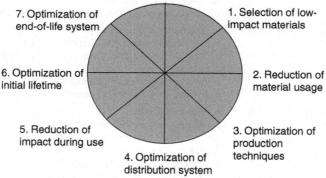

@. New concept development

7. Optimization of end-of-life system

1. Selection of low-impact materials

6. Optimization of initial lifetime

2. Reduction of material usage

5. Reduction of impact during use

3. Optimization of production techniques

4. Optimization of distribution system

FIGURE 6.3
Strategy wheel for design for sustainability.

a basis for decision-making. In most engineering designs, the measure is cost and the goal is to minimize cost. This suggests that one mechanism for incorporating objectives related to sustainability into engineering design is to monetize them. However, some indicators of environmental and societal performance will be difficult, or arguably impossible, to monetize and for these cases other approaches will be needed for incorporating objectives into engineering design. In engineering design, it is most convenient to

TABLE 6.6

Design for Sustainability Strategies and Sub-Strategies

1. Selection of low-impact materials	a. Cleaner b. Renewable c. Have lower energy content d. Recycled e. Recyclable f. Have a positive social impact, (e.g., generate local income)
2. Reduction of materials usage	a. Weight b. Volume (transport)
3. Optimization of production techniques	a. Alternative techniques b. Fewer steps c. Lower/cleaner energy d. Less waste e. Fewer/cleaner materials used to support the production f. Safety and cleanliness of workplace
4. Optimization of distribution system	a. Less/cleaner/reusable packaging b. Energy efficient transport mode c. Energy efficient logistics d. Involve local suppliers
5. Reduction of impact during use	a. Lower energy consumption b. Cleaner energy source c. Fewer consumables needed d. Cleaner consumables e. Health supporting and/or added social added value
6. Optimization of initial lifetime	a. Reliability and durability b. Easier maintenance and repair c. Modular product structure d. Classic design e. Strong product–user relation f. Involve local maintenance and service systems
7. Optimization of end-of-life system	a. Reuse of product b. Remanufacturing/refurbishing c. Recycling of materials d. Safer incineration e. Taking into consideration local (informal) collection/recycling systems

Source: Crul et al. (2009).

discuss sustainability in the context of indicators developed through a Life Cycle Sustainability Assessment (LCSA) framework – for example through an environmental and social LCA – and through the indicators of the Sustainable Development Goals (SDGs) and Global Reporting Initiative (GRI) frameworks. Table 6.7 gives a set of potential criteria and indicators based on these tools and frameworks that can assist in effective decision-making.

TABLE 6.7

Examples of Sustainability Design Criteria and Indicators

Environmental Sustainability Indicators	Economic Sustainability Indicators	Social Sustainability Indicators
Global warming	Capital and operating costs	Direct and indirect job creation within the community
Stratospheric ozone depletion	Economic return over project life cycle	Nuisances to the community (noise, odor, visual impact)
Photochemical smog formation	Life-cycle costs	Public acceptability/emotional ownership of the community in the engineering project
Human carcinogenicity	Direct and indirect investment within the community	Improved health and safety of employees, customers, and citizens
Atmospheric acidification	Environmental liabilities	Enhanced social opportunities for members of the community
Aquatic toxicity	Taxes paid	Compliance with labor standards
Terrestrial toxicity		Community mobility and connectivity/access
Depletion of nonrenewable resources		Enhancement of community aesthetics
Use of land and freshwater resources		
Eutrophication		

The above indicators provide key variables that may be assessed using specific metrics when evaluating the degree of sustainability for a specific engineering project or a product. An LCA software and database will normally be required for the assessment of environmental sustainability. Economic evaluation in traditional design is normally based on the micro economic indicators such as NPV, discounted cash flow analysis, returns on capital investment and so on. In addition to these indicators, life-cycle costs and investments in, for example, pollution prevention and decommissioning should also be considered. Social sustainability indicators can be translated into both quantitative and qualitative indicators. An SLCA enables the identification of the most significant social impacts and the hotspots.

More simple and qualitative sustainability assessment methods can also be employed. Figure 6.4 shows an Impact Matrix which is a qualitative or semi-qualitative method that provides an overview of the environmental inputs and outputs, social aspects, and profit flows at each stage of the product life cycle. The columns correspond to the different product life cycle stages and the rows concentrate on the relevant criteria.

Issue	Raw Materials	Suppliers	In house production	Distribution	Use	E-O-L
Materials						
Energy use						
Solid waste						
Toxic emissions						
Social responsibility						
Human resource management						
Distributed economies						
Water						
CO2						
Costs						

FIGURE 6.4
Example of an impact matrix.

Rows – Environmental criteria usually include material use, energy consumption, solid waste and toxic emissions. Social criteria usually include social responsibility, local or regional economic development and human resource management. More issues can be considered by adding rows. Examples include issues such as specific local problems or sustainability issues like water consumption, biodiversity, CO_2 emissions, costs, and cultural heritage. In addition, rows can be added and linked to the relevant drivers.

Columns – Depending on the life cycle process tree of the product, the stages can be named in different ways and the number of columns can be increased. In Figure 6.4 the life cycle has six stages. Depending on the real situation, the design team can decide to add or leave out stages.

Such a matrix can help a design team to make a quick qualitative assessment of the life cycle by identifying the cells that have major "sustainability" impacts. The next step is to prioritize the impacts which will become the focus for developing improvement options. Such a matrix also enables the identification of information gaps.

6.6 Sustainable Design through Sustainable Procurement Criteria

Procurement is a key part of the project development stage, and it is the process which creates, manages, and fulfils contracts. Innovation for sustainability requires the use of the best modern procurement practice. The use of sustainable procurement criteria is an opportunity for clients/owners to build and operate projects in an environmentally friendly manner. In most cases, the procuring organization will need technical support with specific engineering, environmental and economic knowledge to undertake the whole tender process from initial feasibility studies to the final selection of a contractor. An engineering project will necessarily include a design phase, selection of a contractor followed by the construction as such. The subsequent operational phase will, like the earlier phases, include various environmental concerns, and sustainable procurement criteria therefore cover all these phases. This section covers specifically the procurement of design. The following typical phases can be distinguished in the project development phase:

1. Tendering for consultancy services (engineers, planners, and architects): This is typically based upon the consultant's experience in performing similar projects, the qualification and experience of the consultant's personnel, and the consultant's proposal for performing the services. The selection of the consultant is often based upon an evaluation model consisting of the above requirements and may include the consultant's relevant experience in sustainable design, LCA and LCC calculations for engineering projects.

2. The initial phase: This includes a general outline, feasibility study and to some extent conceptual design. In these phases, several potential solutions to the problem are commonly discussed. The decisions made during the initial phases have great impact on the

economic and environmental performance of the project. Thus, it is very important to incorporate sustainability considerations very early in the process. Relevant SPP criteria for the project are selected and an evaluation model and weighting of the different criteria (economic, technical, and environmental criteria) are determined. Life cycle assessment (LCA) and/or life cycle costing (LCC) calculations are carried out for different options.

3. Preparatory phase: This is also called the preliminary design phase where the more specific technical solutions are considered and decided. The answers to these questions in the preparatory phase can be supported by setting up an *evaluation model* which includes economic, technical, and environmental performance/SPP criteria for the specific project. This evaluation model can further be developed during the detailed design and tender phase and be used as a *contract award model*. The calculation of potential environmental impact can be made based on LCA, and assessment of the total economic impact can be based on LCC calculations. These analyses can help the owner to identify the best sustainable solution to the technical problem.

4. Detailed design and tender document phase: The necessary design, technical specifications, and tender documents for the will be developed ready for issue to the tenderers. The level of detail in the design and technical specifications will depend on the contract form. The type of contract that is most frequently used for the implementation of engineering projects is the FIDIC type developed by the Federation Internationale des Ingenieurs-Conseils, or similar national contract types. The SPP criteria include guidance for contractual performance clauses. This is because requirements for the construction and operation of the infrastructure as such include a number of sustainability aspects that will need to be included in the contract as contractual obligations. Performance clauses are understood here as setting requirements for the manner in which delivery takes place in the construction or operating activities. Together with the specification of what must be delivered, the performance clauses constitute what the constructor/operator must "do" according to the contract.

6.7 Case Studies on Sustainable Engineering Design Process

6.7.1 Sustainable Process Design – Case Study (Source: Azapagic et al., 2006)

One of the key challenges for sustainable development of the chemical industry is designing more sustainable processes. Azapagic et al. (2006) proposes a general methodology based on life cycle thinking for integrating sustainability considerations (technical, economic, environmental, and social criteria) into process design. The approach is illustrated on a design case study of the vinyl chloride monomer process which is an important chemical because it is used in the production of the plastic PVC. The use of PVC is increasing which demands further production of VCM. This makes optimization of the production of VCM interesting and potentially profitable. The methodology for integrating sustainability considerations into process design follows the usual stages in process design, i.e. project initiation, preliminary design, detailed design and final design. As shown in Figure 6.5, each of these four stages consists of a number of steps.

The case study looked at a promising alternative for process design. Compared with an average VMC plant, the LCA results, displayed in Figure 6.6, show that, for most impacts, the proposed VCM design is environmentally more sustainable than an average VCM plant. The exception to this is energy use whereby the proposed plant would use around 3% more energy and generate around 7% more greenhouse gases. A much more significant difference is noted for the water discharge: while the proposed plant is using around 30% less water than the average plant, it is discharging into the environment almost twice as much as the average plant. The proposed plant also has higher carcinogenic potential, through a higher release of toxic substances. Therefore, these are the areas where further improvements could be investigated in the detailed design.

The case study illustrates how sustainability considerations could be integrated into process design from project initiation through preliminary to detailed design. This requires a systems approach whereby sustainability is not considered as an "add on" but is systematically integrated into process design taking into account the whole life cycle of the plant and the product. The methodology enables identification of relevant sustainability criteria and indicators, comparison of alternatives, sustainability assessment of the overall design and identification of "hot spots" in the life cycle of the system. In this way, it is possible to arrive at a design configuration

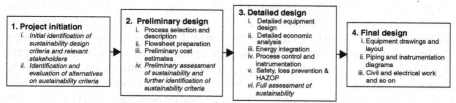

Text in italics—design stages related to sustainability; normal text—stages in traditional design

FIGURE 6.5
Stages in process design for sustainability.

Reprinted with permission from Azapagic et al. (2006).

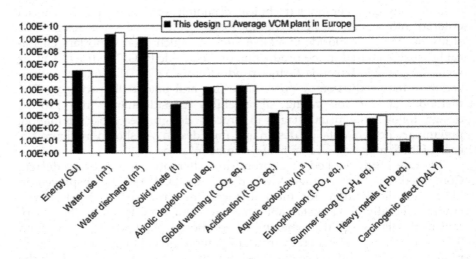

FIGURE 6.6
Comparison of environmental sustainability of the proposed design and an average VCM plant.

Reprinted with permission from Azapagic et al. (2006).

that would ensure the most sustainable performance of the plant and product over their whole life cycles.

6.7.2 Sustainable Building Design – A Case Study (Reprinted with Permission from Santos and Costa, 2016)

Building Information Modeling (BIM), Building Energy Simulation (BES) and Integrated Design Process (IDP) are complementary processes that together contribute to a new process for design of sustainable buildings. IDP is a collaborative process with a multidisciplinary design team that focuses on the design, construction, operation, and occupancy of a building over its complete life cycle, with a clear definition of environmental and economic goals and objectives. BIM is a digital representation of physical and functional characteristics of a facility and serves as a shared source for information for it forming a reliable basis for decisions during its life cycle. Sustainable design processes also usually rely upon a BES software to establish expected energy consumption of building designs. Practitioners have recognized the importance of early design stages when reducing buildings' life cycle environmental impact. Numerous researchers have shown that the earlier decisions are made in the design process and the fewer the changes to these decisions at later stages, the greater is the potential for reducing the building's environmental impact. Integrating BIM software with LCA methodology provides the potential to minimize buildings' environmental impacts during early stage design. BIM supports integrated design and improves information management and cooperation between the different stakeholders, and on the other hand, LCA is a suitable method for assessing environmental performance. Santos and Costa (2016) illustrates the potential of integrating the BIM software with the LCA methodology. Designing environmentally friendly products gets easier with sustainability modeling softwares. Architects and building engineers have been using BIM (building information modeling) software to study energy efficiency and reduce construction costs in their building projects. The pilot case study tests existing tools for energy analysis and environmental performance of buildings. The purpose of the pilot case study was to understand the effect of designer's choices in the environmental impact of the building and to compare the energy consumption and environmental impact of a multi-family house with a single-family house, as shown in Figure 6.7.

Two different solutions representing the residential buildings are analyzed and compared: a single-family building and a multi-family building. Autodesk Revit software is used to create the BIM models, while Revit Energy Analysis tool is used to conduct the energy analysis and Tally tool to perform the LCA study of the above-mentioned solutions. The initial step was to model two simple buildings in Revit both with same envelope solutions, in order to guarantee that the results obtained are independent of material choices. After the development of both BIM models, the authors used the Revit Energy Analysis option to simulate the energy consumption (Table 6.8). The second step of this case study was to analyze

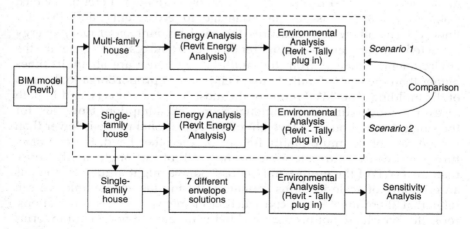

FIGURE 6.7
Pilot case study methodology.

TABLE 6.8

Revit Energy Analysis Results (Multi-Family House vs Single-Family House)

	Multi-family house	Single-family house	Units
Area	1,120	100.00	m^2
Electricity use	88.00	90.00	kWh/m^2/year
LC electricity use	2,956,800.00	270,000.00	kWh
	270.0	24.7	kWh/day
PV low efficiency	46,514.00	11,243.00	kWh/year
PV medium efficiency	93,029.00	22,487.00	kWh/year
PV high efficiency	139,543.00	33,730.00	kWh/year

the environmental impacts due to materials selection, by examining the results of seven different envelope solutions. Unlike the first step, in which the authors' objective was to compare the energy and environmental analysis of different types of buildings, in this step the authors seek to understand the impact of designer's choices of materials, using a single scenario (single-family house).

As it is possible to observe from Table 6.9, Tally provided six different Environmental Impact categories: Acidification Potential, which causes fish mortality, forest decline, and the deterioration of building materials; Eutrophication Potential, which can cause an undesirable shift in species composition; Global Warming Potential (GWP), which causes an increase of the greenhouse effect; Ozone Depletion Potential, which leads to higher levels of UVB ultraviolet rays; Smog Formation Potential, which leads to respiratory issues and damage to ecosystems; and Primary Energy Demand, which measures the total amount of primary energy extracted (non-renewable plus renewable resources). In general, the environmental impacts from the multi-family house are about 10 times higher than the single-family house, being almost proportional to the area of the building. The GWP is one of the most relevant impacts, and in both cases it is mostly due to operational energy consumption. However, for the single-family house, the weight of the operational phase is lower than for the case of the multi-family house. On the other hand, the manufacturing processes contribute more to the GWP of the single-family house than the GWP of the multi-family house. So, we might argue that there is an economy of scale in terms of manufacturing (as an example we can refer that there are several construction elements, such as foundations, roof, etc., which are not proportional to the number of floors) but in terms of operational energy consumption the same logic does not apply, as the multi-family house has higher relative energy consumption (what might

TABLE 6.9

Tally Results (Multi-Family vs Single-Family House)

	Row Labels	Sum of Acidification Potential Total (kgSO$_2$eq)	Sum of Eutrophication Potential Total (kgNeq)	Sum of Global Warming Potential Total (kgCO$_2$eq)	Sum of Ozone Depletion Potential Total (CFC-11eq)	Sum of Smog Formation Potential Total (kgO$_3$eq)	Sum of Primary Energy Demand Total (MJ)	Sum of Non-renewable Energy Demand Total (MJ)	Sum of Renewable Energy Demand Total (MJ)
Multi-family house	End of Life	75	41	174,960	0.00	2,550	575,515	569,443	6,046
	Maintenance and Replacement	3,114	366	319,574	0.01	17,084	3,840,682	3,309,445	531,238
	Manufacturing	5,364	468	1,285,760	0.10	48,229	10,458,306	9,612,019	84,628
	Operations	12,200	630	2,635,708	0.00	142,650	59,812,579	36,338,720	23,473
	Total	**20,754**	**1506**	**4,416,001**	**0.12**	**210,514**	**74,687,083**	**49,829,627**	**2,485,743**
Single-family house	End of Life	−2	4	39,813	0.00	149	42,730	41,394	1329
	Maintenance and Replacement	282	34	26,429	0.00	1,457	339,942	291,080	48,862
	Manufacturing	567	49	192,020	0.02	5,358	1,033,348	940,644	92,704
	Operations	1,100	57	222,181	0.00	13,750	5,201,615	3,060,519	2,141,097
	Total	**1,948**	**143**	**480,442**	**0.03**	**20,715**	**6,617,635**	**4,333,636**	**2,283,992**

be related to the relevant energy consumption of the common/social spaces).

It is also important to understand which kind of materials have higher environmental impact, and if these materials represent a considerable portion of the building's mass. Table 6.10 displays the environmental impact of all materials used in the single-family house, throughout the different project phases. For this purpose, it is meaningless to display both scenarios (multi-family house and single-family house), as our real concern is only to analyze the environmental impacts due to materials selection, not to compare the two scenarios.

As expected, concrete was the material with the highest mass percentage of the building, having a relatively low environmental impact/mass ratio (except in the manufacturing process). On the other hand, metal-based materials and insulation materials have a very high environmental impact/mass ratio, particularly on GWP, Acidification Potential, and Eutrophication Potential, despite their low representativeness in the total mass of the building. If designers are aiming for sustainable solutions, the selection of insulation material and its thickness is an extremely important aspect for the environmental impact of a building. It is relevant to mention that the higher the insulation thickness, the less thermal loss, leading to a decrease of operational energy consumption. So, for a sustainable solution to be reached, it would be advisable to perform a multi-objective optimization of both. However, in order to demonstrate how the designer's choices can profoundly affect a building's environmental impact, the authors decided to use the scenario of a single-family house and select different Revit and Tally solutions for the envelope. As observed in Figure 6.8, the original option (studied earlier) is one of the best sustainable solutions for most environmental impact categories. Also, wood-based solutions (option 3 and option 5) are the ones in which renewable resources can suppress most of the required Primary Energy Demand. Interestingly though, these two solutions also seem to be amongst the least environmentally friendly solutions (option 3 – wood roof: 1st in Acidification Potential, 1st in Eutrophication Potential, and 1st in Primary Energy Demand; and option 5 – timber floor: 2nd in Eutrophication Potential and 2nd in GWP). However, if we examine Figure 6.9, we can conclude that most of those environmental impacts result from the End of Life potential use (recycling/reuse/recovery) of wood-based solutions, assuming that these materials are still in good condition. They are also the only ones with a positive energy return at the end of life (through burning processes). Regarding Smog Formation Potential, Acidification Potential, and demand from non-renewable resources, option 6 (full concrete envelope) comes as the least environmentally friendly solution.

TABLE 6.10

Tally (Single-Family House Material's EI)

	Building Phases	Sum of Acidification Potential Total (kgSO₂eq)	Sum of Eutrophication Potential Total (kgNeq)	Sum of Global Warming Potential Total (kgCO₂eq)	Sum of Ozone Depletion Potential Total (CFC-11eq)	Sum of Smog Formation Potential Total (kgO₃eq)	Sum of Primary Energy Demand Total (MJ)	Sum of Non-renewable Energy Demand Total (MJ)	Sum of Renewable Energy Demand Total, (MJ)
Single-family house	**End of Life**	**−2**	**4**	**39,813**	**0.00**	**149**	**42,730**	**41,394**	**1,329**
	03 – Concrete	17	2	3,676	0.00	304	61,060	58,499	2,561
	04 – Masonry	4	1	792	0.00	65	13,189	12,642	547
	05 – Metals	−19	0	−3,730	0.00	−200	39,564	39,987	423
	07 Thermal and Moisture Protection	−5	0	39,061	0.00	−40	12,243	10,165	2,072
	08 – Openings and Glazing	−1	1	518	0.00	−28	−12,902	−8,244	−4,658
	09 – Finishes	3	0	0	0.00	48	8,704	8,320	384
	Maintenance and Replacement	**282**	**34**	**26,429**	**0.00**	**1,457**	**339,942**	**291,080**	**48,862**
	03 – Concrete	*0*	*0*	*0*	*0.00*	*0*	*0*	*0*	*0*
	04 – Masonry	5	0	654	0.00	29	7244	6854	391
	05 – Metals	235	31	16,334	0.00	990	246,946	211,625	35,321

(Continued)

TABLE 6.10 (Cont.)

Building Phases	Sum of Acidification Potential Total (kgSO₂eq)	Sum of Eutrophication Potential Total (kgNeq)	Sum of Global Warming Potential Total (kgCO₂eq)	Sum of Ozone Depletion Potential Total (CFC-11eq)	Sum of Smog Formation Potential Total (kgO₃eq)	Sum of Primary Energy Demand Total (MJ)	Sum of Non-renewable Energy Demand Total (MJ)	Sum of Renewable Energy Demand Total, (MJ)
07 Thermal and Moisture Protection	2	0	180	0.00	7	1088	770	318
08 – Openings and Glazing	10	1	1,943	0.00	120	35,210	23,277	11,933
09 – Finishes	30	1	7,318	0.00	312	49,453	48,554	899
Manufacturing	**567**	**49**	**192,020**	**0.02**	**5,358**	**1,033,348**	**940,644**	**92,704**
03 – Concrete	181	7	40,938	0.00	2,402	290,957	277,781	13,177
04 – Masonry	19	1	11,139	0.00	295	138,240	130,311	79
05 – Metals	283	33	25,320	0.00	1,725	364,925	324,893	40,031
07 Thermal and Moisture Protection	53	7	105,754	0.02	544	158,259	143,472	14,787
08 – Openings and Glazing	12	1	2,455	0.00	147	47,947	31,358	16,589
09 – Finishes	19	1	6,413	0.00	246	33,021	32,828	192
Operations	**1,100**	**57**	**222,181**	**0.00**	**13,750**	**5,201,615**	**3,060,519**	**2,141,097**
Grand Total	**1,948**	**143**	**480,442**	**0.03**	**20,715**	**6,617,635**	**4,333,636**	**2,283,992**

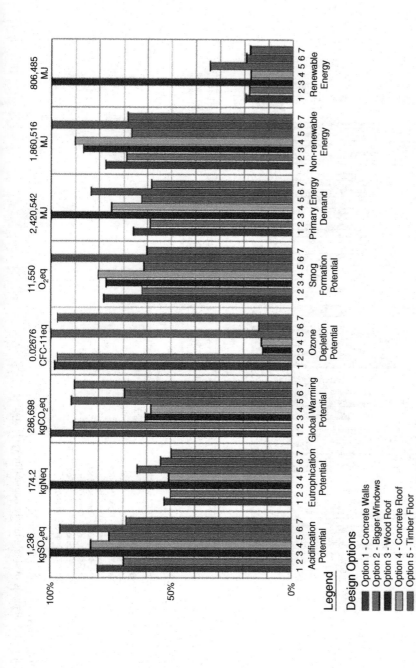

FIGURE 6.8
Tally LCA analysis of single-family house (7 different options).

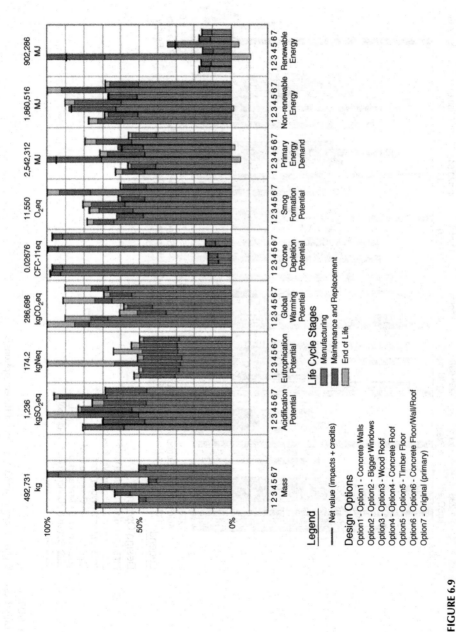

FIGURE 6.9
Tally LCA analysis of single-family house (life cycle stages).

6.7.3 Sustainable Product Design in Mechanical Engineering – A Case Study By P. Ramlogun and K. Elahee, Faculty of Engineering, The University of Mauritius

Environmental concerns generate additional design constraints (Kaebernick et al., 2002), making it difficult to reach to a decision when little information is available at early design stages, for instance at concept development. Hence, it is important to have at least a decision matrix which enables the designer with some freedom in his imagination while designing for the environment. Traditionally, the environmental aspect is included at a later stage during LCC (Ullman, 2010). However, it is possible to integrate it into every stage of the product development which would result into a shift in paradigm (Conteras et al., 2009). The product discussed in this case study used the quality tool called Quality Function Deployment (QFD) along with some qualitative techniques which enabled the integration of the sustainability aspect into the design. The product designed is intended to mechanically harvest *Caricas Papaya* (commonly known as papaya). Traditionally, papaya is hand-picked in many cultivars across the country and the world (Jimenez et al., 2014), (CAC/RCP-53, 2003). Workers often need to climb on the trees or stand on ladders in order to harvest these fruits. Sometimes, they work at a height greater than 3 m above ground level to reach the fruits and this represents a serious trip hazard. A Rapid Entire Body Assessment (REBA) was performed to evaluate the seriousness of the job and a score of 13 was obtained which implies an immediate action of designing because of the high risk (OSHA, 2018). Another drawback of this traditional method is that it tends to bruise the fruits and affecting the fruit quality, thereby reducing the yield (AustralianGov, 2008).

Design for the environment (DFE) provides a framework to minimize these impact in an effort to make the world better (Conteras et al., 2009; Davidson et al., 2007; Dieter and Schmidt, 2009). Environmental impacts of a product may include energy consumption, natural resource depletion, discharges, and solid waste generation. These would broadly fall into two categories – energy and materials. There is also the socio-economic aspect of a product, for instance, in terms of job creation, ease of operation, satisfaction, ethical manufacturing activities and occupational safety. The logic used in the development of the papaya picker is shown in Figure 6.10.

Step 1 of the design process identified the customers of the product and set the design goals toward an environment-friendly orientation. A compilation of some of the design DFE guidelines which were adapted from Telenko et al. (2008) is shown in Table 6.11.

Step 2 involved identifying the customer needs or the voice of the customers (VOC). These were obtained through focus group meetings and off-record interviews with cultivators and then grouped into an

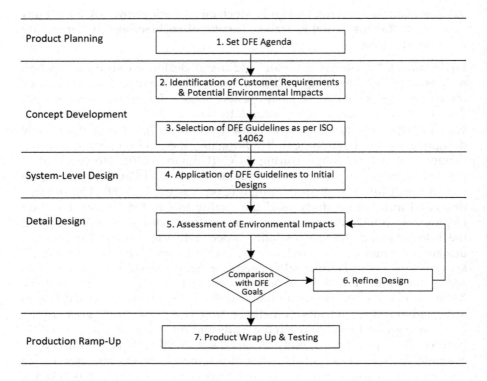

FIGURE 6.10
The DFE process used which shows that it was an iterative process.

affinity diagram based on their hierarchy and category. The customers were then asked to rate the attributes in terms of a relative importance which, when added together, will amount to a score of 100 points. This method was deemed to be more reliable in that it forces the customer(s) to think carefully before allocating the marks (Dixon and Poli, 1995; Ullman, 2010). Focus group meetings and interviews were conducted to obtain necessary information about the expected attributes of the design. The participants were questioned about their preferences and environmental concerns.

To complete Steps 2 and 3, the concept generation broke down the harvesting operations into discrete events. These discretized events were assessed based on their functionality in achieving the desired goal that is safely picking a fruit from the tree. The application of DFE was done right at the concept generation by iteratively referring to the design goals and adapting the concepts along the way. For instance, it was possible to develop a concept and product architecture that reduced the amount of assemblies and other complexities. In Figure 6.11, the concept contained

TABLE 6.11

Design for Environment Guidelines

Life-Cycle Stage		Design for Environment Guidelines
Materials	Sustainability of resources	• Specify recyclable/recycle materials
	Healthy inputs and outputs	• Specify non-hazardous materials
Production	Minimal use of resources in production	• Use few manufacturing steps as possible • Specify materials not requiring surface treatments • Minimize number of assemblies
Distribution	Minimal use of resources in distribution	• Minimize packaging materials • Use structural materials to minimize the total volume of material
Use	Efficiency of resources	• Provide high mechanical advantage • Use of passive components to store energy
	Reliable	• Embed aesthetics and functionality • Ensure minimal maintenance and failure modes
Recovery	Disassembly, separation	• Specify joints which can be undone by hand or using common tools

over 20 components and of varying sizes. Upon reviewing several possibilities and using CAD modeling techniques, a simpler assembly was conceived reducing the total number of parts to 6 (Figure 6.12).

The concept selection was guided by the set of key customer requirements, DFE guidelines and applicable standards. A concept was selected as benchmark for a particular function, and all the other concepts were relatively rated based on the meeting of customer requirements, the applicable standards and sustainability indices. The one having the higher scores was selected for further development.

The final embodiment, shown in Figure 6.13, consists of a grabber assembly which is mounted on a supporting column. The drive system consists of a crank connected to a drive shaft which is simply supported on two ball bearings. Axially connected to the driveshaft is a torque-limiting mechanism which provides slippage when a predetermined torque is exerted on the drive shaft. A winding drum is bolted to the flange of the torque limiter hub. This drum winds the connecting link which is described as a rope. The winding drum is also coupled with a ratcheting wheel which provides free motion in only one direction.

Based on preliminary static analysis and using the maximum contact forces which can be applied on the fruit, a torque of 2 Nm was obtained

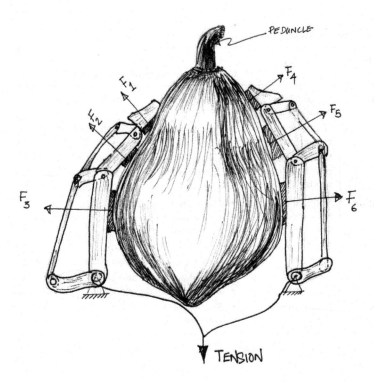

FIGURE 6.11
Grabber inspired by the human hand, hand-drawn by the author.

which could be exerted on the drive shaft having a diameter of 8 mm. This enabled the identification of an adjustable torque limiter on the market and the HPC – SAS 20–6 was chosen as per the manufacturer's datasheet. The overall dimension of this unit helped to size the corresponding drive elements.

The dimensions of the winding drum were guided by the size of the torque-limiting unit overall length. The bore of the drum was obtained using the hub dimensions of the torque limiter. Static analysis was carried out on the drum to evaluate the deformation of the latter under a torque about its rotating axis.

The design of the drive shaft was based on the assumption that the bending stress in the shaft is repeated and reversed as the shaft rotates, but that the torsional shear is nearly uniformed. The tangential force caused by the winching action produces a transverse force on the shaft which resulted in bending. Geometric discontinuities such as key seats, fillet radius and retaining rings grooves were accounted for in the calculations

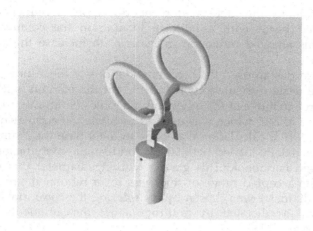

FIGURE 6.12
CAD model for the reviewed and simplified design.

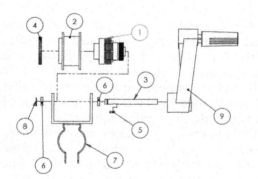

ITEM NO.	PART NUMBER	QTY.
1	HPC___SAS20_6	1
2	Load Drum	1
3	SpindleAISI316	1
4	RatchetBore_6-CW	1
5	JIS B 1301 2x2x6-A	1
6	ISO 15 ABB - 176 - 16,SI,NC,16_68	2
7	Housing Bracket	1
8	Circlip DIN 471 - 6 x 0.7	1
9	Crank Handle	1

FIGURE 6.13
Final assembly of the integrated drive and force control mechanism.

and a shaft diameter of 6 mm was chosen which would fit the torque limiter bore.

The grabber jaw design (see Figure 6.12) was guided by the diameter of locally available papaya species. The maximum diameter of the fruit was found to be 150 mm as per the FAREI (Food and Agricultural Research & Extension Institute). This helped to size the overall dimensions of the jaws. The final jaw opening width was found to be 160 mm. The proposed jaw is a single piece part made out of a composite material which would offer high cleaning abilities and structural rigidity. A finite element analysis was used to evaluate the deformation of the grabber when loaded as a cantilevered beam with a uniform loading in the commercial CAD software. This allowed reduction in material to increase the strength-to-weight ratio.

The column design used the recommendations of the Aluminium Associations. The design calculations assumed that the column is fixed at one end and is free to move at the other end (flagpole configuration). This gave a shape factor of 2.10 which was used to calculate the equivalent length of the column. A trial calculation was done to determine the diameter of the column when axially loaded with 100 N and using a safety factor of 1.95 as per the design handbook. This gave a working diameter that enabled to determine the accepted range of thickness for a tubular member. Finally, the column diameter was chosen by considering the comfortable width of a power grip as dictated by anthropometric measurements (Eastman-Kodak-Company, 1983).

The design of the torsion springs adopted the methodology as described by Carlson (1978). A spring made using music wire having a diameter of 2.8 mm and a single coil is recommended to provide the biasing force to unwind the drum. It has a length of 25 mm extending from each side at an angle of 80°.

The ratcheting mechanism chosen was obtained using the design guide by KHK-USA (www.khkgears.us). The product identified had the catalogue number SRT2/3–50 – J – Bore 10. The ratchet wheel is proposed to be directly mounted onto the drum by means of machine screws (M1.6 × 3). The final part of the design was to design a supporting frame which would contain the drive assembly. This was guided by the overall dimensions of the drive assembly and the diameter of the supporting column.

In the detailed design phase, it was important to select some preliminary materials and manufacturing processes based on the information available from the DFE guidelines and also on the general shape of the parts. References were made to handbooks (Ashby, 2005; Dixon and Poli, 1995; Dieter and Schmidt, 2009; Mott, 2004; Ullman, 2010). Typical design parameters are strength-to-weight ratio (S_y/ρ) and stiffness-to-weight ratio (E/ρ). These are limiting design constraints which enabled the material selection process.

When making a decision on the material, it is desirable to consider the following factors: Material composition (grade of alloy or plastic), cost of material, form of material, size (dimensions and tolerance), heat-treated condition, directionality of mechanical properties (anisotropy), quality level (control of impurities, inclusions, cracks, microstructure, etc.), ease of manufacture (workability, weldability, machinability, etc.) and ease of recycling (Dieter and Schmidt, 2009).

The decision on the manufacturing process is generally based on the following factors: unit cost of manufacture, life-cycle cost per unit, quantity of parts required, and complexity of the part, with respect to shape, features, and size, compatibility of the process for use with candidate materials, ability to consistently make a defect-free part, economically achievable surface finish, economically achievable dimensional accuracy and tolerances, availability of equipment and lead time for delivery of tooling.

Using the DFE guidelines and knowledge from engineering materials and industrial design, it was possible to make the product using environmentally safe and nontoxic materials in that 45% (by weight) was aluminum, 39% recyclable plastics, 15.2% steel and 0.8% of landfill wastes.

The product sustainability assessment was a comprehensive evaluation process with the consideration of the energy, environmental, resource, technical and socio-economic indicators. Given the multi-level and complex relationships among various levels, it was subjectively difficult to obtain accurate pricing mechanism. An attempt was made to determine all the internal and external costs associated with the product throughout its entire journey from manufacturing to an eventual re-integration in the material supply chain as recyclable materials. The cash inflows and outflows were computed using handbook values (Dixon and Poli, 1995) and the net present value (NPV) was determined. Engineering judgement was exercised to compare between concepts and to ensure that the final product was meeting the design goals set.

In conclusion, a mechanical tool to harvest papaya was designed as a result of an ergonomic job assessment done by the authors. This tool had its unique characteristics of being simple and effective to operate while meeting all regulatory standards. It was conceptualized with the help of cultivators and other major stake holders. The use of lean methods like QFD helped to synthesize all information together to help in translating the customer requirements into design variables. Some key consideration in the design for sustainable products were described like for instance the importance of material selection and the manufacturing processes. It was outlined that we need to consider the effects of how the material was made, the exact composition of the candidate material for the application being considered, any processing that may have to be done for shaping the material or fabricating a component, the structure of the material after processing into a component or device, the environment in which the material will be used, and the cost-to-performance ratio. It is hoped that this work will instill interest in potential market segments and also empower local workforce to build up a more sustainable agricultural sector.

6.7.4 Sustainable Product Design in Electronic Engineering (Reprinted with Permission from Andrae et al., 2016)

Annually, it is estimated that about 4 billion units of consumer electronics for mobile communications are produced worldwide. This could lead to various ecological imbalances unless the design and disposal of the products are handled optimally. To illustrate how industry looks at and responds to the increasing social awareness, this case study describes how sustainability is successfully implemented in practice at a large Chinese company, developing and producing various kinds of electronic products used for communication using an Eco-Design Method called EcoSmarT. All companies are continuously investigating what characteristics customers want in the next generation (NG) of products. The features are translated into technical requirements, including environmental requirements, and later the designers come up with different concepts which satisfy these requirements. It is necessary to systematically find a way to integrate eco-design in the usual product development process. The targets for all eco-metrics are set in Step 1 (Concept) and baselined in Step 2 (Plan). All eco-metrics are obtained, improved, and fine-tuned in Step 3 (Development) while actual Field Failure Rate (FFR) and "Rethink, Reduce, Recycle, Refurbish, Resale" (5R) values are obtained in Step 4 (Validation). In the proposed method, the role of LCA in a "Plan, Do, Check, Act" (PDCA) cycle is to "Check" environmental impacts holistically. The features of the design process are as follows:

1. **Eco-Metrics:** The proposed method EcoSmarT includes seven eco-metrics and LCA score.

    (i) Energy efficiency: Energy efficiency is a rather broad concept defined as the quotient between the energy needed to do useful work and total energy actually used. The metric refers to the use stage energy and is product specific. Examples applicable to mobile phones are charger efficiency, absorbed power, charging time, and receiver sensitivity.

    (ii) Packaging materials mass and volume: This metric refers to the mass and volume of the packaging materials, such as cardboard, paper, and plastics. The target is to reduce the volume and mass.

    (iii) Hazardous substances: This metric refers to mass of hazardous substances which are neither regulated nor banned. It can also be qualitative measures such as elimination of substance usage.

    (iv) Precious metals: This metric refers to masses of gold, silver, platinum, palladium, and other valuable metals such as tantalum, indium, rhodium, ruthenium, osmium, and iridium.

    (v) Total mass: This metric refers to the total mass of the designed product including the accessories. The target is to reduce the mass.

    (vi) Recyclability, recoverability, reuse-ability, disassemble-ability: The recyclability rate is typically defined as the amount of materials which can be recycled divided by the total mass of the product. Several new 5R metrics have been defined for products. It is judged from case to case if energy recycling is to be included and which recyclability metrics are to be applied. An example of a 5R metric is time of disassembly. The focus shall generally be on easy and fast disassembly, material identification, fewer material types, and less surface finish, etc., which all help increase the recyclability.

    (vii) Lifetime reliability: This metric usually refers to FFR. FFR is defined as the frequency at which an engineered system fails. The mean time between failure (MTBF) scores can be determined by FFR data.

2. **Collection of Requirements:** The inputs for this stage are, for example, roadmaps for, and LCA performances of, similar products as ED. These inputs form the basis for the collection of technical/functional/ performance requirements, including environmental requirements. The output is the initial specification.

3. **Design Step 1 – Drafting of Design Concepts:** In this step the most promising concepts and optional solutions are listed and drafted based on technical/functional/performance requirements and environmental requirements. The environmental requirement targets can be set based on findings in the customer surveys documented in the initial specification. Additionally, for the Current Generation ED (CGED), in use by the customer, the seven eco-metrics and final LCA score are obtained. For Next Generation ED (NGED) concepts, all eco-metrics are defined and preliminary LCA scores are obtained. The LCA score is calculated by cost-effective LCA methods and LCA tools used by the organization. For CGED, the final LCA score, the real FFR value and the real recyclability are possible to obtain as CGED is a final product used and waste handled in the market. All eco-metrics are estimated and the preliminary LCA scores are calculated for the different NGED concepts which the designers propose. The output from the Concept step is the revised specification.

4. **Design Step 2 – Plan:** In this step, the design is planned. Additionally, the environmental requirement targets in design are documented in the report from Plan. Examples of targets are for power and resource usage, as well as energy efficiency. Revised specifications are developed into a specification baseline to be fine-tuned during the subsequent Development process. The output from Plan is the specification baseline for NGED.

5. **Design Step 3 – Development:** In the Development step, the system architecture and the detailed design for NGED are formed. The detailed NGED design is based on the specification baseline for a NGED concept. Prototypes are created and then fine-tuned repeatedly to meet the technical/functional/performance and environmental requirements. Meanwhile, verification and testing will be repeatedly conducted on Prototypes and their fine-tuning models, including estimations of FFR values. Brainstorming, Theory of Inventive Problem Solving (TRIZ), and guidelines can help generate ideas for fine-tuning. All eco-metrics are quantified and the preliminary LCA scores are calculated for the Prototype and its fine-tuned models of NGED. The designers find ways of fine-tuning the applicable eco-metrics further for the NGED Prototypes resulting in Final product NGED. The new fine-tuned values of the eco-metrics and the LCA score for the Final product NGED design are quantified and calculated, respectively, and put into the report from Development. The Final product NGED is manufactured and goes for sale. The requirements from Step 3.2 Plan are checked (Check) to validate how the eco-metrics and the LCA score were investigated for NGED. It is also checked if the requirements from Step 3.2, Plan, are fulfilled. The Development step is followed by the Customer Validation.

6. **Design Step 4 – Customer Validation:** This step validates data from the use of NGED. Here the final values of the eco-metrics and the LCA score for the final NGED design are quantified and calculated, respectively, and put into the report from Customer Validation. Here the FFR values are based on failure samples returned by customers. As shown in Figure the eco-metrics values from this design step will be used as a benchmark for the next NGED. The actual recyclability rate can also be measured in this step and used as a benchmark for next NGED.

7. **Design Step 5 – Closing Process:** Another LCA score is calculated based on additional data about the life cycle of NGED and the design project for NGED is closed. The LCA score from this design step is usually very close to the previous from Step 4. Design steps 0–5 are then repeated for the next NGED, for example, starting with collecting new customer requirements (from, e.g., roadmaps and LCAs).

The following discusses the application of EcoSmarT to the development of product B to product C.

1. Collection of Requirements: Based on the features obtained from customer surveys and analyses of voluntary trends, several technical/functional/performance requirement targets, and some environmental requirement targets, are set for phone C: increase the stand-by time compared to B; introduce a bio-based plastic; eliminate brominated flame retardants/chlorinated flame retardants from the main body; eliminate polyvinyl chloride (PVC); eliminate phthalates; Introduce Forest Steward Council (FSC) certified color box; introduce soy-ink printing. Note that here environmental requirements refer to customer requirements beyond mandatory legislation. Examples of environmental requirement targets are: reach a certain score for various eco-ratings; fulfill energy star standards; fulfill code of conducts for energy efficiency; fulfill various eco-labels; and remove hazardous substances beyond legislation.

There are several concepts for C which can fulfill the technical/functional/performance and cost requirements. Design for profitability is the key driver and, therefore, it is strived for low-cost and high-performance solutions. Generally, the designers moderate the display, integrated circuits, light emitting diodes, circuit design, battery charging and discharging, and printed circuit board assembly layout. In summary, one of the concepts for C fulfilling the technical/functional/performance requirements, and highly likely meet the eco-design requirement targets, compared to B, is based on the following criteria:

- Uses bio-based plastics for the front shell instead of petro-based plastics.
- Has lower packaging materials volume.
- Has longer talk time and stand-by time.
- Uses soy ink for the packaging box and manual.
- Have no painted mechanical parts.
- Innovated the packaging.

Design Step 1, Drafting of Design Concepts for Phone: In Table 6.12 are shown eco-metrics for phone B and a concept for phone C in step 4.1. Lifetime reliability is outside the scope of this article. Due to confidentiality reasons, the detailed methodology for obtaining the numbers in the tables is not disseminated.

Based on a sensitivity analysis, it is determined that the degree of reuse of C has the strongest effect on the LCA score. Other metrics have weaker correlation with the LCA score' however, gold content has the strongest.

Design Step 2, Plan of Phone: Here the stringent environmental requirement targets in design are documented for C as follows: fine-tune and improve three of seven eco-metrics; has >5% lower packaging volume than B; has at least 10% better autonomy time in stand-by than B; and has better absolute LCA score than B. The place of the precise requirement setting is one of the clearest features of EcoSmarT. The requirements are set on the basis of what can realistically be achieved for C.

TABLE 6.12

Eco-Metric Values for B and a Concept of C in Design Step 4.1

SN	Eco-metric	Value for B	Target value for a C Concept	Unit	Comment
1.	Energy efficiency	422	475	hours	Mobile autonomy time in stand-by mode (charged at 100%).
2.	Packaging materials volume	825	784	cm^3	The metric is also used in Open Eco Rating (OER) [91].
3.	Hazardous substances	Meet the requirements of laws and regulations	PVC, Be, phthalates and triphenyl phosphate eliminated	Quantitative	OER [91] contains similar metrics.
4.	Precious metals (Au, Ag)	240	150	mg	The metric is also used in OER [91].
5.	Total mass	392	454	G	Total mass (mobile phone including battery + charger + accessories + packaging).
6.	Recyclability (metals and polymers)	80	90	%	Can only be measured for C when it has been used by customer.
7.	Lifetime reliability	Not specified	Not specified	%	LIMEv2 weighting method [89].
8.	LCA score	2460 JPY	2510 JPY, (the effect of improved 5R and charging efficiency can be explored)	JPY	For Assembly of C a proxy value is used. Reuse has a relatively strong correlation with the LCA score, whereas the correlation is low for Bio-plastics.

Design Step 3, Development of Phone: The Development step involves fine-tuning of eco-metrics making the Prototypes and Final design of C. Table 6.13 shows the eco-metric values and LCA score which are put in the Development report. These values shall be compared to B values of Table 6.1. The final verification of all requirements is made. In addition, the environmental requirements for C set in 4.2 Plan are checked:

- fine-tuned and improved three of seven eco-metrics (yes, autonomy time in stand-by, packaging volume, and total mass)
- has >5% lower packaging volume than B (yes)

TABLE 6.13

Eco-Metric Values for a Prototype of C and Final Design of C in Design Step 4.3

SN	Eco-metric	Value for B	Target value for a C Concept	Unit	Comment
1.	Energy efficiency	475	480	hours	Mobile autonomy time in stand-by-mode (charged at 100%)
2.	Packaging materials volume	783	780	cm^3	Improvements are made during development
3.	Hazardous substances	PVC, Be, phthalates and triphenyl phosphate eliminated	PVC, Be, phthalates and triphenyl phosphate eliminated	Quantitative	
4.	Precious metals (Au, Ag)	151	154	mg	
5.	Total mass	454	450	G	
6.	Recyclability (metals and polymers)	80	80	%	Recoverability can only be measured for C when it has been used by a customer
7.	Lifetime reliability	Not specified	Not specified	%	Estimations based on similar product to C. LIMEv2 weighting method (Itsubo et al 2012)
8.	LCA score	2510 JPY	2510 JPY, (the effect of improved 5R and charging efficiency can be explored)	JPY	

- has at least 10% better autonomy time in stand-by than B (Yes)
- has better absolute LCA score than B (no, but as more data are collected about actual life cycle performance it could change in step 4.4).

Next, the final product C is sent for final assembly, assembled, and its final LCA value, including the measured impact of the assembly process, is calculated as 2510 JPY (20.3 USD). Next, C is sold.

Design Step 4, Customer Validation of C: Here, the final values of the eco-metrics and the LCA score (Table 6.14) for the final C design are quantified and calculated, respectively, and put into the report from

TABLE 6.14

Eco-Metric Values for Final Design of C in Design Step 4.4

SN	Eco-metric	Value for C Final Design	Unit	Comment
1.	Energy efficiency	480	hours	Mobile autonomy time in stand-by-mode (charged at 100%) + improved charging efficiency when (fast) charging the battery
2.	Packaging materials volume	780	cm^3	Improvements are made during development
3.	Hazardous substances	PVC, Be, phthalates and triphenyl phosphate eliminated	Quantitative	
4.	Precious metals (Au, Ag)	154	mg	
5.	Total mass	450	G	
6.	Recyclability (metals and polymers)	90	%	(5% Reuse, 5% Re-manufacturing, 90% gold recovery, [Huawei, 2014])
7.	Lifetime reliability	Not specified	%	LIMEv2 weighting method (Itsubo et al 2012)
8.	LCA score	2400	JPY	Solar generated power for assembly of C, improved gold recovery, reuse, re-manufacturing and charging efficiency

Customer Validation. These values for eco-metrics and LCA can be used in the design of the next generation of C. As shown in Figure 6.14, the actual improvements of reuse, re-manufacturing, recycling rate, and charging efficiency finally lead to a better overall LCA score for C. B gets more credit from metal recycling as it contains more precious metals.

Design Step 5, Closing the C Design Project: The LCA score from Step 4 is confirmed as 2400 JPY (19.4 USD) based on new data describing the life cycle of C, and next the design project for C is closed. An example of such new data is the use of solar-based electricity for the assembly of C.

6.7.5 Sustainable Product Design in Electrical Engineering – A Case Study (Reprinted with Permission from Gurauskienė and Visvaldas, 2006)

Sustainable electrical engineering is focused on developing the following: renewable electricity sources such as wind and solar power; systems for

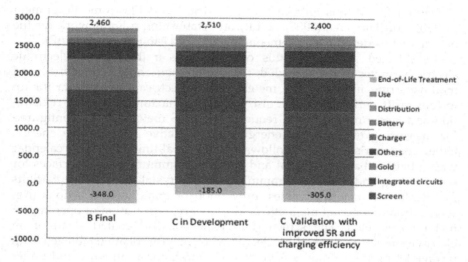

FIGURE 6.14
LCA scores used in product development of B and C.

integrating renewable power into the grid; hybrid and electric vehicles; and energy efficient lights, motors, appliances, and heating and cooling systems. Gurauskienė and Visvaldas (2006) applied the eco-design methodology for electrical and electronic equipment in a company in Lithuania. The company specializes in development and manufacturing of multifunctional static electricity meters. It is a leading manufacturer of electronic electricity meters in the Baltic region. The product range varies from electronic single-phase meters for household applications to three-phase multifunctional electricity meters and devices for industry and utilities. There is a modern line of electronic devices assembly, laboratory for calibration and verification and department of engineers, researchers, and designers, who apply the latest technologies for development of new products and product improvements that satisfy the needs of emerging new markets. The development of products is a permanent process in the company. That is why an eco-design approach has also been implemented. The basic drivers for implementing eco-design are compliance with both environmental legislation and requirements of standards and commitment of the company environmental policy to increase environmental efficiency giving the top priority to the pollution prevention. The main internal drivers are to reduce the costs, to improve the image of the company and products, to improve the quality of products and to make innovations. Compliance with the following environmental legislations is the main external driver: Directive (2002/96/EC) on Waste Electrical and Electronic Equipment (WEEE); Directive (2002/95/EC) on the Restriction of the Use of Certain Hazardous Substances in Electrical and Electronic Equipment (RoHS); and Directive (2005/32/EB) on Establishing a Framework for the Setting of Eco-design Requirements for Energy Using Products (EuP).

The selected product in this case study is a three-phase electronic electricity meter EMS, intended for use in household, commerce, and manufacturing industry. The meters collect electrical energy data for up to four tariff periods per day, also providing maximum demand registration and load profiling upon request. The EMS meters can be configured for measuring either active energy or both active and reactive energy. Other characteristics are as follows: internal real-time clock and calendar with a battery backup; optical and electrical communication interfaces for data reading and meter programming; liquid crystal display with 8 digits for displaying metering values; prevention of unauthorized actions (e.g. registration of external magnetic field influence, registration of openings of meter cover, etc.). The screen (two plates of zinc coated steel) of an electricity meter is covered by clear cowl of polycarbonate (stabilized with ultraviolet ray) – it protects from external mechanical influence and under water. Cowl and cops are fixed with screws and covered by lead. There is a printed circuit board with all electronic components inside of an electricity meter. The measurements of the EE meter are: 328 × 178 × 58 mm. The weight is less than 1.5 kg. This product is classified as an electrical and electronic equipment (EEE), so there is a great potential and importance to implement eco-design tools. The methodology Eco-indicator'99 was used in order to assess an environmental impact of the meter. This LCA methodology is adapted for the eco-design. The Eco-indicator'99 scores are based on an impact assessment methodology that transforms the data of an inventory table into damage scores which can be aggregated to one single score/millipoint (mPt) that expresses the total environmental load of a product or process (weight of materials, emission and electricity energy). The impact on environment is calculated weighing three types of damage: human health, ecosystem quality, resources. The scope of the meter life cycle begins from extraction of raw materials and finishes at end-of-use of the electricity meter. The entire life cycle is approximately 30 years. The functional unit is one electrical electricity meter EMS. The meter is analyzed at all stages of its production, handling and disposal. From the result of environmental impact calculations (Figure 6.15), it is found that all the life cycle consists of 17 points: a production stage – 1.4 Pt, a usage stage – 16Pts (Figure 6.3); and consumption of electricity energy (631 kWh per 30 years) for the equipment operation.

In comparison with other electrical and electronic equipment, this is a very small amount considering the period of 30 years. On the other hand, the function of an electricity meter is to collect the electrical energy data and to economize on energy consumption in time of day and year. Electricity consumption was not considered as an environmental aspect implementing eco-design, because of the long handling period. This

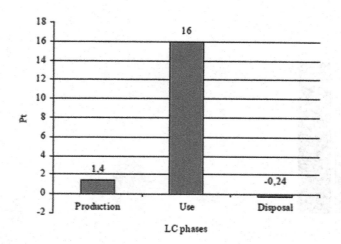

FIGURE 6.15
Environmental impact of all the life cycle (production, use and disposal).

environmental aspect is more positive than negative, because of avoiding the usage of raw materials and energy for production of a new electricity meter. The analysis of the production stage (Figure 6.16) has been made evaluating the environmental impact of all components and assembling processes.

The significant environmental impacts come from the screen of the meter (506 mPt) that is steel production (182 mPt) and zinc coating (324 mPt); the case of the meter (345 mPt): production of PC granules (316 mPt), injection moulding (29 mPt); metal fixing components (188 mPt): lead (109 mPt), screws (79 mPt); electronic components (capacitors, microchips, resistances) (195 mPt); wire (70 mPt); and brazing with lead solder (44 mPt).

As the result of the LCA, the significant environmental impacts from the life-cycle stages were identified as follows: Production (345 mPt) and disposal (2.496 mPt) of the meter body; Production of the screen (506 mPt); Wires (70 mPt); Metal fixing components (188 mPt); and consumption of the electricity at the usage stage (16,000 mPt). Considering the LCIA results of the EE meter, improvement decisions for the electricity meter were made and the new product was improved through the following measures

- Reduction of the case with the amount of plastic used reduced by 22%;
- No steel used in a replaced meter screen;
- Reduction of wires (~3 m);
- No PVC stickers used;
- Metal contacts and screws replaced by wires fixed with chips;

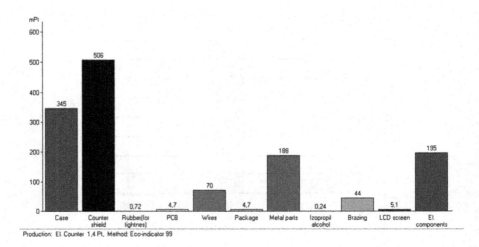

FIGURE 6.16
Environmental impact of production phase.

- Reduced environmental impact at a production stage, by using lead free solder and components;
- New design of PCB with no waste at the assembly stage;
- Optimized end-of-life systems because of the information on the disposal of hazardous substances, plastics marking, and because of the symbol indicating separate collection of electrical and electronic equipment.

The environmental impact of a new EE meter has been analyzed according to the Eco-indicator'99 methodology. Comparison of environmental impact of present and new EE meters is presented in Figure 6.17.

With the new design, the environmental impact in the production stage has been reduced by 2.73 times (Eco-indicator'99 value has been changed from 1,400 to 423.59 mPt). Compliance of legal drivers has been realized; cost reduction and raw materials usage has been optimized and awareness about eco-design of company employees has increased. It indicates that the integration of eco-design into a product development is beneficial for both company and the environment.

6.8 Summary

Design is essentially a complex problem-solving process and the engineering problem-solving methodology is used to carry out the engineering design process for new things or to improve things. To move toward

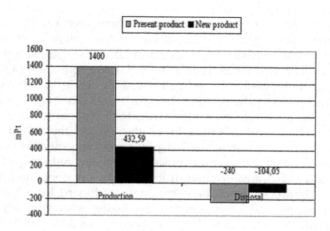

FIGURE 6.17
Environmental impact comparison of present and new EE meters.

a more sustainable practice of engineering, the design process must be modified to enable engineers to tackle sustainability issues in a structured manner. The chapter identifies critical tasks in each design phase deemed necessary to move from the conventional design toward sustainable design. It is crucial to the successful delivery of sustainable development to realize that the problem definition stage is where rigorous consideration of sustainable development issues and in particular life cycle thinking approach will generate the greatest benefits. The use of sustainable procurement criteria is an opportunity for clients/owners to build and operate projects in an environmentally friendly manner. Efforts should be made to be as quantitative as possible when evaluating alternative designs. Indicators and metrics that can be used to measure and quantify environmental sustainability need to be developed to provide a basis for decision-making. Finally, the chapter gives some case studies on the application of the sustainable engineering design process in various engineering disciplines.

SUSTAINABLE ENGINEERING IN FOCUS: SUSTAINABLE PRODUCT REDESIGN AT MAKSS PACKAGING INDUSTRIES LTD. IN KAMPALA, UGANDA

The company has 135 employees, was established in 1994 and produces 2,500,000 kg corrugated cardboard boxes per year. It was one of the first companies to get in touch with the Uganda Cleaner Production Center (UCPC) in 2002 when the UCPC launched a D4S Redesign project. UCPC found that MAKSS had significant potential to improve their production process of corrugated cardboard boxes and to innovate its products. The design of the fruit box, for example, was traditional and had not changed in 20 years. It consisted of two pieces of cardboard which required a separate production process for each. The traditional boxes were formed by using metal staples or tape. Transportation was done first via trucks on rough roads and later via air, mainly to Europe. Corrugated cardboard boxes need to be very robust and lightweight at the same time. Light weight engineering could improve both aspects: reducing the material input for the sake of environmental considerations and reducing costs due to expensive air transportation. Initial ideas were generated, such as reducing the thickness of the corrugated cardboard from five layers to three layers and strengthening the boxes with stiff edges and stiffeners at the same time. Furthermore, the idea of integrating the lid into the box design, which could help reduce the total mass of the box, was found. In the next phase, MAKSS Packaging Industries started intensive discussions with the different customers (flower producers, fruit and vegetable exporters, etc.) to find out their requirements and adapt the design of the boxes accordingly. MAKSS Packaging Industries Ltd. then redesigned the boxes, reduced costs and the environmental impacts at the same time. As a first result, MAKSS launched two redesigned products on the Uganda market in November 2002 – the 5 kg fruit box and the flower box for export. Both were redesigned according to D4S criteria. The redesigned box for flower export has the following advantages: (i) Resource efficiency – a 167 g reduction in weight equal to 12% of the original design. (ii) Improved production process – the production of the box involves one production step less since the bottom is 3 ply instead of 5. The box is self-locking and does not require any tape or staples. (iii) Cost reduction – the box is sold at a cheaper price to the customer; air cargo charges (approximately 1.5 U$/kg to Europe) are less since it is lighter. (iv) Functionality and customer satisfaction – this design offers better ventilation for the flowers, so the product can be better protected and the flowers are in better shape and consequently have higher value. The D4S redesigned box for fruit has the following advantages: (i) Resource efficiency – A 60 g reduction in weight equal to 10.7%. (ii) Improved production process – the production of the box involves one production process less since the MAKSS D4S redesign box is a one-piece box. Offcuts are utilized to make pads for other boxes, (iii) Cost reduction – the box is sold at a cheaper price to the customer; air cargo charges (approximately 1.5 U$/kg to Europe) are less since it is lighter, (iv) Functionality and customer satisfaction – stability and ventilation are excellent. The easy locking system saves time. A one-piece box is easier to handle and less space is needed for packing. Furthermore, there is no problem with imbalance in stocks between tops and bottoms.

Source: Crul and Diehl (2006).

References

Abraham, M. and Nguyen, N. 2004. "Green Engineering: Defining principles" – Results from the Sandestin Conference. *Environmental Progress*, 22, 233–236. DOI: 10.1002/ep.670220410t.

Allen, D., Bakshani, N. and Rosselot, K. 1992. *Pollution Prevention: Homework and Design Problems for Engineering Curricula*. American Institute of Chemical Engineers, New York, American Institute for Pollution Prevention, and Center for Waste Reduction Technologies, Los Angeles. USA.

Anastas, P. and Zimmerman, J. 2003. Design through the Twelve Principles of Green Engineering. *Environmental Science & Technology*, 37, 94A–101A.

Andrae, A., Mengjun, X., Zhang, J. and Xiaoming, T. 2016. Practical Eco-Design and Eco-Innovation of Consumer Electronics – The Case of Mobile Phones. *Challenges*, 7, 3.

Ashby, M. 2005. *Materials Selection in Mechanical Design*, 3rd ed., Butterworth-Heinemann, Burlington.

AustralianGov. 2008. *The Biology of Carica Papaya*, Department of Health and Ageing Office of the Gene Technology Regulator, Canberra.

Azapagic, A., Millington, A. and Collett, A. 2006. A Methodology for Integrating Sustainability Considerations into Process Design. *Trans IChemE, Part A, Chemical Engineering Research and Design*, 84(A6), 439–452.

Azapagic, A. and Pedan, S. 2005. An Integrated Sustainability Decision-Support Framework-Part I: Problem Structuring. *International Journal of Sustainable Development and World Ecology*, 12(2), 98–111.

Azapagic, A. and Perdan, S. 2011. *Sustainable Development in Practice-Case Studies for Engineers and Sxientists*, 2nd ed., John Wiley, New York.

CAC/RCP-53. 2003. *Code of Hygiene Practice for Fresh Fruits and Vegetables*, FAO, s.l.

Carlson, H. 1978. *Sping Designer's Handbook*, 1st ed., Taylor & Francis, s.l.

Conteras, A., Perez, M., Dominguez, E. and Langenhove, H. 2009. Comparative Life Cycle Assessment of Four Alternatives for Using By-Products of Cane Sugar Production. *Journal of Cleaner Production*, 17, 772–779.

Crul, M. and Diehl, J. 2006. *Design for Sustainability: A Practical Approach for Developing Economies*, UNEP&TU Delft, Paris, France.

Crul, M.R., Diehl, J.C. and Ryan, C. (Eds.). 2009. *Design for Sustainability: A Step-by Step Approach*, United Nations Environmental Program, Paris.

Dandy, G., Daniell, T., Foley, B. and Warner, R. 2017. *Planning & Design of Engineering Systems*, 3rd ed., CRC Press, Boca Raton.

Davidson, C.I., Matthews, H.S., Hendrickson, C.T., Bridges, M.W., Allenby, B.R., Crittenden, J.C., Chen, Y., Williams, E., Allen, D.T., Murphy, C.F. and

Austin, S. 2007. Adding Sustainability to the Engineer's Toolbox: A Challenge for Engineering Educators. *Environmental Science & Technology.*

Dieter, J. and Schmidt, L. 2009. *Engineering Design*, 4th ed., McGraw-Hill, New York.

Dixon, J. and Poli, C. 1995. *Engineering Design and Design for Manufacturing: A Structured Approach*, 1st ed., Field Stone Pub, MA.

Eastman-Kodak-Company. 1983. *Ergonomic Design for People at Work*, 1st ed., American Book Company, New York.

Gagnon, B., Leduc, R. and Savard, L. 2012. From a Conventional to a Sustainable Engineering Design Process: Different Shades of Sustainability. *Journal of Engineering Design*, 23, 49–74.

Graedel, T.E. and Allenby, B.R. 2010. *Industrial Ecology and Sustainable Engineering.* Prentice Hall, US.

Gurauskienė, I. and Visvaldas, V. 2006. Eco-design Methodology for Electrical and Electronic Equipment Industry. *Environmental Research, Engineering and Management*, 3, 43–51.

Huawei. Sustainability Report 2014. Available at: http://www-file.huawei.com/~/media/CORPORATE/PDF/Sustainability/2014%20Huawei%20sustainability%20report-final.pdf?la=en (accessed on 18 November 2019).

Itsubo, N., Sakagami, M., Kuriyama, K. and Inaba, A. 2012. Statistical Analysis for the Development of National Average Weighting Factors – Visualization of the Variability between Each Individual's Environmental Thoughts. *The International Journal of Life Cycle Assessment*, 17, 488–498.

Jiménez, M., Víctor, E.M.-N. and Marco, V.G.-S. 2014. Biology of the Papaya Plant. In R. Ming and P.H. Moore (Eds.), *Genetics and Genomics of Papaya, Plant Genetics and Genomics: Crops and Models*, Vol. 10. Springer Science+Business Media, New York. DOI: 10.1007/978-1-4614-8087-7_2.

Kaebernick, H., Anityasari, M. and Kara, S. 2002. Technical and Economic Model for End-of-Life (EOL) Options of Industrial Products. *International Journal on Environmental Sustainable Development*, I(2), 171–183.

Mott, R. 2004. *Machine Elements in Mechanical Design*, 4th ed., Pearson, NJ.

Paper November 2013 Conference: 13 AIChE Annual Meeting, San Francisco, CA. Available at: http://proceedings.dtu.dk/fedora/repository/dtu:2791/OBJ/x142234_GrnDyst_samlet.117.pdf.

Santos, R. and Costa, A.A. (2016). BIM in LCA/LCEA Analysis: Comparative Analysis of Multi-family House and Single. CIB World Building Congress 2016. Tampere, Finland. Available at: www.irbnet.de/daten/iconda/CIB_DC29646.pdf (accessed on 31 July 2019).

Telenko, C., Seepersad, C.C. and Webber, M.E. 2008. A Compilation of Design for Environment Principles and Guidelines. ASME DETC Design for Manufacturing and the Life Cycle Conference, New York.

Ullman, D.G. 2010. *The Mechanical Design Process*, 4th ed., McGraw-Hill, New York.

Exercises

1. Why is it important to take a life cycle approach in design for sustainability?

2. With reference to the Sandestin Green Engineering Principles and through a literature search,

    i. Provide examples of engineering designs or products from these designs that caused unintended damage to ecosystems (Principle 2).

    ii. Provide one or more examples of wastes from key product systems that have found beneficial uses through engineering designs (Principle 6).

    iii. How have innovations in engineering design helped achieve sustainability goals in developing countries, such as providing basic services like drinking water, power, food preservation, and sanitation? Provide some examples and discuss the engineering design aspects (Principle 8).

    iv. Investigate and report on how engineers became involved in community engagement in your chosen field of study. What kinds of community engagement activities were the engineers involved in (Principle 9)?

3. What are the key tasks in the sustainable engineering design process that differentiates it from the conventional one?

4. Design for sustainability requires multidisciplinary teamwork. Which disciplines do you think need to work together in designing a more sustainable product in your field? Explain why and what kind of knowledge and skills these disciplines bring together. What might be the difficulties of working in multidisciplinary teams?

5. Give some creativity tools that can used by a design team to generate ideas and concepts during the design process.

6. Give some examples of sustainability criteria that you think are relevant in the sustainable design of a consumer electronic product. Why is it important to identify such criteria before specifying the design alternatives? What would happen if we did it the other way round?

7. Identify the design for sustainability strategies which have been used in the case study presented in the box: Sustainable Engineering in Focus.

8. Read the following case study on a product redesign regarding a trailer for rural transport of crops in Ghana (Source: Crul and Diehl, 2006) and identify the design for sustainability strategies which have been used:

The company REAL ("Rural Enterprise for Agro Logistics") is an enterprise working for, and partly owned, by farmers and agro-managers in Ghana. The project was carried out in 2003 and had the aim to design a means of rural transport for crops that would reduce post-harvest loss, thus increasing crop market value and improving labor conditions for farmers (men and women) in Ghana. This sustainable transport system was designed with locally available materials and production methods suitable for Ghana. The design took into consideration the influences of local culture and social habits. The concept development phase began with the generation of concepts from a functional point of view. The concepts were made out of combinations of the several functions and were clustered into three design directions. After identifying the available materials, further constraints were encountered and considerations were made. Three design directions resulted in three concepts, the Plain trailer, the Combi trailer, and the Crate trailer. The Plain trailer concept consisted of one loading space, created with wire mesh. The Combi trailer could carry almost any kind of container and had the possibility to create one space, from wire mesh as well. The Crate trailer could only fit plastic crates from an industrial containers producer in Tema. After taking several steps in development a fourth concept, the Multi trailer was generated out of the Combi and Crate trailers. The Multi trailer concept offered the opportunity to use almost any kind of container to transport the crops. It was especially designed for current use as well as future use anticipating the introduction of a complete logistic system based on crates by REAL. After comparing the four concepts on design guidelines, the Crate trailer and the Multi trailer proved to be the two most feasible concepts. Both concepts were constructed in a Ghanaian workshop. Before going into production, models were constructed. This required locating a workshop, a welder, buying the materials, and finding specific components for the construction of details of the trailers. As soon as the models were finished, they were tested in the two regions in the south. After observation and evaluation of the models in use, small changes were made to improve performance. The last test in the north was performed and evaluated as well. It appeared that at least twice the amount of head load could be transported in the same time or even faster. The users of the trailer appreciated the ease of use and maneuvering, the increased amount of products that could be loaded and the width of the total trailer which could be used on small paths. After testing both the Crate trailer and the Multi trailer on the design guidelines, it was evident that the Multi trailer needed to be further defined. The dimensions of the Multi trailer were further optimized in Solidworks, a 3D modeler, in the Netherlands. This model contains the basic construction and some details. The Multi trailer potential results include:

- *Transportability was improved: The trailer was suitable for single path roads, double track grass roads, feeder roads and asphalt roads, except for Abor where it is only suitable for single and double track roads.*
- *Efficiency was increased: Twice the amount of crops could be transported in the same time or faster.*
- *Ergonomics of handling crops was improved: The burden on the farmers (men and women) was lightened.*

- *The trailer could be produced locally: It could be produced with production methods available in Ghana and made from locally available materials. The construction of the trailer could be optimized for weight and weld type and production could be simplified. Promote the use of trailers amongst women, as well as in the transport of the crops.*

In summary, the Multi trailer reduced the post-harvest losses thereby increasing the farmers' income. Labor conditions have improved due to the lighter burden.

9. Team up with another student. You have to design a table to be used in a fruit processing factory. It is to have a steel surface and the surface is to be covered with a soft foam to reduce fruit damage. The foam top and legs of the table will need to be replaced periodically and the factory owner, who expects to purchase several hundred tables, wants the replacement of the components to be quick and effective. With the help of your local hardware store (if needed), design the table for optimum disassembly (adapted from Graedel and Allenby 2010).

Recommended Reading and Websites

- Gagnon, B., Leduc, R. and Savard, L. 2012. From a Conventional to a Sustainable Engineering Design Process: Different Shades of Sustainability. *Journal of Engineering Design*, 23, 49–74. This paper places the conventional design process and the sustainable design process on a continuum along which the engineer can position himself or herself.
- Azapagic, A. and Perdan, S. 2011. *Sustainable Development in Practice: Case Studies for Engineers and Scientists*. Second edition. Wiley-Blackwell. Taking a life cycle approach to addressing economic, environmental, and social issues, the book presents a series of new practical case studies drawn from a range of sectors, including mining, energy, food, buildings, transport, waste, and health.
- Azapagic, A., Millington, A. and Collett, A. 2006. A Methodology for Integrating Sustainability Considerations into Process Design. *Chemical Engineering Research & Design*, 84, 439–452. This paper illustrates how different sustainability criteria can be integrated within a common framework to guide the design of more sustainable processes and chemical plants.
- Crul M.R., Diehl, J.C. and Ryan C. (Eds.). 2009. *Design for Sustainability: A Step-by Step Approach*, United Nations Environmental Program, Paris, www.d4s-sbs.org/d4s_sbs_manual_site.pdf The publication targets

designers and other professionals working in the area of industrial product development. It is useful to those new to ecodesign as well as those interested in breakthrough innovation for sustainability. There is also a sister publication "Design for Sustainability: A Practical Approach for Developing Economies", 2006, www.d4s-de.org/manual/d4stotalmanual.pdf which focuses on the specific needs of small- and medium-sized companies in developing economies

- Anders S.G., Andrae, Mengjun, X., Jianli, Z., and Xiaoming, T. 2016. Practical Eco-Design and Eco-Innovation of Consumer Electronics – the Case of Mobile Phones *Challenges*, 7(3), www.mdpi.com/2078-1547/7/1/3

  This article describes how sustainability is successfully implemented in practice at a large Chinese company, developing and producing various kinds of electronic products used for communication. It also describes how a variety of eco-innovations and business models contribute to reducing the environmental impact, for example, through increased recovery and recycling. A new kind of eco-design procedure is presented along with a new methodology which shows how a mobile phone gradually becomes more sustainable from one generation to the next.

- The text *Pollution Prevention: Homework and Design Problems for Engineering Curricula*, published by the American Institute of Chemical Engineers, contains more than 20 case studies related to life cycles, design of materials, thermodynamics, and transport phenomena (Allen et al., 1992). More case studies, formatted for classroom use, have been developed by participants in faculty workshops sponsored by the Center for Sustainable Engineering (www.csengin.org). The case studies are freely available, without copyright restrictions. Examples include, but are not limited to, estimating the environmental impacts of concrete; water system design, including the use of reclaimed water; methods for incorporating social dimensions of sustainability; landfill power generation; accounting for environmental costs and benefits in a semiconductor facility; assessing environmental product claims; electric power generation; and recycling systems for vehicles. Finally, additional examples can be drawn from the Green Engineering Web site of the U.S. EPA (www.epa.gov/oppt/greenengineering) and the links available at that site.

7

Sustainable Buildings and Sustainable Infrastructure Rating Systems

7.1 Introduction

The shift to a sustainable built environment is being driven by the growing evidence of accelerated destruction of our planetary ecosystems and the increasing demand for natural resources due to population growth and increased consumption. The building sector accounts for nearly 40% of the world's energy consumption, 30% of raw material use, 25% of solid waste, 25% of water use, 12% of land use, and 33% of the related global greenhouse gas (GHG) emissions (UNEP-SBCI, 2009). Yet this sector has great potential to significantly reduce GHG emissions at lower costs, both in developed and developing countries (UN Environment and International Energy Agency, 2017). Furthermore, these cuts can be achieved using readily available technologies, and at the same time there is the opportunity to create green jobs. With proven and commercially available technologies, energy consumption in both new and existing buildings can be cut by an estimated 30–50% with potential net profit during the lifespan of the building (UNEP-SBCI, 2009). Table 7.1 summarizes the key elements and corresponding factors for sustainable built environment design. While governments tend to focus more on increasing the energy efficiency of buildings, there is the need to assess the performance of a building from a broader perspective, taking into account also the environmental, social, and economic impacts of constructions. This has prompted the creation of green building and infrastructure rating systems. The measurement of the sustainability performance of a building through a recognized metric impels people to adjust their behavior to optimize their score on that metric and hence mitigate impacts (Box 7.1). The practice of designing sustainable buildings makes business sense as it leads to improved building performance with increased revenue through higher sales price and lower cost of utilities.

TABLE 7.1

Key Elements and Corresponding Factors for Sustainable Built Environment Design

Factor 1: Material Selection

1. Prescribing low energy materials
2. Use of locally available materials
3. Use of durable/high-performance materials
4. Use of materials with low health risk and pollution
5. Material reuse

Factor 2: Economic Considerations

1. Cost/benefit analysis
2. Life cycle cost analysis
3. Cost efficiency

Factor 3: Policy and Regulations

1. Presence of design sustainability regulatory requirements
2. Presence of sustainability rating systems
3. Inclusion of sustainability requirements in project briefs

Factor 5: Design and Project Management

1. Early contractors involvement at design stage
1. Early suppliers involvement at design stage
2. Selection of appropriate contract/project delivery type
3. Inclusion of sustainability related clauses in contract documents
4. Proper construction quality control procedure

Factor 6: Technical Considerations

1. Exhaustive site survey and ground investigation
2. Considering alternatives prior to proposing a solution
3. Multi-disciplinary integrated design team beginning from feasibility study stage
4. Meeting functional requirements and users Comfort documents
5. Value engineering
8. Harmony with the surrounding environment

Factor 4: Social Considerations

1. Public/beneficiaries participation
2. Client participation
3. Accessibility of the infrastructure to the public including people with specific needs
4. Health and safety consideration for construction workers and the public during construction and operation stages
5. Security consideration during construction and use
6. Satisfaction of the public
7. Protection of cultural heritage
8. Protection of landscape, historical areas and archaeological sites
9. Risk analysis and disaster mitigation

Factor 7: Environmental Considerations

1. Climate resiliency (resistant to climate change)
2. Ensuring efficient energy utilization both during construction and operation phases
3. Optimizing uses of natural resources
4. Optimizing site potentials (land use)
5. Uses of less energy during construction and operation
6. Waste minimization/design optimization

Factor 8: Design Professionals and the Design Process

1. Awareness of clients about sustainability
2. Awareness of designers about sustainability
3. Knowledge, skills and experience of designers
4. Appropriating adequate time for design
7. Presence of design guidelines/ procedures for sustainable infrastructure
8. Proper coordination among designers from different disciplines
9. Willingness of designers to implement sustainability design concept and practices in their designs

Source: Haupt and Nuramo (2017).

BOX 7.1 SUSTAINABLE BUILDINGS IN BRAZIL

Developing tools and strategies for achieving the wide acceptance and adoption of sustainable building practices throughout the world is one of four key goals guiding the work of the UNEP Sustainable Buildings and Climate Initiative (UNEP-SBCI). To achieve this goal, UNEP-SBCI provides policy advice and support for achieving high-efficiency and low-GHG-emission buildings, particularly in developing countries. The Sustainable Urban Housing Initiative (SUSHI), drawing on the research of UNEP-SBCI and its network of expert members, develops approaches for including sustainable building principles in the design and construction phases of social housing. The added value of SUSHI lies in addressing common problems in the development of affordable housing, such as employing standardized design solutions, which are often poorly adapted to local conditions, and the use of low-quality materials (Vanderley, 2010).

SUSHI has already been piloted in Sao Paulo, Brazil, and Bangkok, Thailand (Vanderley, 2010). A range of stakeholders have been engaged in these two locations, including federal and local governments, the Federal Economic Bank, the Community Organization Development Institute, the Brazilian Chamber of Construction Industry, UNEP-SBCI members and private companies. At the Brazilian pilot site (in Cubatão, Sao Paulo), SUSHI has resulted in improved working conditions, higher-quality construction, better waste management and reduced impact of new buildings on their surroundings. It has highlighted some key design features that are easily replicable. These include the use of natural ventilation, lighting and shading to increase energy efficiency and reduce energy demand, and the use of rainwater harvesting and water recycling techniques to minimize the production of wastewater from domestic appliances and activities (SUSHI, 2010). In Brazil, the National Programme on Energy Efficiency provides building ratings by selecting building materials, considering the choices of lighting and air conditioning (SUSHI, 2010). Additionally, the Federal Economic Bank labels the housing projects according to the degree the developer adheres to sustainability practices. To award a Blue House Seal, the Federal Economic Bank analyzes criteria grouped in six categories: urban insertion, design and comfort, energy efficiency, conservation of material resources, good use of water and social practices. The objective is to encourage construction of housing units that respect the environment during construction and operation, while at the same time providing good comfort and health conditions for their users. The Blue House Seal is divided into Gold, Silver and Bronze categories. To receive a Gold ranking, the building must meet at least 24 of the 46 conditions.

Those that meet 19 criteria will receive Silver, and those that meet at least the 14 mandatory criteria will receive Bronze. This system allows developers to properly market projects that have gone beyond the minimum building requirements.

Source: Vanderley (2010).

Rating tools are design checklists, and credit rating calculators developed to assist designers in identifying design criteria and documenting proposed design performance. A common objective of these rating systems is that projects awarded or certified within these programs are designed to reduce the overall impact of the built environment on human health and the natural environment. The benefits of using a sustainability rating system for buildings and infrastructure are as follows:

   (i) Sustainability aspirations can be set.
  (ii) Clear goals can be worked.
 (iii) Sustainability performance can be verified through a third party.
 (iv) Sustainability performance can be demonstrated to third parties.
  (v) Improvement can be measured and demonstrated.
 (vi) Sustainability education can be facilitated and encouraged.
(vii) Positive marketing can be generated.

The use of sustainability rating tools in the built environment began in the 1990s and 2000s with the introduction of rating systems for buildings. BREEAM (Building Research Establishment Environmental Assessment Method) was launched in the UK in 1990, LEED (Leadership in Energy and Environmental Design) in the United States in 2000, and Green Star (Australia and New Zealand) in 2003 and 2007, respectively. The need to look beyond individual buildings or facilities and consider networks and neighborhoods has resulted in the introduction of neighborhood and community tools, e.g. BREEAM Communities and LEED-ND. Whilst a greater emphasis of these sustainability rating tools was on the buildings, there was a lack of similar tools for the "horizontal" infrastructure beyond buildings. However, infrastructure (roads, railways, water supply and wastewater systems, energy generation and distribution systems) plays a massive role in the sustainability of the built environment. Unlike buildings, infrastructure works at a larger scale and requires careful planning to work well and efficiently. It requires a lot of resources and faces a high amount of wear and tear, while facing the expectation of a longer life cycle than other new construction. Furthermore, construction of infrastructure often has a great impact on local residents and must therefore be completed in as quickly a time as possible with many efforts taken to limit disturbances. Because of these unique issues, infrastructure is not easily handled by systems like LEED or BREEAM, which have been successful at the scale of a building. Recently, sustainable infrastructure rating systems have been developed with three of them – Envision in the United States (Envision, 2016), CEEQUAL in the UK (CEEQUAL, 2016), and the Infrastructure Sustainability (IS) Rating scheme in Australia (IS, 2016) – being able to evaluate all types and sizes of civil infrastructures, including ports, airports, highways, dams, bridges, water and wastewater treatment facilities, tunnels, and railways.

As a sustainability tool, sustainability rating systems have the objective of encouraging sustainability practices beyond the regulatory minimum, and to communicate sustainability effectively. Furthermore, rating systems are useful when other means of quantification (e.g. LCA) fail to capture the full range of sustainability best practice impacts. This chapter is intended to provide the engineer with sufficient information to understand the principles of these rating systems so as to use them in future design.

7.2 Sustainable Buildings Rating Systems

The term green building or sustainable building is also known as high performance building. A precise definition is not easy (Kubba, 2012). The USEPA defines it as 'the practice of creating structures and using processes that are environmentally responsible, and resource-efficient throughout a building's life cycle from siting to design, construction, operation, maintenance, renovation and deconstruction' (https://archive.epa.gov/green building/web/html/about.html). This practice expands and complements the classical building design concepts of economy, utility, durability, and comfort. For the purposes of this chapter, a green building can be defined as a structure which employs multiple strategies, such as energy efficiency, water conservation, responsible use of materials and resources, and indoor air quality, in order to alleviate the negative sustainability impacts of the built environment. Green building is essentially a whole-systems approach to improve conventional design and construction practices and standards and we can identify the following fundamental principles for sustainable buildings:

- Energy, material, and water resources are efficiently managed
- Waste is managed according to the hierarchy
- Environmental quality is restored and protected
- Health and indoor environmental quality is restored and protected
- Natural systems are reinforced

- Analysis of the life cycle costs and benefits of materials and methods
- Integration of sustainability in the design decision-making process

CalRecycle (www.calrecycle.ca.gov/greenbuilding/basics) cites the main elements of green buildings and sustainability as follows:

Siting: Start by selecting a site well suited to take advantage of mass transit; Protect and retain existing landscaping and natural features. Select plants that have low water and pesticide needs, and generate minimum plant trimmings. Use compost and mulches. This will save water and time, and recycled content paving materials, furnishings, and mulches help close the recycling loop.

Energy efficiency: Passive design strategies can dramatically affect building energy performance. These measures include building shape and orientation, passive solar design, and the use of natural lighting. Develop strategies to provide natural lighting. Studies have shown that it has a positive impact on productivity and well-being. Install high-efficiency lighting systems with advanced lighting controls. Include motion sensors tied to dimmable lighting controls. Task lighting reduces general overhead light levels. Use a properly sized and energy-efficient heat/cooling system in conjunction with a thermally efficient building shell. Maximize light colors for roofing and wall finish materials; install high R-value wall and ceiling insulation; and use minimal glass on east and west exposures. Minimize the electric loads from lighting, equipment, and appliances. Consider alternative energy sources such as photovoltaics and fuel cells that are now available in new products and applications. Renewable energy sources provide a great symbol of emerging technologies for the future. Computer modeling is an extremely useful tool in optimizing design of electrical and mechanical systems and the building shell.

Materials efficiency: Select sustainable construction materials and products by evaluating several characteristics such as reused and recycled content, zero or low off gassing of harmful air emissions, zero or low toxicity, sustainably harvested materials, high recyclability, durability, longevity, and local production. Such products promote resource conservation and efficiency. Using recycled-content products also helps develop markets for recycled materials that are being diverted from. Use dimensional planning and other material efficiency strategies. These strategies reduce the amount of building materials needed and cut construction costs. Reuse and recycle construction and demolition materials. For example, using inert demolition materials as a base course for a parking lot keeps materials out of landfills and costs less. Require plans for managing materials through deconstruction, demolition, and construction. Design with adequate space to facilitate recycling collection and to incorporate a solid waste management program that prevents waste generation.

Water efficiency: Design for dual plumbing to use recycled water for toilet flushing or a gray water system that recovers rainwater or other non-potable water for site irrigation; minimize wastewater by using ultra low-flush toilets, low-flow shower heads, and other water conserving fixtures; use recirculating systems for centralized hot water distribution; Install point-of-use hot water heating systems for more distant locations; use a water budget approach that schedules irrigation for landscaping; meter the landscape separately from buildings; use micro-irrigation (which excludes sprinklers and high-pressure sprayers) to supply water in non turf areas; use state-of-the-art irrigation controllers and self-closing nozzles on hoses.

Occupant health and safety: Recent studies reveal that buildings with good overall environmental quality can reduce the rate of respiratory disease, allergy, asthma, sick building symptoms, and enhance worker performance. Choose construction materials and interior finish products with zero or low emissions to improve indoor air quality. Many building materials and cleaning/maintenance products emit toxic gases, such as volatile organic compounds (VOC) and formaldehyde. These gases can have a detrimental impact on occupants' health and productivity. Provide adequate ventilation and a high-efficiency, in-duct filtration system: Heating and cooling systems that ensure adequate ventilation and proper filtration can have a dramatic and positive impact on indoor air quality. Prevent indoor microbial contamination through selection of materials resistant to microbial growth, provide effective drainage from the roof and surrounding landscape, install adequate ventilation in bathrooms, allow proper drainage of air-conditioning coils, and design other building systems to control humidity.

Building operation and maintenance: Green building measures cannot achieve their goals unless they work as intended. Building commissioning includes testing and adjusting the mechanical, electrical, and plumbing systems to ensure that all equipment meets design criteria. It also includes instructing the staff on the operation and maintenance of equipment. Over time, building performance can be assured through measurement, adjustment, and upgrading. Proper maintenance ensures that a building continues to perform as designed and commissioned.

Metrics for the "green" benefits derived from the above-named practices are certified through rating systems established by many national green building councils for ambitious performance goals for the built environment in their countries. These councils develop and supervise building assessment systems that provide ratings for buildings based on a holistic evaluation of their performance against a wide array of sustainability requirements. BREEAM (Building Research Establishment Environmental Assessment Method) was the first rating system that was established in the UK in 1990 providing both a standard definition for green building and a means of evaluating its performance against the requirements of the building assessment system(www.breeam.com). BREEAM was a success and Canada and Hong Kong subsequently adopted BREEAM as the platform for their national building assessment systems. In the United States, the USGBC developed an American building rating system with the acronym LEED (Leadership in Energy and Environmental Design) which when launched in 2000 rapidly dominated the market for third party green building certification. Similar systems were developed in other major countries such as CASBEE (Comprehensive Assessment System for Building Environmental Efficiency) in Japan (2004) and the Green Building Council of Australia (GBCA) Green Star Rating Tool (2006). Green Star has been customized with national versions in New Zealand and South Africa. In Germany a building assessment system known as the DGNB was developed and regarded by many as the most advanced evolution of building assessment systems. It is now thought that there are around 600 green certification systems worldwide and the numbers continue to grow. BREEAM, LEED, CASBEE, Green Star and DGNB represent the cutting edge of today's high-performance green building assessment systems. Table 7.2 gives a summary of existing sustainable buildings rating systems.

There are different approaches in the application of rating tools. They can be applied to the planning and design, construction, operation and maintenance, renovation, and eventual demolition phases of a green building. They can also differ in the type of buildings they are applied to, with specific tools or subsets of tools used for different building types such as homes, commercial buildings or even whole neighborhoods.

Leadership in Energy and Environmental Design (LEED) was created in 2000 by the U.S. Green Building Council (USGBC), for rating design and construction practices that would define a green building in the United States. LEED is used throughout North America as well as in more than 30 countries. LEED consists of credits which earn points in seven categories: Site Selection, Water Efficiency, Energy and Atmosphere, Materials and Resources, Indoor Environmental Quality, Regional Priority, and Innovation in Design. One hundred points are available across these categories with mandatory prerequisites such as minimum energy and water-use reduction, recycling collection, and tobacco smoke control. Within each category are credits that pertain to specific strategies for sustainability, such as the use of low-emitting products, reduced water consumption, energy efficiency, access to public transportation, recycled content, renewable energy, and daylighting. Since its inception, LEED standards have become more stringent as the market has changed and expanded to include distinct rating systems that address different building types. Building owners can apply to be certified within four different LEED categories (see Figure 7.1).

The current LEED version is version 4.0. In a nutshell, LEED certification works like this: Projects apply for LEED certification under a particular rating system and projects achieve points by satisfying various

TABLE 7.2

Summary of Existing Sustainable Buildings Rating Systems

System	Type of Standard/ Certification/ Rating Schemes	Managing Organization	Categories/ Areas of Focus/ Issues	Certification Levels
BREEAM (UK, EU, EFTA member states)	• New Construction • Communities • In-use Buildings • EcoHomes • Courts • Education • Healthcare • Industrial • International • Offices • Prisons • Retail • Multi-residential	BRE Global	• Energy and water use • Internal environment • Health and well-being • Pollution • Transport • Materials • Waste • Ecology and land use • Management	• Pass • Good • Very good • Excellent • Outstanding
LEED (United States)	• New construction (NC) • Existing Buildings, Operations & Maintenance (EB O&M) • Commercial Interiors (CI)	U.S. Green Building Council	• Sustainable Sites • Water Efficiency • Energy & Atmosphere • Materials & Resources	• Certified • Silver • Gold • Platinum

(Continued)

TABLE 7.2 (Cont.)

System	Type of Standard/ Certification/ Rating Schemes	Managing Organization	Categories/ Areas of Focus/ Issues	Certification Levels
	• Core & Shell (CS) • Schools (SCH) • Retail • Healthcare (HC) • Homes • Neighborhood Development (ND)		• Indoor Environmental Quality • Locations & Linkages • Awareness & Education • Innovation in Design • Regional Priority through a set of prerequisites and credits	
Green Star (Australia, New Zealand South Africa)	• Design and as built communities: performance, interiors • Legacy rating tools: education, health care, industrial, multiunit residential, office, office interiors, retail center and public building	Green Building Council of Australia, New Zealand or South Africa administers program Independent assessors to assess and score projects	• Management • Indoor Environmental Quality • Energy • Transport • Water • Materials • Land Use & Ecology • Emissions • Innovation • Governance • Design • Livability • Economic prosperity • Environment	• 4 Star • 5 Star • 6 Star For design and as-built communities and interiors • 1–6 Star for performance

| CASBEE (Japan) | • Pre-design
• New Construction
• Existing Building
• Heat island
• Home
• Renovation
• Urban area and buildings
• Urban development | JSBC (Japan Sustainable Building Consortium) and its affiliated subcommittees | • Energy efficiency
• Resource efficiency
• Local environment, and
• Indoor environment | • S (Excellent)
• A
• B+
• B-
• C (Poor) |

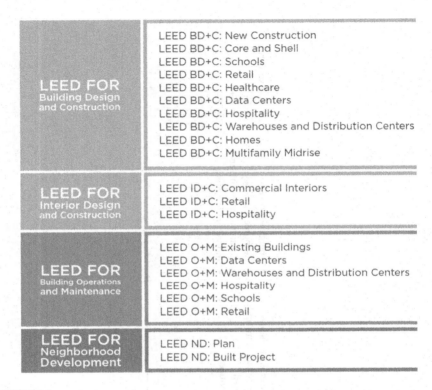

FIGURE 7.1
LEED categories.

Retrieved from https://new.usgbc.org/leed

requirements, which are geared toward different methods of green building practice. Points are distributed as follows:

- Sustainable Sites (26 possible points)
- Water Efficiency (10 possible pints)
- Energy and Atmosphere (35 possible points)
- Materials and Resources (14 possible points)
- Indoor Environmental Quality (15 possible points)
- Innovation in Design (6 possible points)
- Regional Priority (4 possible points)

Points are allocated based on potential for good – more points go for actions with the least environmental impact and most human benefit.

Project teams select the credits that fit best with their project, and that will help them reach their desired LEED certification. If a project earns enough points, it gets a ranking (certified, silver, gold, or platinum) and is then dubbed a "LEED certified" project. An applicant for a LEED certification for a project must submit documentation that verifies compliance with the requirements of the rating system. The Green Building Council responsible for issuing the LEED system used on the project is the entity that grants certification.

Many sustainable rating systems also now allow to assess the performance of neighborhood and cities (see Box 7.2). There are three well-known tools: BREEAM Communities, CASBEE for Urban Development and LEED for Neighborhood Development. The most important category within CASBEE for Urban Development and LEED for Neighborhood Development is "Infrastructure," (LEED for Neighbourhood Development, 2009) while in BREEAM Communities the most stressed categories are "Infrastructure" and "Transportation."

BOX 7.2 URRIÐAHOLT (URRIDAHOLT) SUSTAINABLE NEIGHBORHOOD

The Urriðaholt (Urridaholt) urban master plan in Garðabær, Iceland, covers an environmentally sensitive area, with a pristine lake and panoramic views. It is a walkable neighborhood in a beautiful natural setting. Community participation was a key element in the preparation of the master plan. Special attention was given to the relationship between the urban character and the natural environment. SUDS and BREEAM Communities certification were introduced here for the first time in Iceland. In Iceland, land is comparatively cheap and plentiful. This has fueled urban sprawl and low-density residential suburbs where distances become too far to walk, pavements disappear, and bus routes become unviable. One objective was to show how this trend could be reversed, using Urriðaholt as a model, and recalling the walkable character of the central core of downtown Reykjavík. The aim was to create a compact and diverse mixed-use, walkable neighborhood, with local amenities and opportunities to run good public transport. Integrated with sustainable urban design strategies, this will enhance the well-being of the people that will be living and working in Urriðaholt in the future.

The site encompasses a hill that rises some 50 m above a lava field, wetlands, and a portion of a pristine lake, in total 100 ha. The site is surrounded by a beautiful, unspoiled landscape and has spectacular views of mountains, volcanoes and the sea.

Another objective was to design a neighborhood that would fit into, and work with, the unspoiled surrounding landscape, and to integrate the built environment with nature. The pristine lake was also to be protected, both in terms of the cleanliness of the water and the water level. The aim was to use a Sustainable Urban Drainage System (SUDS) in the neighborhood, as traditional drainage solutions would result in a loss of water catchment area and shrinking of the lake itself. The SUDS maintains the natural hydrology of the site, and ensures that rain falling within the catchment area will feed the lake naturally.

The project serves as a bridge from the seaside community in Gardabaer to the inland nature preserve, providing new connections to the Green Scarf – a continuous natural landscape wrapping the Reykjavik capital area. The master plan was created with respect for the environment and an emphasis on sustainability – aiming for a BREEAM assessment and certification. When the planning process started, BREEAM communities' certification did not yet exist. To ensure the development met with the original sustainable vision, the planners therefore used existing guidelines and checklist. The master plan served as an excellent base for the BREEAM communities 2012 criteria and guidelines. The BREEAM community guidelines assisted the development of local plans in Urridaholt.

The project began with substantial collaborative activities, following Charrette methodology, with municipality council politicians and officers. This included a site visit, briefings, dialogue workshops and hands-on planning groups. The result was a Vision for a walkable, climatically responsive, mixed-use neighborhood, where the lake and the lava fields were protected as much as possible. After this, the local residents were invited to a Community Planning Weekend, where they were first introduced to the development concepts, and then invited to take part in workshops, walkabouts and hands-on planning sessions to consider challenges and opportunities for the site. What emerged was the desire to create a highly sustainable community. This meant traffic-calmed streets and green links to the protected lake and the wider natural environment, and a detailed Vision for the site, incorporating ideas from the Winter Cities movement, SUDS, etc.

Urridaholt was designed to consist of a mixture of some 1,600 dwellings, 90,000 m^2 for office and retail, and 65,000 m^2 for public uses. Up to 9,000 people will be living and working in Urridaholt when fully built. Mixed-use principles are utilized. The public buildings hosting most activities are sited to minimize walking distances – around the top of the hill at the geographic center of Urridaholt. This hub is surrounded by a series of residential areas, each with their own identity. The hillside's own "Green Scarfs" lead through the landscape to surrounding natural areas. A range of housing types is planned to encourage development of a mixed community. Pedestrians are given priority: streets are designed to reduce traffic speed through changes in direction and use of landscaping. Solutions encouraging an active outdoor life throughout the year were incorporated to enhance the community's health and well-being during the dark winter months. Urridaholt features the first Sustainable Urban Drainage System (SUDS), a large-scale Blue-Green drainage solution. Impervious materials are kept to a minimum, and the area design forms a network of swales, placed to collect water from roads and roofs. Enabling walking, biking and use of public transports promotes resource efficiency. Guidance for the use of sustainable materials in the area also encourages sustainable building practices. Educational material on environmental practices is provided for residents and the elementary school has a special environmental focus.

Urridaholt has received international awards and certification:

- Recipient of the 2007 citation from the Urban Design Committee of the Boston Society of Architects
- International Award for Livable Communities (LivCom), silver level, for the category "Environmentally Sustainable Projects"
- First international project to achieve a certification under BREEAM Communities 2012, and the first urban master plan in Iceland to receive a BREEAM Communities certification. The local plan for the North side phase 2 is the first phase to achieve a final certification – with a "Very Good" rating.

Buildings are sited and designed to take advantage of daylight from the low-angled sun to reduce energy consumption. Sustainable design guidelines for designers promote the use of local and sustainable materials. Encouraging cycling and the use of public transport through safe streets and cycle paths reduces fossil fuel consumption. There is also bike parking at regular intervals in the streets and requirements to include shelters for bikes in all private houses and apartment buildings. Safe disposal of hazardous materials is encouraged. Service stations for recycled material (glass, paper, plastics and metal) are provided within a walking distance. Sustainable drainage systems are used in the area. This protects the lake and the surrounding environment. Permeable surfaces exist where possible, to get the water naturally into the ground. The rain from the roofs drains into the soil in raingardens and there are grassy water-channels, (swales) by the roadsides, into which storm water can drain. A network of swales collects water from the roads and allows infiltration, whilst the rainwater runs along the contours to the lake. The SUDS render the area a Low Impact Development, by the use of Blue-Green drainage solutions, which preserve the local hydrology and protect the environmental quality of lake Urridavatn.

The construction of the area and the buildings is somewhat more costly than traditional construction. However, it is expected that the buyers of the flats and houses will appreciate the area's environmental profile and the green solutions, and be willing to pay the somewhat higher price.

Source: One Planet Network. (Retrieved with permission from One Planet Network. www.oneplanetnetwork.org/)

7.3 Sustainable Infrastructure Rating Systems

Infrastructure is the underlying system of structures, facilities and services that are essential to the functioning of an economy, encompassing energy (including fuel) generation and distribution, transportation, telecommunications, and water and sanitation (including waste) (Bhattacharya et al., 2015). In addition to human-built infrastructure, the term can also be interpreted to include nature-based solutions, recognizing that ecosystems can provide biodiversity and other essential goods and services – for example, through agriculture, forestry, wetlands – that offer alternatives to those provided by human-built infrastructure (Silva and Wheeler, 2017).

In addition to its effects on society and the economy, long operational lifetimes make infrastructure potentially impactful on the environment as well as vulnerable to climate variations over the many decades of its use. Both the long lifespan of infrastructure and the fact that approximately 75% of the infrastructure that will be in place in 2050 does not exist today (UNEP, 2016) represent a huge opportunity for many countries to ensure their infrastructure "leapfrogs" the inefficient, sprawling and polluting systems of the past to an infrastructure that is sustainable, resource efficient, low-carbon and climate resilient. In the absence of an agreed definition in the literature, this chapter uses the following practical definition of Sustainable Infrastructure (SI) by the Inter-American Development Bank based on a number of frameworks, principles and standards and which encompasses four major pillars of sustainability:

> SI refers to infrastructure projects that are planned, designed, constructed, operated and decommissioned in a manner to ensure economic and financial, social, environmental (including climate resilience) and institutional sustainability over the entire life cycle of the project.
>
> (IDB, 2018)

The four dimensions of sustainability for infrastructure are discussed below:

Socially sustainable: Sustainable infrastructure is inclusive and respects human rights. Such infrastructure meets the needs of the poor by increasing infrastructure access to basic services, supporting general poverty reduction, and reducing vulnerability to climate change risks. For example, infrastructure such as distributed renewable power generation in previously un-electrified rural areas can increase household income and improve gender equality by reducing the time needed for basic household chores. Socially sustainable natural infrastructure will protect the resources that communities depend on and build resilience to natural hazards and climate change.

Economically sustainable: Sustainable infrastructure is also economically sustainable. It positively impacts GDP per capita and job outcomes. Sustainable infrastructure does not burden governments with debt they cannot repay, or end-users – especially the poor – with tariffs they cannot afford. Economically sustainable infrastructure may also include opportunities to build local developer capacity. Sustainable infrastructure is not only good for the planet, it's good for investors' profit margins too. Financial and economic benefits can result from reduced use of materials, improved pollution prevention, reduced carbon emissions, payment for environmental services, and better labor and community relations. Sustainable infrastructure can trigger a host of positive externalities, including industrial expansion, employment, productivity, and technological innovation.

Environmentally sustainable: Sustainable infrastructure is also environmentally sustainable. This includes infrastructure that establishes the foundation for a transition to a low-carbon, resource-efficient economy. Environmentally sustainable infrastructure mitigates pollution and carbon emissions during construction and operation (e.g., high-energy efficiency standards). It includes investments in systems that improve resource efficiency and demand-side management, such as energy and water efficiency measures. Sustainable infrastructure is also resilient to climate change risks (e.g., by building public transport systems in less fragile places or to different specifications due to climate change risks). Tackling climate risks through infrastructure design, maintenance and operation will benefit all and is essential to reduce poverty and protect the most vulnerable populations.

Institutionally sustainable: Institutionally, SI is aligned with national and international commitments and is based on transparent and consistent governance systems over the project cycle. Robust institutional capacity and clearly defined procedures for project planning, procurement, and operation are enablers of institutional sustainability. The development of local capacity is critical to enhance sustainability and promote systemic change. SI must develop technical and engineering capacities as well as systems for data collection, monitoring and evaluation to generate empirical evidence and quantify impacts or benefits.

One approach to evaluating infrastructure projects for meeting the above sustainability goals involves a rating system. Today, many sustainability-oriented rating systems exist for infrastructure and their use is expanding; some are applicable across sectors (e.g., Envision (USA) (>, Civil Engineering Environmental Quality (CEEQUAL) assessment (UK) (www.ceequal. com) and Infrastructure Sustainability (IS) Rating tool (Australia)) and others are sector specific (e.g., GreenroadsTM). Table 7.3 gives a summary of existing Sustainable Infrastructure rating tools.

The CEEQUAL and IS rating tools focus on the design and construction phases of the project development while the Envision tool focuses on the planning and design phases (Griffiths et al., 2015). The following section presents two of the most widely used infrastructure rating systems: CEEQUAL and ENVISION.

CEEQUAL (Civil Engineering Environmental Quality) is an international evidence-based sustainability assessment, rating and awards scheme for civil engineering, infrastructure, landscaping and works in public spaces. The Institution of Civil Engineers (ICE) led to the development of CEEQUAL with financial support from the UK government between 1999 and 2003. CEEQUAL was developed with the involvement of many stakeholders like the UK government departments and agencies, civil engineering consultants, major contractors and professional and industry associations. The tool was launched in September 2003 and became public in June 2004 after publishing Version 3 of the Assessment Manual for Projects. Since then, CEEQUAL has been updated until the latest Version 6 was launched in June 2019. CEEQUAL for International Projects has been developed for adoption by other countries other than the UK and Ireland with application of weightings to take on board a particular country's cultural influence on sustainable performance and environmental conditions. CEEQUAL works on a scheme that uses a points-scoring-based assessment, applied to any civil engineering or public realm project. This scheme is made up of around 200 questions revolving around a specifically designed CEEQUAL Manual. The two types of CEEQUAL Manuals used are for maintenance works and construction. The Manual contains background information and references, guidance on scoring and scoping out, and examples of what is considered appropriate evidence.

The question set is split into nine sections (as listed hereunder) and 48 sub-sections from the perspective of the three key stakeholders (Clients, Designers and Contractors) involved in the project.

TABLE 7.3

Summary of Existing Sustainable Infrastructure Rating Tools

Characteristics	Civil Engineering Environment Quality (CEEQUAL) (Version 5)	Infrastructure Sustainability (IS) (Version 1.0)	Envision (Version 2.0 Stage 2)
Supporting Institution	CEEQUAL Ltd.	Infrastructure Sustainability Council of Australia (ISCA)	Institute for Sustainable Development (ISI)
Geographical Context	The UK and Ireland/ International	Australia and New Zealand	The United States and Canada
Year of launching	2003	2012	2012
Manuals	CEEQUAL for Projects/ CEEQUAL for Term Contracts	Infrastructure Sustainability (IS)	Envision
Life cycle phases covered	Design and construction	Design, constructions and operations	Planning and design
Categories	9	6	5
Sub-categories	48	15	60
Levels of Achievement	4 (Pass, Good, Very Good, Excellent)	3 (Commended, Excellent, Leading)	4 (Bronze, Silver, Gold, Platinum)
Awards	6 (CEEQUAL for Projects) and 2 (CEEQUAL for Term Contracts)	3 (Design, As Built, Operation)	1 (Planning and Design)
Verification Agents	Independent CEEQUAL-trained verifiers	Independent ISCA-trained verifiers	ISI independent third-party verifiers

Source: Diaz-Sarachaga et al. (2016).

1. Project/Contract Strategy
2. Project/Contract Management
3. People & Communities
4. Land Use & Landscape
5. The Historic Environment
6. Ecology and Biodiversity
7. Water Environment
8. Physical Resources Use and Management
9. Transport

Project assessments are self-assessments carried out by an assessor from within any part of a project team, or contracted in, who must have been trained by CEEQUAL. CEEQUAL also appoints a trained verifier to the project, who is independent of the project team and acts to support the assessor and provide validation of the completed assessment and subsequent Award. There are several different CEEQUAL Award levels that a project can achieve, depending on the percentage number of points scored against the question set.

Envision sustainable infrastructure rating system was created in the United States in 2010 by a strategic alliance of the Zofnass Program for Sustainable Infrastructure at the Harvard University Graduate School of Design and the Institute for Sustainable Infrastructure (ISI). Envision is a holistic sustainability rating system that can help project teams identify sustainable approaches during planning, design, construction, and operation of civil infrastructure (e.g., roads, bridges, airports, dams, landfills, water treatment systems, etc.). It is a framework that guides users to initiate a systemic change in the planning, design and delivery of sustainable and resilient infrastructure. Envision is a decision-making guide, not a set of prescriptive measures and provides industry-wide sustainability metrics for all types and sizes of infrastructure to help users assess and measure the extent to which their project contributes to conditions of sustainability across the full range of social, economic, and environmental indicators. Fundamentally, Envision is about supporting higher performance through more sustainable choices in infrastructure development. The framework provides a flexible system of criteria and performance objectives to aid decision makers and help project teams identify sustainable approaches during planning, design, and construction that will carry forward throughout the project's operations and maintenance and end-of-life phases. Using Envision as a guidance tool, owners, communities, designers, contractors, and other stakeholders are able to collaborate to make more informed decisions about the sustainability of infrastructure.

ISI launched the Envision Version 2.0 in 2012 until recently in 2018 Version 3 has been launched. Just like CEEQUAL, Envision has a rating system which includes a guidance manual and online scoring system. Envision Version 3.0 has 64 sustainability credits arranged in five categories that address major impact areas in terms of the Triple Bottom Line pillars namely: (1) Quality of Life; (2) Leadership; (3) Resource Allocation; (4) Natural World; and (5) Climate and Risk/Resilience. See Figure 7.2. Envision provides innovation points for projects with advanced sustainable infrastructure practices or exceptional performance beyond expectations. Five levels of achievement are defined by Envision to assess performance and foster project improvement: Improved (performance is above conventional); Enhanced (sustainable performance adheres to Envision principles); Superior (sustainable performance is note-worthy); Conserving (performance results in zero impact); and Restorative (performance

ENVISION POINTS TABLE

			IMPROVED	ENHANCED	SUPERIOR	CONSERVING	RESTORATIVE
QUALITY OF LIFE	PURPOSE	QL1.1 Improve community quality of life	2	5	10	20	25
		QL1.2 Stimulate sustainable growth and development	1	2	5	13	16
		QL1.3 Develop local skills and capabilities	1	2	5	12	15
	WELLBEING	QL2.1 Enhance public health and safety	2			16	
		QL2.2 Minimize noise and vibration	1			8	11
		QL2.3 Minimize light pollution	1	2	4	8	11
		QL2.4 Improve community mobility and access	1	4	7	14	
		QL2.5 Encourage alternative modes of transportation	1	3	6	12	15
		QL2.6 Improve site accessibility, safety and wayfinding		3	6	12	15
	COMMUNITY	QL3.1 Preserve historic and cultural resources	1		7	13	16
		QL3.2 Preserve views and local character	1	3	6	11	14
		QL3.3 Enhance public space	1	3	6	11	13
		Maximum QL Points:					**181**
LEADERSHIP	COLLABORATION	LD1.1 Provide effective leadership and commitment	2	4	9	17	
		LD1.2 Establish a sustainability management system	1	4	7	14	
		LD1.3 Foster collaboration and teamwork	1	4	8	15	
		LD1.4 Provide for stakeholder involvement	1	5	9	14	
	MANAGEMENT	LD2.1 Pursue by-product synergy opportunities	1	3	6	12	15
		LD2.2 Improve infrastructure integration	1	3	7	13	16
	PLANNING	LD3.1 Plan for long-term monitoring and maintenance	1	3		10	
		LD3.2 Address conflicting regulations and policies	1	2	4	8	
		LD3.3 Extend useful life	1	3	6	12	
		Maximum LD Points:					**121**
RESOURCE ALLOCATION	MATERIALS	RA1.1 Reduce net embodied energy	2	6	12	18	
		RA1.2 Support sustainable procurement practices	2	3	6	9	
		RA1.3 Use recycled materials	2	5	11	14	
		RA1.4 Use regional materials	3	6	9	10	
		RA1.5 Divert waste from landfills	3	6	8	11	
		RA1.6 Reduce excavated materials taken off site	2	4	5	6	
		RA1.7 Provide for deconstruction and recycling	1	4	8	12	
	ENERGY	RA2.1 Reduce energy consumption	3	7	12	18	
		RA2.2 Use renewable energy	4	6	13	16	20
		RA2.3 Commission and monitor energy systems		3		11	
	WATER	RA3.1 Protect fresh water availability	2	4	9	17	21
		RA3.2 Reduce potable water consumption	4	9	13	17	21
		RA3.3 Monitor water systems	1	3	6	11	
		Maximum RA Points:					**162**
NATURAL WORLD	SITING	NW1.1 Preserve prime habitat			9	14	18
		NW1.2 Protect wetlands and surface water	1	4	9	14	18
		NW1.3 Preserve prime farmland			6	12	15
		NW1.4 Avoid adverse geology	1	2	3	5	
		NW1.5 Preserve floodplain functions	2	5	8	14	
		NW1.6 Avoid unsuitable development on steep slopes	1		4	6	
		NW1.7 Preserve greenfields	3	6	10	15	23
	LAND & WATER	NW2.1 Manage stormwater		4	9	17	21
		NW2.2 Reduce pesticide and fertilizer impacts	1	2	5	9	
		NW2.3 Prevent surface and groundwater contamination	1	4	9	14	18
	BIODIVERSITY	NW3.1 Preserve species biodiversity	2			13	16
		NW3.2 Control invasive species			5	9	11
		NW3.3 Restore disturbed soils				8	10
		NW3.4 Maintain wetland and surface water functions	3	6	9	15	19
		Maximum NW Points:					**203**
CLIMATE & RISK	EMISSIONS	CR1.1 Reduce greenhouse gas emissions	4	7	13	18	25
		CR1.2 Reduce air pollutant emissions	2	6		12	15
	RESILIENCE	CR2.1 Assess climate threat				15	
		CR2.2 Avoid traps and vulnerabilities	2	6	12	16	20
		CR2.3 Prepare for long-term adaptability				16	20
		CR2.4 Prepare for short-term hazards	3		10	17	21
		CR2.5 Manage heat islands effects	1	2	4	6	
		Maximum CR Points:					**122**
		Maximum TOTAL Points:					**809**

*Not every credit has a restorative level. Therefore tables include the maximum possible points for each credit whether conserving or restorative.

FIGURE 7.2
Ratings in the five categories of Envision.

Retrieved from http://sustainableinfrastructure.org/login.cfm

restores natural or social systems). There are four Envision award levels (bronze, silver, gold and platinum) according to the percentage of credits obtained. The projects that have received awards can be found at: http://sustainableinfrastructure.org/awards/index.cfm.

A checklist is also available which consists of a series of yes or/no questions based on the sustainability rating criteria, that which can help engineers perform quick self-assessments and also familiarize them with different sustainability aspects included in the rating system. The Envision Rating System and the checklist can be downloaded at no cost from: http://sustainableinfrastructure.org/login.cfm.

7.4 Role of the Engineer in Rating Systems

Limited coordination between different stakeholders throughout a building's life span is a major constraint for the sustainable development of the building and construction sector. Rating systems require an integrated design process to create projects that are environmentally responsible and resource-efficient throughout the lifecycle: from siting to design, construction, operation, maintenance, renovation, and demolition. As such, they create the conditions and incentives that encourage all stakeholders to promote jointly sustainable building or infrastructure practices. As project managers or members in the project team, engineers can use rating systems to encourage greater dialogue and teamwork to consider the issues in the question set at the most appropriate time. Rating tools are most useful to the engineer during the design stage when measures can be incorporated to minimize impacts. It is important to note that a building or an infrastructure does not have to be certified to be sustainable and well built. Sometimes project sponsors may want to evaluate the sustainability of the project but do not want to formally register or formally qualify the project. Or, the client may not want to comply with all the requirements. Some organizations, like the Institute for Sustainable Infrastructure (ISI), provide tools for use that can be used in a self-assessment mode to increase the environmental performance of a project regardless of whether the project will ever go through the verification process. Envision in its simplest form can be used as a preliminary self-assessment tool. Such checklists are good educational tools that help engineers familiarize themselves with the sustainability aspects of project design. They can be used as a stand-alone assessment to quickly compare project alternatives or to prepare for a more detailed assessment. The engineer may have to deal with resistance to rating systems due to either the third-party assessment aspects or a perception of extra work and effort. However, the rating systems are very powerful planning tools that can be used to embed sustainable thinking into each and every decision point in the development of a project. When green thinking becomes an integral part of the initial plan, it is easier to design and incorporate green elements into the project. The use of these tools also fosters innovation, encouraging the creation of new environmentally friendly products by engineers.

As building rating and certification systems are continuously being refined to reflect new standards and goals for higher levels of sustainability, it is essential for the engineer to investigate the most current versions of these programs. Also, there is a gradual shift away from a prescriptive approach to sustainable design towards the scientific evaluation of actual performance through Life Cycle Assessments (LCA). While LCAs are not yet a consistent requirement of green building rating systems and codes, there is a trend toward requiring LCAs and improving the methods for conducting them. Building and infrastructure rating systems promote the development and diffusion of sustainable construction practices. Their use encourages building owners and the construction industry to strive for higher levels of sustainability and in turn elevate the ambition of government building codes and regulation, workforce training, and corporate strategies. The unique difference between codes and building rating systems is that codes are mandatory. If green codes become adopted on a widespread basis, their impact can change the building environment rapidly and extensively. When undertaking a project, whether it is new construction or a renovation, the engineer should check to see if there is a local green code that will dictate the direction and scope the project must take.

7.5 Summary

Given the scale of the impact of buildings and infrastructure, there is the need to assess their performance from a broad perspective, taking into account also the environmental, social, and economic impacts of constructions. This has prompted the creation of green building and infrastructure standards, certifications, and rating systems aimed at mitigating the impacts through the measurement and recognition of sustainability performance. The essential purpose of most sustainability rating systems is a tool to encourage sustainability practices beyond the regulatory minimum, and to communicate sustainability in a comprehensible manner. Buildings rating systems (such as BREEAM or LEED) and infrastructure rating systems (such as CEEQUAL or ENVISION) promote the development and diffusion of sustainable construction practices. The checklist in the various rating systems contains good educational tools that help engineers familiarize themselves with the sustainability aspects of project design. The checklist can be used as a stand-alone assessment to quickly compare project alternatives or to prepare for a more detailed assessment. Rating systems are very powerful planning tools that can be used to embed sustainable thinking into each and every decision point in the development of a project.

SUSTAINABLE ENGINEERING IN FOCUS: CASE STUDIES OF RATING SYSTEMS

1. *BREEAM Case Study: University of California, Davis, USA: BREEAM In-Use Excellent rating unlocks flexibility (Retrieved with permission from www.breeam.com/case-studies/education/university-of-california-davis/)*

As one of the most sustainable universities in the world, UC Davis saw great value in lowering carbon emissions using this science-based standard. Built in 2001 and located in Davis, CA, Plant and Environmental Sciences is a three-story, 126,651-square-feet building that provides state-of-the-art facilities for research. Laboratories make up 80% of the building with the rest consisting of faculty offices from the Departments of Plant Sciences and Land, Air and Water Resources. The Green Building Team chose this building to pilot the BREEAM In-Use program because the building had undergone recent upgrades to increase energy efficiency, reduce water use, and improve indoor air quality and wanted to recognize the completed work.

The BREEAM In-Use Assessment:

BREEAM In-Use	Part 1 (Asset Performance)	Part 2 (Building Management)
Rating	Excellent	Good
Star Rating	5 Stars	3 Stars

UC Davis is one of the most sustainable universities in the world. According to the University of Indonesia GreenMetric World University Ranking, it ranks No. 1 in the 2016 assessment of 516 colleges and universities in 74 countries for environmentally friendly campus operations and policies, and research and education on sustainability. UC Davis is also a Top 10 "Cool School" in Sierra magazine's ranking of America's greenest colleges and universities. As a global leader in sustainability, UC Davis understands the impact that buildings have in the big picture of sustainability and are eager to make a difference by lowering carbon emissions across campus.

Although UC Davis relies on the Solar Farm to lower their carbon footprint, they also realize how important it is not to "solarize" your inefficiencies. The facilities team has been outstanding in improving building operations. Recent upgrades include:

- In 2014, the Facilities Management Energy Conservation Office performed an HVAC retrofit on the laboratories, which resulted in 36% total energy savings. The following activities contributed to major energy savings:

- Temperature setbacks.
- Reduced exhaust velocities established through wind tunnel studies.
- Reduced air flow rates when building was unoccupied.
- Replaced 10 existing exhaust fan motors with new high efficiency motors.
- Upgraded to new relays and direct digital controls (DDC).
- Established occupancy-based lighting and HVAC control for spaces and fume hoods.
- Low-flow fixtures.
- Energy Star computers.
- New LED lighting with HVAC control protocols in use to minimize energy consumption and lower carbon footprint.

UC Davis Key Sustainability Facts:

- 14,000 metric tonnes (9%) of campus's carbon footprint reduced by the 62-acre Solar Farm – the largest installation in the UC system and largest "behind-the-meter" solar plant on a US college campus offsetting electricity demand.
- 14% of campus's power supply comes from the solar installation.
- US$5 million (15–20%) of savings annually at maturity from optimizing HVAC control systems for all campus buildings.
- 15,500 tonnes of material reduced, reused, recycled, and composted in 2015–2016.
- 61 million gallons of potable water (9% of campus's total potable water use for a year) saved annually by switching to recycled water in four cooling towers at UC Davis.
- The campus aims to be carbon neutral by 2025.

2. LEED Case Study: Infosys Hyderabad Building Awarded Highest LEED Rating www.infosys.com/newsroom/press-releases/Documents/2012/LEED-india-platinum-rating-hyderabad.pdf

Infosys, a global Consulting and Technology leader, was awarded the LEED (Leadership in Energy and Environmental Design) India "Platinum" rating by Indian Green Building Council (IGBC) for its Software Development Block 1 (SDB 1) at its Pocharam campus in Hyderabad, India. LEED Green Building Rating System is a nationally and internationally accepted benchmark for design, construction and operation of high-performance green buildings.

The SDB 1 uses the innovative Radiant-cooling technology – the first in a commercial building in India, setting out higher standards for energy efficiency in building systems design. It has been built keeping in mind a holistic approach to sustainability in five key areas, including sustainable site development, water savings, energy efficiency, materials selection, and indoor environmental quality. This is the fourth Infosys building that has won a Platinum rating, taking the total Platinum certified building area at Infosys to over 1 million sq. ft.

Given the high-standards in terms of building design achieved at the SDB1 in Hyderabad, it has now been showcased in the "Best Practices Guide for High Performance Indian Office Buildings" by Lawrence Berkeley National Lab, a U.S. Department of Energy (DoE) National Laboratory. Researchers from the Technical University of Braunschweig, Germany have also acknowledged the high energy efficiency and comfort parameters in this building after detailed monitoring and study.

Key features of this Platinum-rated building include:

- Water Efficiency: A 48% reduction in overall water consumption has been achieved in the building through the use of efficient plumbing fixtures and by water recycling. Also, 100% of waste water from the campus will be treated on site, helping in the reduction of potable water consumption.

- Energy Efficiency: The building is 40% more efficient than the globally accepted ASHRAE standard. This has been achieved through an efficient building envelope including high performance glazing and adequate shading, radiant cooling system, efficient chillers, pumps and fans, efficient lighting system and smart building automation.

- Day lighting: Over 90% of the office space has natural light, reducing the need for artificial lighting during daytime. The design includes light shelves along all windows to ensure that the natural light travels as deep into the building as possible.

- Efficient Material Selection and Management: Recycled materials account for 18% of the total value of materials in the building; these include aluminum, glass, steel, plywood and tiles among others. Thirty-eight percent of the total project material by cost was manufactured regionally, thereby reducing pollution due to transportation.

3. Envision Case Study: New York City Department of Environmental Protection – 26th Ward Wastewater Treatment Plant – Preliminary Treatment and Reliability Improvements

Retrieved with permission from www.wef.org/globalassets/assets-wef/ direct-download-library/public/03—resources/wsec-2017-fs-018-envision- for-wastewater-infrastructure—final—9.19.17.pdf

The City of New York (NYC) is the single largest municipal and regional economy in the United States. Home to over 8.5 million people, NYC is the most densely populated city in the country which swells daily to 9.5 million people with commuters, visitors and tourists. NYC recognizes that a sustainable city is one that is grounded in the recognition that people, economic development, and the environment are interconnected, and for any to thrive, all must thrive together. NYC Department of Environmental Protection (NYCDEP) supplies and distributes more than 1 billion gallons of high-quality drinking water daily to nine million New Yorkers and visitors, and treats 1.3 billion gallons of wastewater to achieve the smallest possible impact on water quality in New York Harbor. This monumental task is accomplished through the ownership and maintenance of 19 reservoirs and three controlled lakes, and the capture and delivery of wastewater via 7,000 miles of water mains and 7,400 miles of sewers. In-city infrastructure alone includes 14 wastewater treatment plants and 96 pumping stations. These critical structures are in need of constant maintenance, and are often located in densely populated locations by virtue of the necessary coastal locations. NYCDEP aspires to stand at the forefront of sustainability in New York City, the nation, and the world. NYCDEP's mission is "to be the safest, most efficient, costeffective, resilient, and transparent water utility in the nation." As the agency strives to meet this goal, the need for a systemic and quantifiable approach to integrate the triple bottom line criteria of social, economic, and environmental impacts becomes ever more acute. The triple bottom line is currently considered in NYCDEP projects based primarily on economic value, with environmental impacts rapidly gaining second, and social impacts largely underrepresented in critical design decisions. The incentive for meaningful sustainability integration is being rapidly applied on a city-wide level. Living in dense spaces with a legacy of environmental issues, compounded with the continual sprawl and associated demographic shift of urban areas and NYC in particular, creates an environment where environmental and social issues are inextricably linked. The current landscape increasingly motivates designers to place the triple bottom line at the forefront of design. The challenges with identifying a universally appropriate triple bottom line analysis were numerous. NYCDEP infrastructure is by design unseen. Pump stations, wastewater treatment plants, and drinking water purification operations are not open to the public for safety and security. The community should not interact tactilely with this infrastructure and therefore it cannot be subject to traditional measurements of social impact. Designers as well are challenged to convey the indirect community benefit that can be gained from NYCDEP projects.

Implementing Envision into NYCDEP practice was a process of incremental change. The process began in 2012, when the Sustainability Division was created at NYCDEP within the Bureau of Engineering Design & Construction (BEDC). Projects are evaluated for their sustainable attributes early in design, allowing for flexibility in the addition of features with lower costs and effort than if these methods were implemented retroactively. The BEDC sustainability program key performance indicators are updated at each major project milestone, starting with a series of workshops which occur at project kickoff, 30% design, and continuing through 100% design completion. Within the program, Envision is embedded as a tool for evaluating the sustainable components inherent in the project design and to identify areas in need of further consideration. Thus far, NYCDEP has trained over 300 employees to use the system, and currently has over 80 registered ENV SPs on staff. The use of the Envision has aided NYCDEP designers to push the boundaries of traditionally gray infrastructure projects, therefore incentivizing the inclusion of sustainable stormwater management solutions and low impact development techniques. When placed in the context of greater community and city resiliency, and even further defined through an accredited rating system, these concepts become less abstract. Implementing Envision on NYCDEP projects can level the field of understanding and better illustrate the sustainable design decisions consciously inserted into a project.

The 26thWard Wastewater Treatment Plant is located on a 57.3 acre site in Canarsie, Brooklyn, New York City, USA. The plant is located in Brooklyn Community District 5, which covers the east-central portion of Brooklyn. The plant currently serves a population of approximately 283,400 from a drainage area of 5,907 acres. The current design dry weather flow is 85 million gallons per day (mgd). The treated and chlorinated effluent is discharged into the Hendrix Street Canal that borders the plant to the east and is tributary to Jamaica Bay. The 26W-20 Preliminary Treatment and Reliability Improvements project has the overall goal of increasing the reliability of preliminary treatment at the plant, and to improve flow, solids, and grit distribution to the primary settling tanks. The project will include improvements to the Primary Treatment Facilities, Pump and Blower House, Sludge Degritting Wing, and Biological Nutrient Removal Building, as well as the installation of the first green roof installed on an in-city wastewater treatment plant. This contract was estimated at $134 million dollars. The project design was completed in December 2014 and construction completion is targeted at October 2021. 26W-20 received an Envision Silver Rating in August of 2015. This project was the first wastewater treatment plant to receive an Envision certification in the United States, and the first NYCDEP project to achieve full third-party accreditation. Goals to incorporate TBL initiatives into this traditionally gray infrastructure project included increased durability and energy efficiency, sustainable procurement and reused materials, strategic use of landscaping techniques, installation of a green roof, reduction of climate threat, and meaningful stakeholder involvement. Defining the level of public engagement as a key indicator of project success solidified its importance in design and implementation. Envision was used as a platform to guide designers in the creation of a project incorporating all aspects of the triple bottom line.

References

Bhattacharya, A., Oppenheim, J. and Stern, N. 2015. *Driving Sustainable Development through Better Infrastructure: Key Elements of a Transformation Program*, Global Economy and Development Working Paper 92, Brookings Institution, Washington, DC.

Diaz-Sarachaga, J.M., Jato-Espino, D., Badr, A. and Fresno, D. 2016. Evaluation of Existing Sustainable Infrastructure Rating Systems for Their Application in Developing Countries. *Ecological Indicators*, 71, 491–502.

Griffiths, K., Boyle, C., and Henning, T.F.P. 2015. Infrastructure Sustainability Rating Tools – How They Have Developed and What We Might Expect to See in the Future. In IPWEA International Conference 2015: Sustainable Communities Sharing Knowledge.

Haupt, T. and Nuramo, D. 2017. Key Elements for Sustainable Infrastructure Design in Developing Countries. 7th West Africa Built Environment Research (WABER) Conference, At Accra, Ghana, Volume: 994–1008.

IDB. 2018. *What Is Sustainable Infrastructure?: A Framework to Guide Sustainability across the Project Cycle*, Inter-American Development Bank. IDB Invest, Washington, DC.

IS (Infrastructure Sustainability Rating Tool). www.isca.org.au.

Kibert, J.C. 2013. *Sustainable Construction: Green Building Design and Delivery*, 3rd edn., Wiley & Sons, NJ.

Kubba, S. 2012. *Handbook of Green Building Design and Construction: LEED, BREEAM and Green Globes*, Butterworth-Heinemann, London.

LEED for Neighbourhood Development. 2009. *Manual*. The US Green Building Council. LEED (Leadership in Energy and Environmental Design). Available at: www.usgbc.org/leed.

Silva, J.M.C. and Wheeler, E. 2017. Ecosystems as Infrastructure. Perspectives in Ecology and Conservation. *Perspectives in Ecology and Conservation*, 15(1), 32–35. doi:10.1016/j.pecon.2016.11.005.

UN Environment and International Energy Agency. 2017. Towards a Zero-emission, Efficient, and Resilient Buildings and Construction Sector. Global Status Report 2017.

UNEP (United Nations Environment Programme). 2016. *Sustainable Infrastructure and Finance*, Paris, France. June 2016.

UNEP-SBCI. 2009. *Buildings and Climate Change: Summary for Decision-Makers*. Available at: www.unep.org/sbci/pdfs/SBCI-BCCSummary.pdf (Accessed 21st August 2019).

Vanderley, J. 2010, March. *UNEP – SUSHI: Stakeholder Roundtables: Outcomes*, UNEP-SUSHI, Sao Paulo, Brazil. Available from www.unepsbci.org/newSite/SBCIActivities/SUSHI/Documents/SUSHI_Roundtables_Outcomes_1.pdf (Accessed in July 2019).

Exercises

1. The United Nations includes the potential of infrastructure in the SDGs by directly mentioning sustainable and resilient infrastructure in SDG 9: Build resilient infrastructure, promote inclusive and sustainable industrialization and foster innovation. Infrastructure is also related somehow to other SDGs. Achieving SDG 9 will have positive multiplier effects for all other Goals. Give some examples of linkages between SDG 9 and other SDGs.

2. What are the benefits of green buildings rating systems?

3. Give the main differences between infrastructure and buildings, explaining why there are distinct sustainability rating systems for each one.

4. Consider the BREEAM for New Construction rating system. Give a breakdown of the issues covered by each of the environmental sections in this rating system. What are the minimum standards that must be met?

5. Consider the LEED V4 rating system for New Construction and Renovation (LEED-NC). Information on the points and categories can be found online on the USGBC website. Look at the Water Efficiency (WE) category. Give the pre-requisite and for each credit give possible options and/or strategies available to earn points for each credit. You can refer to the LEED case study in the box "Sustainable Engineering in Focus: Case Studies of Rating Systems" for ideas.

6. The design team of a 2,600 m^2 (approx 28,000 sq. ft.) public school hopes to achieve SS Credit 6.2, Stormwater Design: Quality Control, in the LEED-NC Rating System. Give at least three green elements

that can be incorporated into the design that would aid the design team in achieving this credit?

7. Give the key features of the ENVISION rating system for infrastructure. Considering the Natural World category, give possible options and/or strategies that can be implemented in the design to earn points for each sub-category.

___

Recommended Reading and Websites

1. The benefits of green buildings by the World Green Building Council. (www.worldgbc.org/benefits-green-buildings)

2. Green Building Standards and Certification Systems by Stephanie Vierra. www.wbdg.org/resources/green-building-standards-and-cer tification-systems (Accessed 6th May 2019).

3. List of green buildings rating tools by the World Green Building Council. www.worldgbc.org/rating-tools

4. Sustainability rating tools – how they have developed and what we might expect to see in the future. June 2015. Conference: IPWEA 2015: Sustaining Communities, Sharing Knowledge. Rotorua, New Zealand by Kerry Griffiths; Carol Boyle; Theuns Henning.

5. Sustainable building case studies. www.mfe.govt.nz/publications/ sustainability/value-case-sustainable-building-new-zealand/3-sustain able-building-case

6. Websites of Rating Systems
   https://new.usgbc.org/leed
   www.bregroup.com/products/breeam/
   www.ibec.or.jp/CASBEE/english/
   https://new.gbca.org.au/green-star/certification-process/
   www.dgnb-system.de/en/system/certification_system/
   https://sustainableinfrastructure.org/
   www.ceequal.com/www.isca.org.au/www.greenroads.org/
   publications

8

Policies and Instruments for Implementing Sustainable Development from a Life Cycle Perspective

8.1 Introduction

Most sustainability problems originate from a combination of two broad sets of factors, namely (1) "pressure" and (2) "enabling" factors, which make it possible for pressures to cause them, as shown in Figure 8.1.

Sustainability problems occur when the pressure factors are not properly managed, usually due to one or more of these three general types of "enabling" factors: market failures, government failures, and institutional failures.

(i) *Market failures:* Pollution and resources degradation result from individuals' or enterprises' lack of concern towards the potential negative effects of their production on third parties (externalities). For example, industrialists may dispose of waste products in a way that is harmful to the atmosphere or to water systems, as they believe that these resources are limitless, and hence "free". There is market failure when there is a dichotomy between what a private person does given market prices and what society might want him or her to do to protect the environment. This failure implies wastefulness or economic inefficiency.

(ii) *Government failures:* Sound economic policies are normally beneficial for the environment. However, macroeconomic reforms can sometimes lead to environmental degradation. For instance, serious distortions in resource use and allocation are the result of incorrect pricing of environmental resources.

(iii) *Property rights failures:* A lack of definition over property rights can lead to numerous natural resources management problems. For instance, non-private ownership of resources has regularly led to open access, resulting in the overexploitation or degradation of tropical rainforests, mangroves, rivers and fisheries. Other factors leading to environmental degradation are *inadequate awareness and commitment* and *institutional* weaknesses.

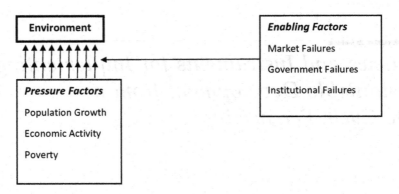

FIGURE 8.1
Causes of sustainability problems.

This chapter introduces some categories of policy tools that governments have at their disposal to correct the above-named failures. It presents the policies which can be designed within each element of life-cycle stages and within key production-consumption areas. It is based on Bengtsson et al. (2010) and on SWITCH-Asia Network Facility (2011) and gives a categorization of some of the more widely applied approaches in sustainability from a life-cycle perspective.

8.2 Policies and Instruments for Sustainable Development

Policies include strategies, action plans and programmes devised by government ministries. Policies will generally comprise a strategic vision along with objectives and potentially concrete targets, as well as identify several instruments that will be employed to achieve these objectives and may include indicators to assess progress toward objectives/targets (SWITCH-Asia Network Facility, 2011). Examples include energy or water or waste management policies and action plan on sustainable production and consumption. Instruments are the practical means for implementing the policies and include a number of different types including regulatory instruments, economic instruments, and information-based and voluntary agreements.

Regulatory instruments such as laws, standards and enforcement measures are sometimes referred to as "command and control" instruments. These tools are often used worldwide in policy making together with mechanisms for monitoring and sanctioning in order to ensure compliance. Regulations that are of relevance for sustainable engineering can be divided into the following three general categories:

(i) Environmental quality standards specify a minimum desired level of environmental quality, or the maximum level of pollution of a certain medium. For example, a typical standard for the Biochemical Oxygen Demand of a treated wastewater discharged into a river is 20 mg/L.

(ii) Technical/emissions standards specify either mandatory technical equipment to be used in certain applications, or maximum levels of emissions or resource consumption for specific products or systems. For example, many countries require automobiles to be equipped with catalytic converters (a technical standard) and, in addition, they regulate maximum emission values for certain pollutants for vehicles (an emission standard).

(iii) Restrictions and bans refer to the direct limitation of an undesirable behavior or technology, or restrictions on the sale or use of certain products/substances with negative environmental and health impacts. An example is the European Community's regulation that deals with the Registration, Evaluation, Authorization and Restriction of Chemicals (REACH). Another example is the prohibition of lead additives in gasoline or a ban on dumping end-of-life vehicles in nature.

Economic instruments are covered in Chapter 5 and their aim is to encourage or discourage certain behaviors and practices, ranging from internalizing external costs to promoting specific technologies. These include instruments such as charges, taxes, subsidies, feed-in tariffs for renewable energy installations, tradable permits, deposit-refund systems, and so on.

The aim of information-based instruments such as ecolabeling or sustainability reporting is to provide knowledge on the environmental performance of certain goods, services, or systems in a standardized manner so that stakeholders, such as consumers and investors, can make better-informed choices. Such instruments have become more popular in recent years, partly because of the Internet revolution which has decreased the costs of information dissemination. Two basic categories can be distinguished:

(i) The government provides information to various actors, a targeted group or society at large. This can range from general information on increasing energy efficiency over the next five years, to highly specific and targeted information, such as technical training for industry on how to improve energy efficiency;

(ii) The government requires certain stakeholders to provide certain information, otherwise known as information disclosure, such as data on emissions of toxic substances from production facilities or on energy consumption of products during the use phase.

Voluntary agreements aim to promote environmental improvements through voluntary action. This usually implies that companies and organizations make commitments that go beyond legal requirements.

For example, the Organization for Economic Cooperation and Development (OECD, 2003) distinguishes four types of voluntary agreements:

    (i) Unilateral commitments made by polluters or resource users.

   (ii) Private agreements between polluters or resource users and those who are negatively affected.

(iii) Negotiated agreements between industry and a public authority, which can include legally binding obligations.

(iv) Voluntary programmes developed by public authorities, to which individual firms are invited to participate. Participating firms agree to standards (related to their performance, their technology or their management) that have been developed by public bodies such as environmental agencies. This can include an agreement to support transparency in sustainability reporting.

Most sustainability policies around the world are based on regulatory tools which have been quite effective to mitigate local environmental impacts as well as global environmental impacts such as the phasing out of ozone depleting substances and for reducing the use of hazardous substances in electronics. However, regulations have a number of limitations and for enhanced effectiveness they need to be combined with other tools. Such policy mixes can for example combine an economic incentive, such as an environmental tax or a user charge, with an information-based policy tool, which makes it easier for households or companies to change their behavior. An example related to households' energy consumption could be a combination of a carbon tax on electricity use (economic tool providing incentive) with mandatory energy consumption labeling of electrical appliances (an information-based tool helping consumers to select better products and cut their energy bills and carbon emissions).

8.3 A Policy Framework for Translating Sustainable Development into Action

One of the key elements for achieving sustainable development is the transition towards Sustainable Consumption and Production (SCP). *SCP is a holistic approach to minimizing the negative environmental impacts from consumption and production systems while promoting quality of life for all* (UNEP, 2012). It is a practical means to achieve sustainable development and requires a holistic approach. At its core is the life-cycle perspective. Its focus on the sustainable and efficient management of resources at all stages of value chains of goods and services encourages the development of processes that use fewer resources and generate less waste, including hazardous substances, while yielding environmental benefits and frequently productivity and economic gains. Such improvements can also increase the competitiveness of enterprises, turning solutions for sustainability challenge into business, employment and export opportunities. SCP also encourages capturing and reusing or recycling valuable resources, thereby turning waste streams into value streams.

Figure 8.2 presents a framework for policies on Sustainable Consumption and Production (SCP). The framework is split into three overall clusters:

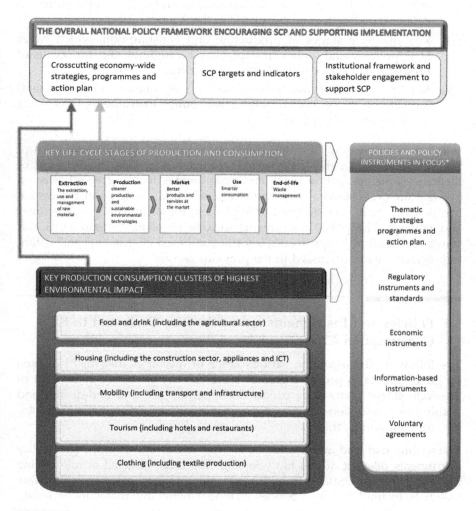

FIGURE 8.2

A sustainable consumption and production policy framework.

(Reprinted with permission from SWITCH-Asia Network Facility, 2011)

A. The overall national policy framework supporting implementation of SCP, which provides the overall policy agenda focuses on economy wide strategies, action plans, targets and indicators and on the national institutional framework for supporting and promoting SCP. It affects all economic stakeholders. Examples include national sustainable development strategies, environmental action plans, Intended Nationally Determined Contributions (INDCs) and other climate change strategies. Since the adoption of the 2030 Agenda for Sustainable Development in 2015, countries have continuously updated institutional arrangements in order to enable government institutions and other stakeholders to deliver the objectives encapsulated in the Agenda.

B. Key life-cycle stages of production and consumption, which cover SCP policies and policy instruments throughout the life cycle of products and services from cradle to grave.

C. Key production–consumption clusters of highest environmental impact, which focuses on policies and policy instruments targeted at high-impact consumption areas.

The lower right box describes elements potentially found under each of the headline area in the left boxes. These include policies in the form of thematic strategies, action plans and programmes focusing specifically on a single life-cycle stage or to a single consumption cluster, and the instruments which have been adopted to implement these strategies, programmes and policies. The policy instruments include regulatory instruments and standards, economic instruments, information-based instruments and voluntary agreements as discussed in the previous section.

8.4 Policies and Instruments across the Life Cycle and in Key Consumption Clusters

The following discusses the various kinds of measures that may be taken by governments to promote greater sustainability within each element of life-cycle stages (extraction, production, market, use and end-of-life) and within key production–consumption areas (food and drink, housing, mobility, tourism and clothing).

Extraction, use and management of raw materials: Policies and policy instruments relevant at this lifecycle stage are those targeted at minimizing the environmental impacts from the extraction, use and management of raw materials. Examples of such policies include national raw materials strategies, renewable materials strategies, water management strategies and taxes on raw materials. Examples of specific policies and policy instruments within the field of extraction, use and management of raw materials are presented in Figure 8.3.

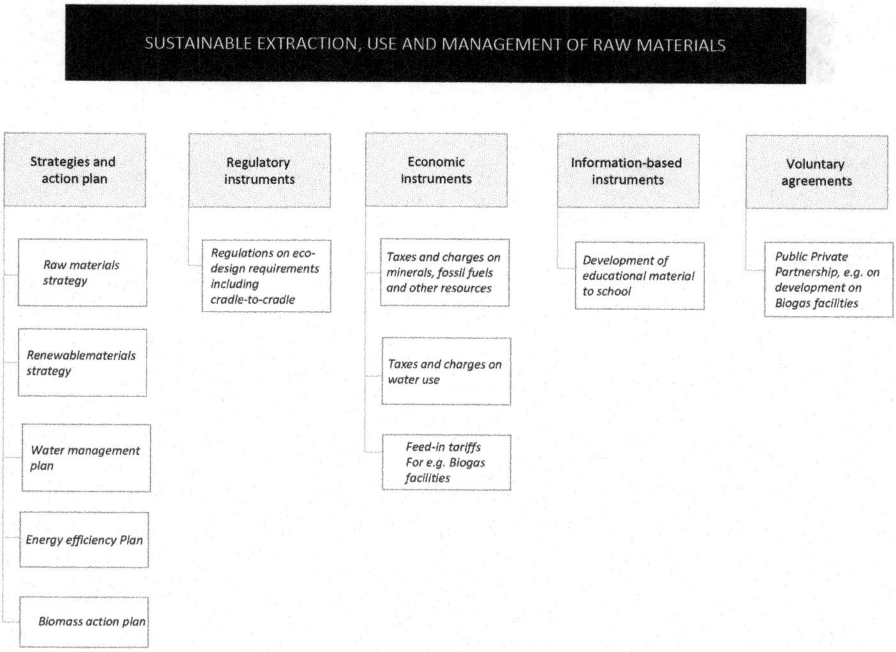

SUSTAINABLE EXTRACTION, USE AND MANAGEMENT OF RAW MATERIALS

Strategies and action plan	Regulatory instruments	Economic Instruments	Information-based instruments	Voluntary agreements
Raw materials strategy	*Regulations on eco-design requirements including cradle-to-cradle*	*Taxes and charges on minerals, fossil fuels and other resources*	*Development of educational material to school*	*Public Private Partnership, e.g. on development on Biogas facilities*
Renewablematerials strategy		*Taxes and charges on water use*		
Water management plan		*Feed-in tariffs For e.g. Biogas facilities*		
Energy efficiency Plan				
Biomass action plan				

FIGURE 8.3

Instruments for sustainable extraction, use and management of raw materials.

(Reprinted with permission from SWITCH-Asia Network Facility, 2011)

Cleaner production and sustainable environmental technologies: This refers to policies aimed at greening production and promoting environmental technologies and include policies promoting application of cleaner production, use of environmental management systems in business, greening of supply chains, corporate social responsibility, environmental accounting and reporting as well as environmental technologies, including renewable energy. Examples include national cleaner production strategies, life-cycle assessment (LCA) programmes, environmental technology action plans, and industrial ecology initiatives. Examples of specific policies and policy instruments within the field of leaner and cleaner production and the environmental technologies sector are presented in Figure 8.4.

Better products and services and the marketplace: This element covers policies and related policy instruments aimed at promoting the supply and sale of greener/more sustainable products and services. Examples include Integrated Product Policy (IPP) strategies in the EU, eco-design policies,

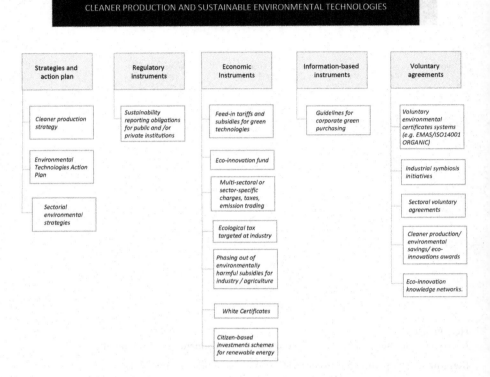

FIGURE 8.4
Instruments for cleaner production.

(Reprinted with permission from SWITCH-Asia Network Facility, 2011)

ecolabel programmes, policies addressing the retail sector, and policies supporting fair trade. Examples of specific policies and policy instruments within the field of better products and services and the marketplace are presented in Figure 8.5.

Smarter consumption (household consumption and government consumption): This includes policies and related instruments aimed at exerting direct influence on the decision-making of private consumers and policies aimed at changing or adjusting the "framework conditions", as well as policies on more sustainable procurement by the public sector. Examples include consumer policies, Sustainable Public Procurement (SPP) policies, consumer campaigns, and green taxes aimed at consumers (see Figure 8.6). Most policies aimed at promoting more sustainable household consumption have been using soft policy instruments, such as campaigns,

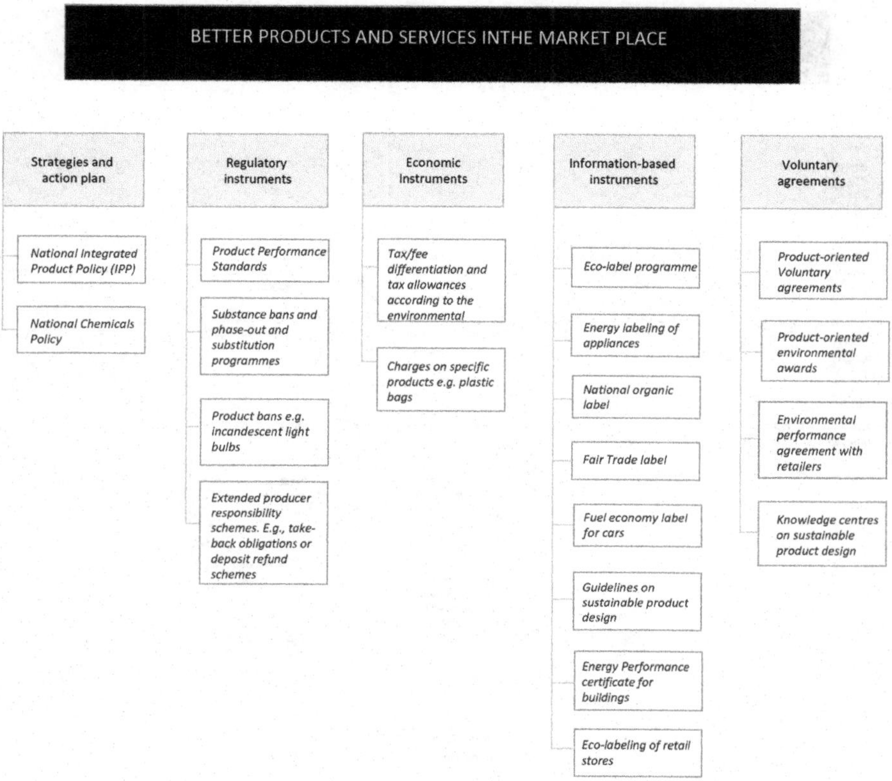

FIGURE 8.5
Instruments for better product and services and the marketplace.

(Reprinted with permission from SWITCH-Asia Network Facility, 2011)

eco-labels, and other information-based instruments. Regulatory measures and extensive use of economic instruments to reduce impacts from private consumption are politically sensitive and consequently rare. Prominent exceptions include increasing taxes on supply of water, electricity, and other energy services to households and measures introduced for traffic management including increasing fuel taxes, increased parking restrictions, and road pricing such as in the City of London.

End-of-life management: This component covers policies aimed at waste prevention and promoting sustainable waste management practices. Examples are presented in Figure 8.7 and include waste management plans, landfill taxes and extended producer responsibility schemes. The most predominant actions in this area concern end-of-life and waste management, including recycling. Regulatory and economic measures

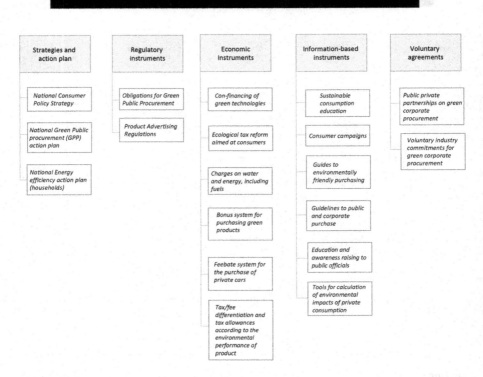

FIGURE 8.6
Instruments for smarter consumption.

(Reprinted with permission from SWITCH-Asia Network Facility, 2011)

are commonly employed to ensure that different waste types are appropriately handled. Waste prevention has seen fewer and generally softer initiatives, employing mainly information-based tools to reduce waste generated from both for individual households and businesses. Examples of specific policies and policy instruments within the field of end-of-life management are presented below.

Policies focused on the full lifecycles of specific production–consumption clusters associated with particularly high environmental burdens have the potential to achieve large SCP gains. Food and Drink, Mobility and Housing have been repeatedly identified as areas of consumption that create a significant portion of overall impacts in Europe and elsewhere (Tukker et al., 2006). Under each production–consumption cluster, the same elements as for life-cycle stages can be identified, i.e. Strategies and action

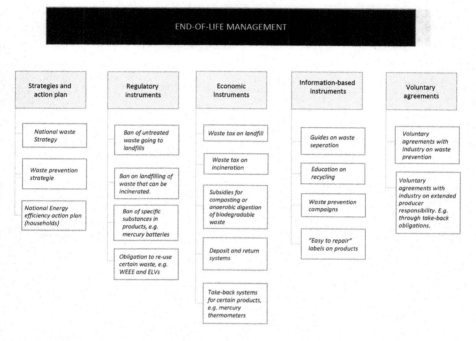

FIGURE 8.7
Instruments for end-of-life management.

(Reprinted with permission from SWITCH-Asia Network Facility, 2011)

plans, Regulatory instruments and standards, Economic instruments, Information-based instruments and Voluntary agreements that are directed at and/or affect that particular production–consumption cluster. Types of policies and instruments relevant for two clusters – housing and mobility – are described below:

Housing. This category covers policies and policy instruments aimed at more sustainable use of construction materials, promotion of low energy and zero energy housing, reduction of electricity use for appliances, green ICT and prevention of construction waste. Examples include sustainable construction strategies, housing policies and strategies on reducing energy consumption in buildings. Housing, including energy consumption during the lifetime of buildings – for space heating, water heating and use of electric appliances in the buildings – is a key cause of environmental impacts, making up roughly 25–30% of overall environmental pressures, including greenhouse gas emissions and use of raw materials. The majority of these environmental pressures are caused by energy use during the use phase of houses. Figure 8.8 provides an

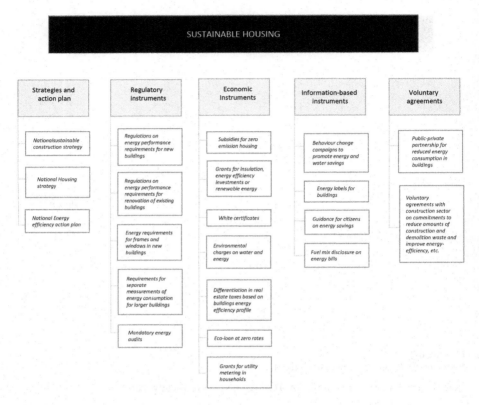

FIGURE 8.8
Instruments for sustainable housing.

(Reprinted with permission from SWITCH-Asia Network Facility, 2011)

overview with examples of policy instruments used to reduce environmental impacts from housing.

Mobility. This category covers policies and policy instruments aimed at promoting more sustainable modes of transportation, at reducing transport demand, and at improving the eco-efficiency of transportation. Examples include transport policies and nonmotorized mobility (cycling, walking) strategies. Mobility contribute significantly to overall environmental pressures – not least greenhouse gas emissions and air pollutants – in all regions of the world and the impacts are increasing as people are moving further and faster than ever before. Figure 8.9 below provides an overview with examples of policy instruments used to reduce environmental impacts from mobility.

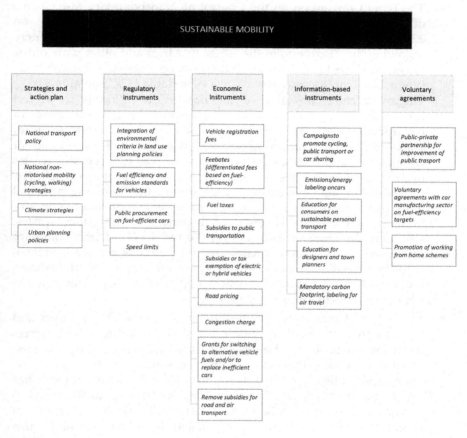

FIGURE 8.9
Instruments for sustainable mobility.

(Reprinted with permission from SWITCH-Asia Network Facility, 2011)

8.5 Multilateral Environmental Agreements

One way in which governments have promoted sustainable development has been through the negotiation and implementation of international environmental agreements also called multilateral environmental agreements (MEAs) in their national legislation. The conclusion of agreements rests on reaching consensus around a strong body of scientific evidence and its implications. A list of all MEAs is available on the following website: www.informea.org/(accessed on 6 August 2019). The objectives of some key multilateral agreements in engineering work are given below:

- The Basel Convention on the Control of Transboundary Movements of Hazardous Wastes and Their Disposal was adopted on 22 March 1989 in Basel, Switzerland, in response to a public outcry following the discovery, in the 1980s, in Africa and other parts of the developing world of deposits of toxic wastes imported from abroad. (www.basel.int/theconvention/overview/tabid/1271/default.aspx) The overarching objective of the Basel Convention is to protect human health and the environment against the adverse effects of hazardous wastes. Its scope of application covers a wide range of wastes defined as "hazardous wastes" based on their origin and/or composition and their characteristics, as well as two types of wastes defined as "other wastes" – household waste and incinerator ash. The provisions of the Convention center around the following principal aims: (i) the reduction of hazardous waste generation and the promotion of environmentally sound management of hazardous wastes, wherever the place of disposal; (ii) the restriction of transboundary movements of hazardous wastes except where it is perceived to be in accordance with the principles of environmentally sound management; and (iii) a regulatory system applying to cases where transboundary movements are permissible.

- The Vienna Convention for the Protection of the Ozone Layer and the Montreal Protocol, negotiated in 1985, is a framework agreement for which legally binding measures were set out by the 1987 Montreal Protocol that entered into force in 1989. (https://ozone.unep.org/treaties/vienna-convention). The Montreal Protocol has 196 parties and is legally binding. The Montreal Protocol agreed to phasing out the production and consumption of entire groups of Ozone Depleting Substances (ODSs), setting out provisions for this to take place more gradually in developing countries.

- The UN Framework Convention on Climate Change (UNFCCC) entered into force on 21 March 1994 (https://unfccc.int/). Today, it has near-universal membership. The objective of the treaty is to "stabilize greenhouse gas concentrations in the atmosphere at a level that would prevent dangerous anthropogenic interference with the climate system." The treaty itself set no binding limits on greenhouse gas emissions for individual countries and contains no enforcement mechanisms. In that sense, the treaty is considered legally non-binding. Instead, the treaty provides a framework for negotiating specific international treaties (called "protocols") that may set binding limits on greenhouse gases. For example, legally binding measures were set out by the 1997 Kyoto Protocol, which entered into force in 2005. The Paris Agreement of 2015 is the latest development in this process.

- The Stockholm Convention on Persistent Organic Pollutants (POPs) is a global treaty adopted in 2001 and which entered into force in

2004 with the aim to protect human health and the environment from chemicals that remain in the environment for long periods of time (www.pops.int/). It is concerned with promoting more sustainable production, as POPs are often inputs or by-products of production processes, and consumption, as POPs may be released into the environment when chemicals are used. The Stockholm Convention lists pollutants that parties have to eliminate, restrict, or reduce unintentional releases of. It also provides a standardized toolkit to identify POPs and to deal with them, and has a global monitoring plan, which includes an effectiveness evaluation of measures taken.

8.6 Summary

This chapter introduces some of the groups of policy tools that governments have at their disposal to correct market, government and institutional failures. It presents the policies which are designed to address specific stages in the economy as seen from the perspective of the lifecycle of materials and product groups. These stages comprise: the extraction of resources; production (including delivery of services); use; and end-of-life (i.e. waste management). Instruments such as regulatory instruments, economic instruments, information-based and voluntary agreements are the practical means for implementing policies. A policy framework based on the concept of Sustainable Consumption and Production (SCP) is helpful to understand the role of the various policy instruments. Within each element of life-cycle stages (extraction, production, market, use and end-of-life) and within production-consumption areas (food and drink, housing, mobility, tourism and clothing), various policy instruments can be implemented by governments to promote greater sustainability.

SUSTAINABLE ENGINEERING IN FOCUS: REDUCTION OF ENERGY USE IN MUNICIPAL BUILDINGS BY AN ENERGY SERVICE COMPANY (ESCO)

The City of Vantaa in Finland decided to improve the energy efficiency of 14 municipal buildings by using the services of an Energy Saving Company (ESCO). The justification for this approach was that the energy saving measures and associated cost savings would be realized more rapidly than would otherwise be possible if making the investments from the City's budget alone. The City decided to pilot ESCO services for 14 buildings over an eight-year time period. These properties are representative in terms of their age, size, design, etc.

The long-term plan is to roll out EPCs for further municipal buildings if energy saving measures will prove successful.

Before publishing the actual procurement notice, Vantaa carried out a stakeholder engagement exercise on the use of ESCOs to carry out such work. Procurement objectives were defined and the City of Vantaa's climate targets were taken into account as part of the process. The City aimed to make energy-saving targets both clear and realistic. After consulting legal advice, Vantaa chose to use a negotiated procedure as the best option to utilize market potential and explore available options. Subject matter was a contract for an ESCO to improve the energy efficiency of 14 municipal buildings.

Procurement criteria:

- Selection criteria: Previous experience of undertaking EPC services was a pre-requisite. However, Vantaa decided not to restrict this to municipal EPCs, as experience in this area is relatively limited in Finland and the city wanted to encourage as much competition as possible. The value of the contract was relatively low – under the EC procurement threshold of 4.85 million Euros (threshold for construction in 2011) – so it was most likely to be Finnish ESCOs that would be interested.

- Technical criteria: The energy savings proposed by the ESCO had to be 100% guaranteed. The maximum repayment period was set at 10 years and indoor air quality must remain similar to the current standard. To promote innovative solutions from the participating companies, the procurement notice didn't define what energy-saving models should be used nor were any boundary conditions presented. Instead, the participants were given detailed initial information about the energy audits for four buildings enclosed with the notice, on the basis of which they were able to select and suggest energy-saving measures. The energy savings proposed by the winning ESCO in its bid were then guaranteed as part of the contract.

- Award criteria: Four negotiations were carried out with each bidder in order to guide Vantaa's approach to the procurement. Aspects discussed included the targets, procurement principles, contract model, and final invitation to tender. The award criteria and their focus areas were selected so that they supported the targets set for the project. The award criteria set for procurement were:

  - Euros saved/year: 20%
  - MWh savings/year: 20%

- Savings/year tCO2e: 30%
- Savings that can still be made in 2023 (MWh): 30%

– Contract management and monitoring: The winning ESCO gave a 100% guarantee of the energy savings they proposed, as required by the City of Vantaa. Any savings above the proposed amount are split between the ESCO and the City and any shortfalls are made up solely by the ESCO. The consumption figures are verified from electricity and district heat meters and the implementation of savings is monitored in follow-up meetings held once a year. In the follow-up, the entire project, and not individual buildings, is examined. This means that if in one building the savings target is not reached, this can be compensated for if it is exceeded in another.

The City of Vantaa was committed to the procurement of ESCO services and developed a model for including both the savings and the investments in the same procurement process. Analytical discussions with research parties and cooperation with legal services were also critical success factors. All in all the procurement process was successful as the end result was in line with the original plan. Four companies responded to the procurement notice and the first investments in energy saving measures began in summer 2014. The ESCO's proposals of the amount of electricity savings were significantly higher than the internal audit estimation of the City of Vantaa. In terms of heat, the ESCO's proposals also tended to be higher, although the ESCO guaranteed significantly less water savings than the internal audits estimated feasible. The initiative is ongoing and final results will be available in 2023.

In this pilot project, the City of Vantaa wished to encourage rather than stifle innovation through its procurement approach, i.e. using stakeholder engagement and a negotiated procedure. However, due to the fact that the City of Vantaa required proposed energy savings to be 100% guaranteed by the ESCO, the companies tended to be risk averse in terms of using innovative energy saving solutions. One of the measures proposed that is predicted to result in the most significant energy savings is LED lighting. It is foreseen that the procurement budget will still have some room for developing new solutions in the implementation stage. The number of different measures was not determined in advance, either. This will give the supplier freedom to choose the measures that are worth investing in, whilst ensuring that the measures specified by the City of Vantaa will be carried out. The investments have now been made in all 14 buildings and follow-up period starts in June 2015.

Sustainability Impacts: Using effective and appropriate procurement procedures and energy performance criteria can reduce energy use significantly, which in turn lowers GHG emissions such as CO_2. Buildings and equipment must be effectively maintained in order to optimize efficiency. Transferring responsibility for energy efficiency to private entities and using incentives and penalties to encourage high performance levels means that the contractor has a clear incentive for ensuring that newly installed equipment is performing correctly and achieving energy savings. Over the lifetime of this energy service contract, the aim is to realize a total of 30,100 MWh energy savings in heat and electricity. According to an average EU household electricity consumption of 2500 kWh (According to European Commission: EuroStat), this is an equivalent amount of energy needed to power over 12,000 homes for a year. It is estimated that in total GHG emissions will be reduced over the eight-year contract period by 7500 tonnes CO_2 equivalent emissions.

Cost-effectiveness: The combined energy cost of the 14 municipal buildings is €1.3 million per year. The total project investment will be €1.5 million over the 8 years. The target of the contract period is to achieve savings of up to 30,100 MWh in heat and electric energy. This will cut 7,500 tonnes CO_2 eq. emissions. This means annual savings of over €200,000 in the energy costs of the City of Vantaa, which corresponds to savings above 15% of total project investments.

(Retrieved with permission from One Planet Network. www.oneplanetnet work.org)

References

Bengtsson, M., Hotta, Y., Hayashi, S. and Akenji, L. 2010. Policy Tools for Sustainable Materials Management: Applications in Asia, IGES Working Paper, Institute for Global Environmental Strategies. http://enviroscope.iges.or.jp/modules/envirolib/upload/2885/attach/policy_tools_smm_dp_final.pdf. (Accessed 23rd July 2019).
OECD. 2003. *Voluntary Approaches for Environmental Policy: Effectiveness, Efficiency and Usage in Policy Mixes*, OECD, Paris.
SWITCH-Asia Network Facility. 2011. *SWITCH Asia, Sustainable Consumption and Production Policies – A Policy Toolbox for Practical Use*, Collaborating Centre on Sustainable Consumption and Production (CSCP), Wuppertal, Germany.
Tukker, A.G., Huppes, J., Guinée, R., Heijungs, A., de Koning, L., van Oers, Suh, S., Geerken, T., van Holder-beke, M., Jansen, B. and Nielsen, P. 2006. *Environmental Impacts of Products (EIPRO). Analysis of the Life-cycle Environmental Impacts*

Related to the Total Final Consumption of the EU -25, Institute for Prospective Technological Studies. (EUR 22284 EN), Sevilla. http://ec.europa.eu/environ ment/ipp/pdf/eipro_report.pdf.

UNEP. 2012. *Global Outlook on Sustainable Consumption and Production Policies: Taking Action Together*, United Nations Environment Programme, Paris https://sustai nabledevelopment.un.org/content/documents/559Global%20Outlook%20on %20SCP%20Policies_full_final.pdf.

Exercises

1. Classify the following policy instruments used in the clothing sector: *Mandatory sustainability reporting for large companies in the textile sector; taxes on pesticide use in cotton production; effluent standards for discharge from textile dyeing factories; fair trade cotton labeling; awareness raising of consumers on sustainable textiles; voluntary commitments from retailers on sales of organic cotton/sustainable textiles.*

2. Classify the following policy instruments used for sustainable tourism: *Limiting access to especially vulnerable areas; establishment of national parks to protect wildlife; grants for ecofriendly Hotels; entrance fees to protection areas; tourism tax earmarked for nature protection; eco-tourism labeling scheme for hotels; awareness raising for tourists; promotion of EMS for the accommodation sector.*

3. List some policy instruments currently being used or which can be used in the future to promote the SCP concept in the energy and water sectors in your country.

4. Through a literature search, give examples of regulatory, economic, information-based, and voluntary instruments used in climate policy.

5. Legislation in the form of regulations and directives for dealing with waste such as waste electrical and electronic equipment (WEEE) is just one of a number of tools the European Union has used to try and manage specific waste streams. Since EEE, and therefore WEEE, may contain hazardous chemicals, it is subject to the requirements of not only the WEEE directive but also the Restriction on Hazardous Substances (RoHS) Directive and the Directive on Establishing a Framework for the Setting of Eco-design Requirements for Energy Using Products (EuP). Through a literature review, give the objectives of the WEEE, ROHS and EuP directives as well as the Registration, Evaluation, Authorization and Restriction of Chemicals (REACH) regulations pertaining to electronics.

6. Search for examples of the implementation of multilateral environmental agreements in the legal system of the state/country in which you reside.

Recommended Reading and Websites

SWITCH Asia Network Facility, 2011. *Sustainable Consumption and Production Policies: a policy toolbox for practical use*, United Nations Environment Programme, Paris.

It provides practical explanations of SCP policies and policy instruments. It further recommends instruments that could be applied to increase the positive environmental and social impacts of specific projects. Special attention is paid to small- and medium-sized enterprises (SMEs), which often form the majority of business operations in developing countries.

OECD 2007, *Instrument Mixes for Environmental Policy*, Organization for Economic Cooperation and Development, Paris.

It is a comprehensive publication that provides descriptions and analysis of how various policy instruments can be combined to achieve higher impact. Chapters are organized to demonstrate combinations of policy instruments for different sectors and different points in the production-consumption system.

UNEP (2012), Global Outlook on SCP Policies: Taking Action Together. Paris: United Nations Environment Programme. www.unep.fr/shared/publications/pdf/DTIx1498xPA-GlobalOutlookonSCPPolicies.pdf

Chapters 1 & 2 exposes the interlinked crises that the world is facing with the current unsustainable patterns of production and consumption; explores the evolution of the SCP concept and its approach, with the life cycle perspective at its core; and provides an outline of international efforts to promote SCP, including intergovernmental policies and business and civil society initiatives.

GTZ (2006). Policy Instruments for Resource Efficiency: Towards Sustainable Consumption and Production. GTZ: Germany. www.uns.ethz.ch/edu/teach/bachelor/autumn/energmob/GTZ_et_al_2006_policy-instruments_resource_efficiency.pdf

It is a good starting point to understanding different types of policy tools, based on real examples, their advantages, disadvantages and where they can be applied towards SCP objectives.

Global Reporting Initiative (2010).

Homepage. Available from www.globalreporting.org.

9

Business and Sustainability

9.1 The Business Case for Sustainability

With increasing concern about the environmental and social impacts of economic development, there are initiatives from businesses to integrate sustainability into their long-term strategy. A balance needs to be struck between adjusting existing practices to become more sustainable and ensuring the changes make good business sense such as new-product development, new markets, resources efficiency, and reputation building. Sustainability tends to be understood in most companies as the management of environmental issues such as greenhouse gas emissions, energy efficiency, and waste management. But sustainability should also include the management of governance issues (such as complying with regulations, maintaining ethical practices, and meeting accepted industry standards), and also the management of social issues (for instance, working conditions and labor standards).

Traditionally, the main focus of stakeholder interest has been upon the financial performance of the company. The phrase "triple bottom line," coined by John Elkington, co-founder of the business consultancy Sustainability, in his 1998 book *Cannibals with Forks: the Triple Bottom Line of 21st Century Business*, has gained recognition as a framework for measuring business performance. In its broadest sense, "triple bottom line" captures the dimensions that organizations must embrace – economic, environmental, and social. There are a number of stakeholders who have interest in the performance of a company, and the range of stakeholders that demand sustainability standards is displayed in Figure 9.1.

The major triggers for companies to consider sustainability as a strategic issue are as follows:

* *Legislation and enforcement*:
  - There is an increasing volume of policies, laws and regulations, and their enforcement. Environmental protection regulations become stricter over time and implementing sustainable business strategies helps organizations stay ahead of the game. Furthermore, regulatory authorities will more readily approve new business activities if the

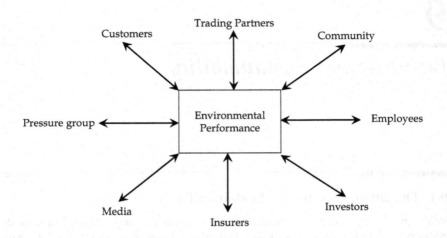

FIGURE 9.1
Stakeholders pressure and sustainability performance.

(Source: Welford and Gouldson (1993))

company shows a history of responsible corporate behavior and contributes to reduced costs associated with compliance and permitting.

- *Stakeholder pressure*:

    - The increasing pressure of third parties such as financial institutions, insurance, and companies;

    - Pressures from shareholders and employees;

    - Attention from environmental interest groups, consumers and their organizations, and the general public (in the locality). There is evidence that the ethical conduct of companies exerts a growing influence on the purchasing decisions of customers.

- *Awareness, image, and reputation*:

    - Growing awareness in the business community concerning the environment (responsible care);

    - Media publicity and improved image and reputation;

    - Some of the impacts on business of accidents and failures in environmental management controls, which leads to negative publicity and damage to corporate image;

    - There is a growing demand for corporate disclosure from stakeholders, including customers, suppliers, employees, communities, investors, and activist organizations.

- *Competitiveness*:

  - There are economic benefits for companies to demonstrate environmental and social responsibility, as in some markets and for certain products, such differentiation can provide the opportunity for higher margin and/or higher sales volumes.

  - There is growing awareness that the environmental aspects of products and processes may play a role in international competitiveness.

  - Just as society as a whole is prioritizing sustainability, potential employees and investors are interested in working with companies that have sustainability initiatives in place. Talented and committed employees are increasingly attracted to sustainability initiatives.

  - In the supply chain, increasingly, larger companies, through their procurement policies, require that suppliers demonstrate socially and environmentally responsible behavior.

  - With increasing resource scarcity and increasing prices of key natural resources such as water, fossil fuels or metals, a number of forward-looking companies are seeking opportunities associated with the development of renewable energy sources or resources efficiency in key consumption areas such as food or mobility or buildings.

- *Finance*:

  - The introduction of economic (financial) instruments, such as taxes or levies on emissions (e.g., waste), to stimulate a decrease in pollution levels;

  - Some of the impacts on business of accidents and failures in environmental management controls (liability issues, cost of remediation, business interruption, and inefficient production processes);

  - Incentives from government (licensing), banking (more attractive credit facilities), and insurance companies (more attractive premiums);

  - Investors are changing the way they assess companies' performance, and are making decisions based on criteria that include ethical concerns. Several investors today use environmental, social, and governance (ESG) metrics to analyze an organization's ethical impact and sustainability practices;

  - Cost savings through cleaner production and eco-efficiency associated with reduced waste and pollution disposal costs, and lower input and resource expenses;

- There is an improved access to capital. The investment community sees the provision for environmental and social concerns in corporate strategy as a proxy for sound financial and risk management. The recent growth in socially responsible investment funds, the increasing number of sustainability indices (such as FTSE4Good and the Dow Jones Sustainability Index), and the development of initiatives such as the Equator Principles. The Extractive Industries' Transparency Initiative and the Carbon Disclosure Project is indicative of the growing interest of the financial community in sustainability issues.

- *Life cycle thinking*:

  - Greening of the supply chain with formal requirements from large companies for certified EMS (e.g., ISO 14001) or codes of conduct from suppliers.
  - Extended Producers' Responsibility (extension of the responsibility of producers along the life cycle – in particular to the stage of disposal – even if the production side is abroad.

Sustainable business development can therefore be defined as doing business by taking into account the company's economic, environmental and social impacts, risks and opportunities with the aim to reinforce its strategy and impose its competitiveness. Figure 9.2 shows how companies pursing a sustainability strategy will benefit in a number of ways.

9.2 The Business Sustainability Journey

Figure 9.3 shows the sustainability journey for an organization in tiers with associated strategies. Normally, sustainability should be implemented in this succession (Modak, 2014). At the minimum, the journey toward sustainability must start with compliance with the laws of the land. This may require complying with procedures, doing associated documentation; investing on mitigation measures, deploying qualified staff and conducting required monitoring and reporting. However, while being compliant the organization is more reactive than pro-active. Sustainability is seen as a cost and more as a philanthropic activity. The next tier on sustainability is to go "beyond compliance" with the organization undertaking sustainability-related initiatives to reduce its footprint as driven by its vision/mission and goals/objectives and not solely directed by needs of compliance. For example, following an objective of resource efficiency, the organization may install rainwater harvesting units or implement solar water heaters to reduce consumption of fossil fuels. Many of these initiatives also lead to cost savings and help improve the resource

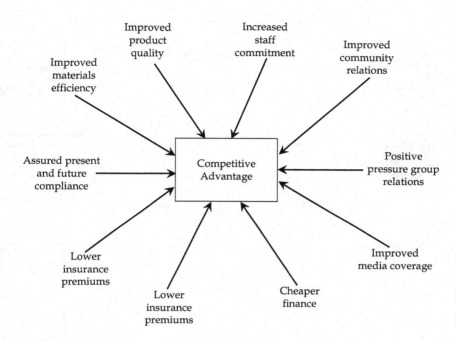

FIGURE 9.2
The constituents of competitive advantage.

security. Strategies and tools such as "cleaner production" and EMS help in this endeavor. Companies begin to see that governance structures allow them to better manage their risks and start identifying opportunities that sustainability can present. The third tier on sustainability is to expand the "boundary beyond the factory gates" and address wider set of stakeholders across the supply chain or even the life cycle. Top management is fully involved and there is a strategic approach with the company bringing all the individual activities across the organization together as part of an overall strategy. Suppliers are engaged to improve resource efficiency and waste recycling. Further, vendor selection criteria may be developed and imposed asking for a code of practice on sustainability. Accordingly, technical/financial assistance may be provided to SMEs. On the consumption side, the organization can show "extended producer responsibility" and, for example, have a "take back policy" and promote recycling of packaging and used goods by setting up collection centers. This strategy of practicing sustainability often adds a brand value to the organization reduces risks and adds to its competitiveness over a long run. Sustainability reporting is also usually practiced at this tier, which improves internal

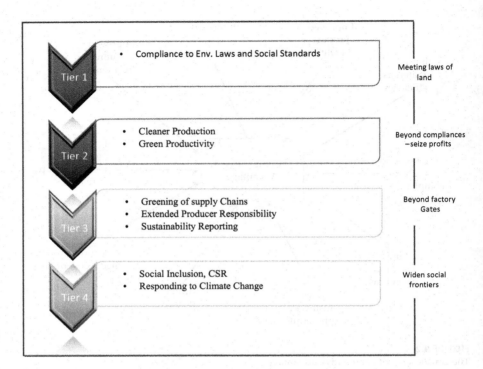

FIGURE 9.3
The four tiers of sustainability.

and external communication. The fourth tier on sustainability is to widen the "social frontiers" of the organization by addressing the "catchment area", that is, "going beyond project boundaries" but without any direct commercial motives. For example, the organization addresses needs of its neighborhood (or catchment area) and works with interested partners and implements projects that help improve the livelihoods and natural resources of the region. Supporting school education and primary health centers, carrying out afforestation and watershed development programmes are some examples. These projects usually form part of organizations efforts toward "Corporate Social Responsibility" (CSR). The returns on these investments are however over a long term. Addressing of global issues, such as climate change, are also considered at this tier – especially financing projects that assist the communities toward adaptation. On mitigation, the organizations expand sustainability reporting of the earlier tier into specialized reporting such as on GHGs, for example, by subscribing to projects such as "carbon disclosure."

Most organizations today in the developing economies belong to Tier 1. Tier 2 is generally practiced by medium- to large-scale organizations that have technical capacities, financial resources and enjoy support of top management. Multinationals with global supply chains are leaders on Tier 3 and their sustainability driver is the market and competition. Tier 4 is practiced by medium to large industries. These companies have sustainability as a top priority in their CEOs' agendas, that it is formally embedded in business practices. Sustainability is a part of the fabric of these companies.

CSR refers to a growing practice of for-profit organizations aligning with relevant causes and social good programs. It promotes a vision of business accountability to a wide range of stakeholders, besides shareholders and investors. Many companies now focus on a few broad CSR categories

1. *Environmental protection*: Reducing carbon and other environmental footprints are considered good for both the company and society.

2. *Philanthropy*: Businesses can practice social responsibility by donating resources to social causes and nonprofits.

3. *Ethical labor practices*: by treating employees fairly and ethically, companies can demonstrate their social responsibility.

4. *Volunteering*: by doing good deeds without expecting anything in return, companies can express their concern for specific issues and commitment to certain organizations.

The Harvard Kennedy School of Government's definition of corporate responsibility captures well the above practices:

> Corporate Responsibility goes beyond philanthropy and compliance to address the manner in which companies manage their economic, social and environmental impacts and their stakeholder relationships in all their spheres of influence: the workplace, the marketplace, the supply chain, the community and the public policy realm.
>
> (Kytle and Ruggie, 2005)

Box 9.1 gives a few examples of companies doing CSR taken from information available on their websites. These firms have all made strong commitments to sustainability, in large part through transparency and addressing material issues. They are practicing CSR on a strategic level and following a three-pillared approach that consists of economic, social, and environmental factors. The business strategies are transformative, attempting to change mindsets and business culture overall. The treatment of stakeholders in an ethically and socially responsible manner is at the core of CSR.

As a support to CSR, nonfinancial reporting on sustainability has expanded over the past three decades. The Global Reporting Initiative (GRI) set up in

1997 jointly by UNEP and CERES (a US-based national coalition of investors, environmental organizations and public interest groups working with companies to address sustainability challenges) produces one of the world's most prevalent standards for sustainability reporting. The core goals of the GRI include the mainstreaming of disclosure on environmental, social, and governance performance, across both the private and public sectors. The GRI Reporting Framework sets out principles and performance indicators that organizations can employ to measure and report their economic, environmental, and social performance. The cornerstone of the GRI Reporting Framework is the *Sustainability Reporting Guidelines*.

BOX 9.1 EXAMPLES OF CSR IN ACTION

Johnson & Johnson: Their initiatives range from leveraging the power of the wind to providing safe water to communities around the world. The company continues to seek out renewable energy options with the goal to procure 35% of their energy needs from renewable sources.

The Walt Disney Company: They committed to reducing their carbon footprint with goals for zero net greenhouse gas emissions and zero-waste, and have committed to conserve water. They are actively ensuring that they set strict international labor policies to protect the safety and rights of their employees. They are also active in the community and encourage employees to do the same. They also have healthy living initiatives to promote healthy eating habits amongst employees. They developed an informative online sustainability report, available in six languages and that is connected to the global reporting initiative (GRI), showing that the company is truly committed to CSR and has an integrated strategy to manage it. Disney also has a well-structured philanthropy organization that donates to nonprofits all over the globe.

Patagonia: They set out to be as transparent as possible and tackle different environmental and social issues at every step of their supply chain. For example, they switched to using organic cotton all the way back in 1994 and are forging deep relationships down to the farmers of the cotton to ensure quality. They are also well known for their "Traceable Down Standard" that also works with farmers to make sure that the ducks used to supply their down jackets are protected against force-feeding and live-plucking. Patagonia works with many factories to produce their clothing and is working on finding ways to ensure that the employees not only receive minimum wage but also a living wage. Apart from making sure that they use renewable energy, work out of green buildings, and measure their carbon footprint, Patagonia also supports grassroots activists through what they call an "Earth Tax." They donate 1% of their sales to conservation efforts.

Starbucks: The acronym for Starbuck's CSR initiative stands for Coffee and Farmer Equity (C.A.F.E.). In its simplest form, C.A.F.E. is a set of buying guidelines that ensure that all coffee purchased by the company meets their standards of quality and it must be sustainably grown. The company requires suppliers to submit evidence of payments made throughout the coffee supply chain to demonstrate how much of the price Starbucks pays for green (unroasted) coffee gets to the farmer. They ensure safe, fair, and humane working conditions for workers and providing adequate living conditions.

Google: Google Green is a corporate effort to use resources efficiently and support renewable power. But recycling and turning off the lights does more for Google than lower costs. Investments in these efforts have real-world effects on the bottom line. Google has seen an overall drop in power requirements for their data centers by an average of 50%. These savings can then be redirected to other areas of the business or to investors. They have been investing in green energy programs to power their data centers with 100% renewable electricity. At the same time, Google.org, Google's philanthropic foundation, is also helping to fight poverty and grow sustainable development.

Xerox: The printing giant offers many programs supporting corporate social responsibility. Their Community Involvement Program encourages it by directly involving employees. Since 1974, more than half a million Xerox employees have participated in the program. In 2013 alone, Xerox earmarked more than $1.3 million to facilitate 13,000 employees to participate in community-focused causes. The return for Xerox comes not only in community recognition, but also in the commitment employees feel when causes they care for are supported by their employers.

Nike and Adidas: Nike has focused on reducing waste and minimizing its footprint, whereas Adidas has created a greener supply chain and targeted specific issues like dyeing and eliminating plastic bags.

Unilever and Nestlé: They have both taken on major commitments; Unilever notably on organic palm oil and its overall waste and resource footprint, and Nestlé in areas such as product life cycle, climate, water efficiency, and waste.

Walmart, IKEA, and H&M: They have moved toward more sustainable retailing, largely by leading collaboration across their supply chains to reduce waste, increase resource productivity, and optimize material usage. It also has taken steps to address local labor conditions with suppliers from emerging markets.

9.3 Sustainability by Departmental Functions

Implementing sustainability initiatives requires participation from individuals and departments throughout an organization, from senior leadership to front-line workers. Figure 9.4 illustrates the importance of all departments in implementing sustainability in an organization. As per Hitchcock, D and Willard M (2009), their roles can be described as follows:

Corporate management: Corporate management plays a role in setting organizational priorities and in making decisions to manage risks and pursue opportunities. Management support for the development and implementation of a business sustainability strategy can create significant business value. Investors are also looking for ethics and good governance.

Plant management: Plant managers have an important role in protecting operations from potential business constraints. Plant manager's attentiveness to environmental issues and support for environmental initiatives can pave the way for numerous projects that reduce costs, lower business risks, enhance community relations, and create new business opportunities.

FIGURE 9.4
Bringing all departments together for a sustainability strategy.

Operations: Operations personnel can affect many significant resource uses and impacts through their activities, work practices, equipment management and maintenance, process design and operation, and other activities. Operations personnel are often the source of creative ideas for improving resources efficiency and reducing environmental impacts, and they are often the first line of defense against problems. For environmental strategies to be effective, the involvement of operations personnel is often essential. Implementing quality-control measures earlier in the production process and incorporating these into the design process will improve overall efficiency and reduce waste.

Marketing and public relations: Marketing personnel provide an important link to the organization's customers, enabling the business to understand customer sustainability-related needs and expectations. Marketing activities can also be used to inform customers and others about the sustainability-related attributes or products or services, as well as the proper use or disposal of products. Marketing personnel can also play a key role in assessing the viability of market opportunities. Public relations personnel play a critical role in creating channels for communication between the organization and its neighbors, local communities, government, and the broader public. They enable the company to understand the sustainability-related concerns and needs of these groups, while facilitating opportunities to engage the public in sustainability-related activities.

Product development: The design of new products and services can impact sustainability in powerful ways throughout the value chain, from input selection to final product disposition.

Purchasing: Determine what to buy through a sustainable procurement policy and strategy and how to work with suppliers.

Environmental management: Sustainability/environmental managers can often provide the cohesion and coordination that is necessary to get a business sustainability strategy off the ground. Since organizational responsibilities for addressing sustainability challenges and opportunities are often not well defined, sustainability/environmental managers are frequently best positioned to launch and coordinate initial efforts to develop and implement a business sustainability strategy.

Accounting: The role of the accountant in sustainability is primarily to collect information to assist internal decision-makers (management accounting); to prepare financial and sustainability information for external stakeholders (financial accounting); and to provide assurance on published results.

HR and organizational behavior: The HR department plays a key role in promoting positive behavior, creating an engaging workforce and creating an environment where sustainability is embedded in every aspect of the employee's life cycle from recruitment to retirement.

Figure 9.5 summarizes the key functions of departments in an organization pursuing sustainability.

9.4 Putting Sustainability into Business Practice

An ideal company practicing sustainability does the following to drive sustainability into the fabric of the organization such that it becomes systemic:

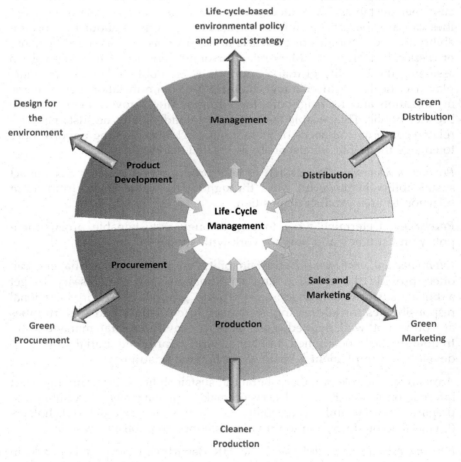

FIGURE 9.5
Key functions of departments in organization for sustainability.

- Integrates environmental and social issues into its core strategy
- Quantifies the social and environmental costs of its activities
- Displays innovation throughout the full life cycle of its products and services
- Implements sound corporate governance practices
- Commits to transparency and accountability
- Promotes meaningful change amongst its peers, within its neighboring communities, and throughout its supply chain

Section 3.2 in Chapter 3 provides an overview of the tools available for putting sustainability into practice. Specifically, for business, we can distinguish four categories of tools (UNGC, 2011):

1. Management tools (EMS, LCM, SEA, sustainable procurement, strategic CSR, etc.)
2. Assessment tools (LCA, carbon footprinting, LCC, LCSA, MFA, EIA, ERA, MCA, etc.)
3. Monitoring and auditing tools (sustainability audits, performance indicators, social and ethical codes of conduct)
4. Reporting and communication tools (sustainability reporting, stakeholder engagement, CSR, eco-labels)

Consider a hypothetical company whose management has decided that sustainability is now a top priority and is considering which tools it should adopt and their sequence for implementation. The company could first decide to manage its environmental risks and opportunities by implementing an environmental management system (EMS), which includes a strategy of cleaner production. A cleaner production opportunity assessment (CPOA) will then be undertaken, and this will encompass or lead to one or more of the following: environmental auditing, pollution and waste audits, and supply chain audits and assessments. Environmental performance indicators will be generated and options for improved environmental responsibility will be identified. LCA and CBA can then be used for more detailed assessments of the options to assist decision makers on determining their feasibility and long-term sustainability. A sustainability report will then communicate the carbon or ecological footprint and other environmental performance indicators together with the identified options for improved environmental responsibility. Similarly, a new design or business idea can be developed using tools such as Design for Sustainability (D4S), and can be assessed using EIAs, LCAs, ERAs, and CBAs. The new business can be operated under an EMS and according to the strategies of CP/eco-efficiency/industrial ecology and with regular audits being undertaken to ensure the business remains environmentally responsible and to identify further opportunities for continuous improvement. The EMS will help tie all the different aspects together by providing a structured approach.

Implementing sustainable business practices takes commitment and buy-in from the top down. Executives and leadership can implement a comprehensive sustainability plan, while employees help ensure its success. As organizations commit to CSR policies and a sustainability agenda, companies need to decide on which management structure will best ensure implementation of the sustainability strategy. We can distinguish between a centralized approach, which involves naming a chief sustainability officer (CSO) as the leader responsible for integrating environmental concerns into core business strategies, and a decentralized approach whereby CEOs decentralize sustainability by spreading particular sustainability efforts across the organization. Each approach has its advantages and drawbacks.

The engineering profession has a variety of ethical responsibilities to society and the environment. But these professional social responsibilities may be in tension with the business side of engineering. The majority of engineers work for businesses, whose primary motivation is often profit and corporate stockholders, rather than societal impacts. Luckily, this has begun to change based on movement toward CSR and realizations that companies can thrive economically while considering social and environmental impacts (the triple bottom line). Developing a business strategy is the first step in planning, at the very "front end" of the project delivery process. This where owners define what service they must deliver to the public by reference to their business model or strategy. Putting sustainability principles into practice in the business strategy will help engineers create the potential for the most sustainable outcomes. The business strategy can set sustainability objectives and targets, extend the boundaries and consider all life cycle stages, apply a visioning approach and plan long term, choose the business model, have a multidisciplinary approach, and engage with stakeholders.

9.5 Summary

This chapter discusses the reasons why businesses should nowadays care about environment and sustainable development issues. The guiding principles and the main tools for sustainability management in business are presented. Sustainability can help companies to approach environmental, social, and economic issues systematically and to integrate environmental and social care as a normal part of their operations and business strategy. Sustainability helps businesses see relationships between different issues and more accurately forecast what may occur in the future – a framework for making sense of what is happening. It examines the world as a whole system, revealing threats and opportunities. It forces the business to see relationships between social, economic, and environmental trends. It challenges the business to make decisions that simultaneously improve the economy, the community, and the environment

SUSTAINABLE ENGINEERING IN FOCUS: SUSTAINABILITY REPORTING FOR INFORMING, BENCHMARKING AND IMPROVING PRODUCTS AND PROCESSES

The Global Reporting Initiative (GRI) is a multi-stakeholder platform and network-based organization that launched the first-ever sustainability reporting framework – its *Sustainability Reporting Guidelines* in 2000. The GRI's core goal is to mainstream disclosure on environmental, social, and governance performance across all organizations and sectors. In 2010, over 1,800 reports following the GRI framework were issued worldwide. The guidelines facilitate comparability for companies and reduce the costs of sustainability reporting. The GRI's main product is its *Sustainability Reporting Guidelines,* which set out the reporting principles and generic performance indicators that organizations can use to measure and report their economic, environmental, and social performance. By reporting on the sustainability of their operations, businesses create the information needed to improve their sustainability, as well as raise awareness and encourage other businesses to do the same. The GRI Guidelines are developed through consensus seeking, multi-stakeholder processes. Participants stem from global business, civil society, labor, academic, and professional institutions. To complement the third version *(G3) Guidelines,* which are applicable to all organizations, the GRI also engages in more specific activities:

- Sector Supplements contain industry-specific performance indicators and have so far been developed for following sectors: electricity, mining and metals, food processing, financial services, automotive, logistics, telecommunications, apparel and footwear, and even public agency and NGO sectors. Efforts are currently underway to develop Sector Supplements for the airport, construction and real estate, event organization, media, and oil and gas sectors.

- National Annexes contain country-level information of relevance to "triple-bottom-line" reporting (covering environmental, social, and economic sustainability). The first pilot project to develop a National Annex was launched in 2010 for Brazil. The experiences from this project are to guide the development of other National Annexes around the world.

The GRI also recognizes the need for integrated reporting, which presents an organization's financial performance together with its environmental, social, and governance (ESG) performance. It is recognized today that, for such holistic reporting to be viable, it must be underpinned by standardized financial and ESG reporting frameworks. ESG reporting frameworks, in particular the GRI's *G3*, together with financial reporting standards – such as the International Financial Reporting Standards or the United States Generally Accepted Accounting Principles – can form the base for new, integrated reporting frameworks. With the aim of creating such an integrated reporting framework that is, moreover, globally accepted, the GRI and the Prince's Accounting for Sustainability Project announced the formation of the International Integrated Reporting Committee in August 2010. The objective of the International Integrated Reporting Committee is an integrated framework that brings together financial, environmental, social and governance information in a clear, concise, consistent, and comparable format. The GRI intends the next generation of its guidelines, the *G4*, to boost the robustness of the integrated reporting framework, enabling better analysis and assurance of integrated reports.

(Source: UNEP (2012))

References

Hitchcock, D. and Willard, M. 2009. *The Business Guide to Sustainability: Practical Strategies and Tools for Organizations*, 2nd edn., Earthscan, London.

Kytle, B. and Ruggie, J.G. 2005. Corporate Social Responsibility as Risk Management: A Model for Multinationals. Corporate Social Responsibility Initiative Working Paper, No. 10. John F Kennedy School of Government, Cambridge, MA.

Modak, P. 2014. https://prasadmodakblog.com/2014/08/12/four-tiers-of-sustainability/ (accessed on 6th August 2019).

UNEP. 2012. *Global Outlook on Sustainable Consumption and Production Policies: Taking Action Together*, United Nations Environment Programme, Paris. https://sustainabledevelopment.un.org/content/documents/559Global%20Outlook%20on%20SCP%20Policies_full_final.pdf.

UNGC. 2011. United Nations Global Compact. United Nations. www.unglobalcompact.org/.

Welford, R. and Gouldson, A. 1993. *Environmental Management and Business Strategy*, Pitman Publishing, London.

Exercises

1. Individual businesses interact with a number of stakeholders, all of whom have an interest in the performance of that company. Give the range of stakeholders who demand sustainability standards.

2. The International Chamber of Commerce (ICC) is a nongovernmental organization serving world business. In response to the Brundtland report, ICC developed a "Business charter for sustainable development," which sets 16 principles for environmental management. Through a literature search, summarize the 16 principles set out in the charter.

3. The Vision 2050 report of the World Business Council for Sustainable Development lays out a pathway leading to a global population of some 9 billion people living well within the resource limits of the planet by 2050. Download the report (www.wbcsd.org/Overview/About-us/Vision2050/Resources/Vision-2050-The-new-agenda-for-business) and read the Executive summary to summarize the critical pathway to attain the vision.

4. List the competitive advantages of a business which has made sustainability a core strategy. What are the threats that a business could face by not integrating sustainability in its core activity?

5. What do you think could be some of the barriers to embed sustainability within a company's activities?

6. Improving business performance through improving environmental performance Browse the Web site of the World Business Council for Sustainable Development (www.wbcsd.ch) and identify a case study of a company improving business performance through improving environmental performance. Write a one-page summary of the case study.

7. Elaborate a business strategy incorporating sustainability principles that engineers can propose to their senior management for project delivery. You can choose a specific field of engineering.

8. Compare the sustainability reports of two large companies in the same sector related to your field of engineering. How do these companies differ? Which company contributes most to sustainable development in your view?

Recommended Reading and Websites

- The Ecology of Commerce by Paul Hawken (1994).
- Cannibals with Forks: The Triple Bottom Line of 21st Century Business. John Elkington

- Natural capitalism by Paul Hawken and Amory and Hunter Lovins. Little Brown and Co. (available to download free of charge at www. natcap.org).
- Mc Donough, W. and Braugart, M. (2002). *Cradle to Cradle: Remaking the Way we Make Things*. New York. North Point Press.
- The Sustainability Advantage (2002) and the Next Sustainability Wave (2005) by Bob Willard.
- United Nations Global Compact (www.unglobalcompact.org/library). It is a voluntary initiative based on CEO commitments to implement universal sustainability principles and to take steps to support UN goals.
- Global Reporting Initiative (GRI) (www.globalreporting.org/Pages/default.aspx)

"The Global Reporting Initiative (GRI) is a leading organization in the sustainability field. GRI promotes the use of sustainability reporting as a way for organizations to become more sustainable and contribute to sustainable development."

- UNEP – Resources for business and industry. In partnership with business and industry, the United Nations Environment Programme (UNEP) promotes informed decision-making and industrial practices to encourage and facilitate environmental sustainability. A selection of UNEP resources is outlined in this brochure: www.unepfi.org/fileadmin/documents/DTIE_Brochure.pdf
- WBCSD (World Business Council for Sustainable Development) www.wbcsd.org/Overview/Resources

The World Business Council for Sustainable Development (WBCSD) is a CEO-led organization of forward-thinking companies that galvanizes the global business community to create a sustainable future for business, society, and the environment.

- The Global 100 Most Sustainable Corporations in the World, announced each year at the World Economic Forum in Davos (www.global100.com)

10

The Contribution of Engineers to Sustainability

> The difference between what we are doing and what we are capable of doing would solve most of the world's problems.
>
> Mahatma Gandhi

10.1 Introduction

The engineering community contributes directly to many of the sustainable development goals (SDGs), including clean water and sanitation; affordable and clean energy; industry, innovation, and infrastructure; sustainable cities and communities; responsible consumption and production; and climate action. To achieve the goals of sustainability resources, efficiency need to be improved significantly, by factors of 4- to 10-fold, and this reduction will only occur through cleaner production, recycling, renewable energy development, and sustainable engineering design. Engineers need to innovate for more sustainable services. This chapter discusses the critical roles of engineers for innovation for sustainability in all the stages of project delivery. In addition, it presents the individual principles that engineers must nurture for action on sustainability during their professional career.

10.2 Innovation for Sustainability

Innovation refers to creating new activities, products, processes, and services; improving existing processes and functions; disseminating new activities or ideas; and adopting things that have been successfully tried elsewhere. Innovation can be diverse; covering minor quality improvements to "cutting edge" products and services (Engineers Australia, 2012). Box 10.1 gives some examples of innovations for sustainability. There is increasing awareness among nations and firms of the importance of being ahead of the next so-called "waves" of innovation, for both prosperity and maintaining economic growth (Figure 10.1).

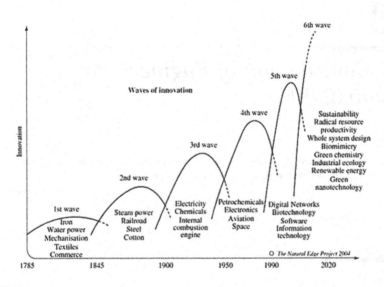

FIGURE 10.1
Waves of innovation.

(Reprinted with permission from Hargroves, K and Smith, M.H. 2005)

BOX 10.1 NINE SUSTAINABLE INNOVATIONS SET TO IMPROVE OUR FUTURES

As the battle on climate change increases, more and more overwhelming statistics and foreboding scientific data fill our feeds. Feeling despondent to it all is an understandable side effect. However, on the flip side of potential environmental catastrophe lies the innovation that could bring it, and therefore us, back from the brink. Here are nine sustainable innovations in food waste, ocean pollution, water-scarcity, and plastic manufacture set to shape our lives and improve our futures.

Cleaning Up Our Oceanic Act of Unkindness

It is estimated that close on 13 tons of plastic pollution is dumped into our oceans each year, causing harm to not only our own health, but that of marine biodiversity too. These ocean-geared innovations are combatting the plastic soup.

1. *Boat-fixed turbines collect plastic en-route*
   By far the biggest portion of plastic waste is dumped in rivers. A young Ecuadorian innovator, Inty Grønneberg, has designed turbines capable of filtering and collecting plastic waste before it ends up in the ocean. Installed on any boat, they are able to collect

up to 80 tons of plastic during navigation. His innovation is focused on not requiring new infrastructure and vessels but rather to make the most use of as many existing boats as possible.

2. *Bins for emptying ocean trash*
   Surfers, Andrew Turton and Pete Ceglinski, invented the Seabin, submersible bins with fitted pumps capable of sifting the water and trapping the floating pollutants such as plastic, detergent, and oil. The bins only need to be emptied once per month.

Food Industry Transformations through AI and Intelligent Plants
Half the world's food harvest is wasted due to fruit and vegetables being so susceptible to pathogens. Take into consideration the greatly reduced pollinating powers of our declining bee population and you could see where, without innovative thinking like this, we would be facing an increased severity of food shortages and waste.

1. *Droid bees doing the heavy lifting pollination*
   Researchers at the University of Warsaw have created robotic bees, or B-Droids, who could help in boosting the bee population by taking up the tasks of low-nutritional and high-labor pollinating. Their concept is a managing platform and "swarm" of autonomous and semi-autonomous robotic bees, who could identify crops and pollinate them effectively.

2. *Blockchain eliminating food waste*
   Part of IBM's research arm, "5 in 5 innovations," works on 5 AI technologies that can be implemented in the next 5 years. One of these innovations calls for a future where blockchain, AI and IoT devices work together to eliminate the costly wastage and losses in the food supply chain. By collecting and sharing these data within the grocery supply chain, planting, ordering, and shipment can be calculated to exact quantities, resulting in reduced waste and fresher food for the consumer.

3. *Plant-based preservative cuts out cold storage*
   Demetra, founded by Italian start-up Green Code, is a 100% plant-based preservative that not only improves the shelf life of fresh produce but could also eliminate cold-storage during transit, thereby simultaneously saving on energy.

Intelligent Plastics Made from Waste and Wasteful Emissions
Scientists estimate that there may be around 5.25 trillion macro- and microplastic floating in the vast oceans. While arising technologies that sifon out this harmful plastic solve one half of the problem,

other innovators are creating better plastics that will prevent it from occurring again in the future.

1. *Bio-plastic from the humble avo-pip*
   Mexican chemical engineer, Scott Munguia, has engineered a plastic from a biopolymer discovered inside the pip of an avocado. The single-use cutlery was made from his bioplastic biodegrades in 240 days. Most notably, this bioplastic is not only manufactured from a previously discarded food item – an estimated 300,000 tons are discarded annually in Mexico alone, but also not manufactured from a food source crop such as corn, cassava, or sugarcane.

2. *Greenhouse gases become profitable plastics*
   Carbon capture and storage (CCS) technology makes it possible for carbon to be captured and compressed at the source, where it can then be moved to storage. While this technology can significantly reduce greenhouse gasses, it is now also being used profitably to make products.

   AirCarbon is a verified carbon-negative material that can be used to manufacture everyday products, which would have otherwise used fossil fuels and oil, by combining the sequestered carbon with oxygen. The entire manufacturing process is carbon negative.

Water-wiser Engineering
30–40% of the world will face water scarcity by 2020. As a climate-change exacerbated reality, 1.8 billion people will live in regions of water scarcity by 2025. This not only will affect fresh drinking water but also the ability to grow food. This innovative thinking could help.

1. *Growing plants in the desert*
   The Groasis Waterboxx makes growing crops in an intelligent plant device that germinates, incubates, and waters saplings possible. It also has a 90% less water requirement than traditional methods.

2. *Drinking water from clouds*
   For coastal and mountainous areas with less access to clean drinking water, a 3D mesh innovated by CloudFisher can convert the fog into safe drinking water while withstanding winds of 120 km/h. The design is an improved-upon innovation on the first fog collectors installed in Eritrea in 2007.

(Retrieved with permission from: Fedder A. (n.d.). Sustainability Management School (SUMAS, Switzerland) Retrieved from https://sumas.ch/nine-sustainable-innovations-set-to-improve-our-futures/)

Porter and Stern (2003) conveyed the critical role that engineers play in innovation by showing the strong correlation between the proportion of scientists and engineers available in a country and the National Innovative Capacity Index (NICI) of the country. As they put it, "The foundation of a nation's common innovation infrastructure is its pool of scientists and engineers available to contribute to innovation throughout the economy." Sustainability is a key driver of innovation to improve productivity (doing more with less) and at the same time significantly reducing the impacts on the environment. Innovation is not only an explicit focus of SDG Goal 9 (build resilient infrastructure, promote inclusive and sustainable industrialization and foster innovation) but also a key enabler of most – if not all – of the SDGs. Achieving the SDGs requires social and institutional innovation just as much as technological innovation. There has been an evolution of design approaches for sustainability in the past decades, from product design to design for system innovations and transition. As per Ceschin and Gaziulusoy (2016), the design for sustainability approaches can be categorized in four different innovation levels with each category being progressively more significant and more far-reaching:

(1) *Product innovation level*: These are eco-design approaches focusing on improving existing or developing completely new products by lowering environmental impacts through the whole life cycle of products from extraction of raw materials to final disposal. However, although the design approaches at this level used by engineers are crucial to reduce environmental impacts of products and production processes, they are not on their own sufficient to obtain the radical improvements required to achieve sustainability. In fact, they can even generate unwanted rebound effects through an increase in consumption level. There is thus a need to move focus from that on product improvements-alone toward a wider approach focused on producing structural changes in the way production and consumption systems are organized.

(2) *Product–service system (PSS) innovation level*: Here the focus is beyond individual products toward development of new business models that integrate combinations of products and services. A PSS is the result of an innovative strategy that shifts the center of the business design and sale of products only (physical) to systems offering products and services that are jointly capable of satisfying a given application (UNEP, 2002). There is a shift from a consumption based on ownership to a consumption based on access and sharing. Examples are from selling heating systems to providing thermal comfort services or from selling cars to offering mobility services. Further examples are given in Box 10.2. Engineers need to consider such new business models at the early stages of engineering designs.

BOX 10.2 EXAMPLES OF PRODUCT–SERVICE SYSTEMS

1. *Car sharing, StattAuto, Germany Company (Source: d4s-sbs.org/ MC.pdf)*

StattAuto was the first car-sharing corporation in Germany and the second worldwide. In 1988, the first German car-sharing company started with one car and an answering machine as part of a scientific research project. Within two years, the model had been increased to four cars and approximately 50 participants. As the results had been promising, the researchers decided to set up the StattAuto company in 1990 (the name StattAuto was chosen as a pun on "city-car" and "instead-of car"). System Users can book a car by phone or via the website 24 hours a day, 7 days a week, and if they call with their mobile phone from any one of parking lots of StattAuto, they can drive the car off immediately. Also users can use a car as long and much as they would like (can afford) to. The cars are stationed at 100 distribution points. Each distribution point has between two and seven parking spaces. The distribution system for cars is spread over the city with a concentration in densely populated inner city districts. On average, a StattAuto member reaches a distribution point in 10 minutes. Most types of car equipment and accessories such as snow chains, children's seats, ski racks, and bike racks are available at no extra cost at the central office. The group has a "moonshine rate" for women. Between midnight and 8 a.m., women drive free to their destination and return the car in the morning, avoiding a potentially dangerous walk in the dark.

Goals: The philosophy of car-sharing companies is to organize the mobility of their customers on the one hand, and to keep the overall cost down on the other hand. Car sharing can be regarded as a supplementary means of transport.

Results: The combined mobility public transport and car sharing is assumed by StattAuto as the most economical and ecological traffic concept for urban agglomerations in the future. Now the service contributes to the urban eco-balance with a reduction of 510,000 car kilometers and an annual decrease in CO_2 emissions of 80.32 tonnes. A shared car from StattAuto is driven approximately 30,000 km per year compared to the national 14,500 km per year with private cars. On average, two persons travel in a StattAuto car compared to the national average of 1.3 persons in private car use.

2. *PSS: Carpet Lease, Interface Inc. (Source: d4s-sbs.org/MC.pdf)*

Specializing in modular floor-covering with carpet tiles, Interface is the world leader in the design, production, and sales of modular carpet and commercial fabrics. Ray Anderson, the founder of Interface Inc., fully devoted his life to sustainability and Mission Zero was the name given to his vision articulated in 1997 that, for most outside the company, seemed audacious at the time: "To be the first company that, by its deeds, shows the entire industrial world what sustainability is in all its dimensions: People, process, product, place and profits – by 2020 – and in doing so we will become restorative through the power of influence."

System: Among many innovative initiatives, Interface sought to introduce a business strategy of leasing – rather than selling – its carpets, known as "Evergreen Lease," so as to be able to take them back to recycle. Customers can choose dematerialized or recycled content Interface products for their installation. Through offering ongoing maintenance as part of the lease, the first quality appearance of the product is maintained for longer and hence its length of time on site is increased. Comprehensive maintenance prevents degradation of the fibers and prolongs the product's life: for example, carpet tiles can be moved from areas of heavy wear such as corridors to areas of lighter wear such as under desks, increasing the useful first life of the carpet. Interface also controls the installation of the product, meaning that flooring contractors will use the Interface Factor Four adhesive system – an immediate improvement in both the amount used and the environmental impact of the adhesives.

Goals: The Evergreen Lease is a revolutionary way of meeting customers' floor-covering needs. It offers a complete package that increases resource productivity at each stage of the life cycle.

Results: The Evergreen Lease program prolongs the life of the carpets while proving end of life recycling. By making the carpeting process more efficient and recycling, Interface is greatly reducing the environmental impact of commercial carpeting.

3. Distributed Solar Energy and Electrical Devices as An All-inclusive Package, Brazil. (Source: Vezzoli et al., 2014)

In 2001, Fabio Rosa began exploring a new business model to provide Brazil's rural people with what they needed: energy services, not just solar energy. Rosa founded both a for-profit corporation, Agroelectric System of Appropriate Technology (STA), and a not-for profit organization, the Institute for Development of Natural Energy and Sustainability (IDEAAS). To that end, TSSFA developed a leasing structure whereby customers pay a monthly fee for the use of cost-effective solar energy packages, a basic photovoltaic solar home system that could be rented for USD10/month plus an initial installation fee. This not only fits with the traditional way people pay for energy, it also saves its customers from paying the 50% sales tax that would be required if they were to purchase the systems instead of renting them. Solar home kits, as TSSFA calls them, include the hardware needed to generate energy, while also providing the installation service and products that use the electricity generated by the solar home system, such as lighting and electrical outlets. All of the tangible inputs are owned by STA and only the services provided by these materials are leased to customer.

(3) *Spatio-social innovation level*: Here the context of innovation is on human settlements and this can be addressed on different scales, from neighborhoods to cities. To achieve sustainability, there is a need for technological innovations to be complemented by social innovations (Geels, 2005). Examples are community car-pooling systems or networks linking consumers directly with producers. Manzini (2014) defines design for social innovation as "a constellation of design initiatives geared toward making social innovation more probable, effective, long-lasting, and apt to spread" and points that it can be part of top-down (driven by experts, decision makers, and political activists), bottom-up (driven by local communities), or hybrid (a combination of both) approaches. Even if social innovations are often driven by nonprofessional designers, professional designers can play a significant role in promoting and supporting them (Manzini, 2015).

(4) *Socio-technical system innovation level*: This innovation level focuses on the transformation of socio-technical systems through technological, social, organizational, and institutional innovations. In other words, system innovation occurs when the societal system functions differently and thus there is a requirement for fundamental structural change (Frantzeskaki and de Haan, 2009). Design approaches focus on promoting radical changes on how societal needs, such as nutrition and transport/mobility, are fulfilled, and thus on supporting transitions to new socio-technical systems. This is an emerging research and practice area. Some historical examples of system innovation are the transition from sailing ships to steam ships or the transition from horse-and-carriage to automobiles (Geels, 2005). The agricultural and industrial revolutions, both of which fundamentally changed how the society operates, are much more profound examples of system innovation. Our society is currently experiencing another profound system innovation through the rapid development and diffusion of information and communication technologies.

In order to achieve sustainability, design and innovation efforts of engineers should intervene at these different levels; the higher the levels of innovation, the more significant will be the contribution to "Factor X" reductions in environmental impacts (Figure 10.2).

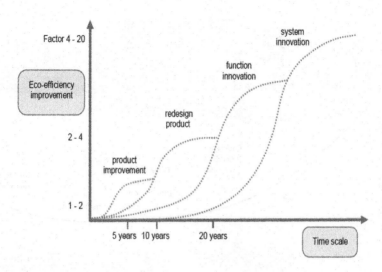

FIGURE 10.2
Four levels of environmental innovation towards reducing environmental impact.

(Wilt, 1997)

10.3 Role of Engineers across Stages of Project Delivery

To innovate for sustainability, a shift must be achieved in the way engineers design and implement projects. Every engineering project contributes to or detracts from sustainability and engineers have to take responsibility. Sustainability is about a new set of questions to be asked at the right time throughout project delivery, which enables innovation that saves energy, money, and resources. Issues beyond the traditional criteria of time, cost, and quality need to be considered such as ethics, boundaries, social impacts, and future generations.

Figure 10.3 shows a diagram of the typical stages of engineering project planning and delivery.

Within any one, the main generic stages of project delivery – planning, design, construction, operation, and decommissioning – sustainability can be improved by what we do and by what questions are asked. As the value management curve shows, the biggest opportunities are often in the earlier stages of the process and also at the very start of each stage (Ainger and Fenner, 2014). The best opportunities to deliver more sustainable engineering projects come by asking radical questions right at the start of each stage. Table 10.1 summarizes what engineers can specifically do at each stage of project delivery. In addition to these actions, engineers have to engage with stakeholders throughout the project delivery process and also

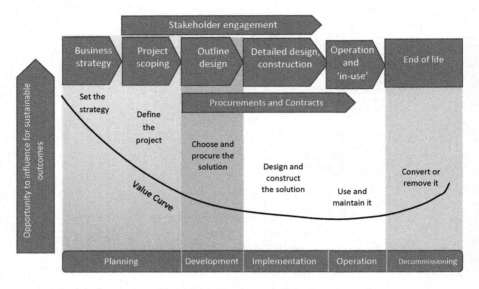

FIGURE 10.3
Opportunities for more sustainable outcomes at typical stages of project delivery.

(Reprinted with permission from Ainger and Fenner, 2014)

use the procurement process in the project development phase (design and construction) to provide sustainability benefits. Problem framing and problem scoping tend to use participative and qualitative methods, while planning and design tend to use more strategic and analytic methods. Implementation, delivery, and operations use more managerial and quantitative methods.

10.4 The Sustainability Competent Engineer

The IEA Graduate Attributes and Professional Competencies model includes a high expectation for sustainability. Engineers should understand their important potential role in the sustainable development of communities and recognize the impacts of an engineering project on communities, global, or local, and consider the views of the community. Engineers need to recognize this professional responsibility and build the necessary competencies during their education and career to take a leadership role in the whole debate about sustainability.

Dealing with complex systems, the three dimensions of sustainability, long time scales, and future uncertainty require some additions to the

TABLE 10.1

Actions of Engineers for Sustainability at Project Delivery Stages (Adapted from Ainger and Fenner, 2014)

Stages of engineering projects delivery	What engineers can do
Business strategy (set the strategy)	• Set sustainability objectives and targets and choose wider boundaries; consider all life cycle stages • Plan for the long term; integrate disciplines and consult with key stakeholders
Project scoping (define the project)	• Ensure planning is anchored in strategic sustainability objectives • Demand management before traditional supply to meet new demand project. Consider new business models such as PSS • Involve local communities and supply chains • Seek new opportunities in the assets. Consider integrated needs • Apply a scoping hierarchy, consider the right design life and apply cost effectiveness for cost evaluation • Ensure that the project scope defines the required performance and not a pre-judged solution • Sustainability must be part of the problem framing. Put sustainability principles into practice at this early stage, demonstrating to nonengineering professionals that they have thought holistically about wider problems
Outline design (feasibility study)	• Work in a multidisciplinary team • Widen the scope so that sustainability is reflected in the performance and scope to consider real sustainable alternatives • Include sustainability criteria in the evaluation and use decision-making processes that reflect sustainability principles • Consider all life cycle stages • Consider DfX strategies • Respect people and human rights • Engage with ideas and opinions of stakeholders in their local context
Detailed design	• Ensure all sustainability objectives identified in the previous stages are consistently applied now to the project detail • Keep sustainability metrics in mind • Challenge traditional approaches and design standards
Construction	• Deliver on the sustainability features of the designs already decided on but try to further minimize the impacts • Source all materials and inputs sustainably
Use	• Deliver and improve on the desired sustainability performance through measurement and follow-up action

(Continued)

TABLE 10.1 (Cont.)

Stages of engineering projects delivery	What engineers can do
	• Incentivize O&M teams and user behavior for sustainable performance • Add environmental performance and resilience into effective maintenance
End of life	• Apply the end-of-life hierarchy of approaches, in order of preference: directly reuse, dismantle; reclaim and reuse; demolish; reclaim and recycle; demolish and dispose

mindsets and skill sets of engineers (Ainger and Fenner, 2014). They require lateral thinking and knowledge transfer between the social, life and physical sciences as well as between engineering disciplines. To meet the sustainability challenges, the engineering community has to work closely with experts from fields that they may not have interacted with previously, and to overcome inherent and fundamental barriers to their cooperation (Rahimifard and Trollman, 2018). Sustainable development with its future-oriented approach also requires professionals who are creative and who have the capacity to be innovative. On the other hand, CSR requires staff who have a sense of responsibility and a social conscience. Every engineer, in addition to disciplinary competences for sustainable development, should develop the following general professional competencies related to sustainability (adapted from Roorda, 2012):

Responsibility: The sustainably competent engineer should bear responsibility for his or her own work, that is, can create a stakeholder analysis, take personal responsibility, and be held personally accountable with respect to society.

System thinking: The sustainably competent engineer thinks and acts from a systemic perspective and apply life cycle thinking tools in the decision making.

Emotional intelligence: A sustainably competent engineer empathizes with the values and emotions of others, that is, recognize and respect his or her own values and those of other people and cultures and cooperate on an interdisciplinary basis.

Future orientation: A sustainably competent engineer thinks and acts on the basis of a perspective of the future, that is, think on different time scales and think innovatively, creatively, and out of the box.

Action skills: A sustainably competent engineer is decisive and capable of acting, that is, deals with uncertainties and weighs all criteria in the decision making.

> *Personal involvement*: A sustainably competent engineer has a personal involvement in sustainable development, that is, consistently involve sustainable development in his or her own work as a professional (sustainable attitude) and employ his or her own conscience as the ultimate yardstick.

Personal values and ethics play an important role in the pursuit of sustainability. Professional ethics now require all engineers to take some responsibility for delivering on sustainability. In its model codes of ethics for engineers globally, the World Federation of Engineering Organization (WFEO) gives the core reasons of why sustainable development matters so much to engineers:

> Because of the rapid advancements in technology and the increasing ability of engineering activities to impact on the environment, engineers have an obligation to be mindful of the effect that their decisions will have on the environment and the well-being of society, and to report any concerns of this nature. With the rapid advancement of technology in today's world and the possible social impacts on large populations of people, engineers must endeavor to foster the public's understanding of technical issues and the role of engineering more than ever before. As population and growth place increasing pressures on our social and biophysical environment, engineers must accept increased responsibilities to develop sustainable solutions to meet community needs, overcome extreme poverty and prevent segregation of people. The education of engineers needs to inculcate an understanding of sustainability and cultural and social sensitivities as well. The engineering code of ethics must reflect a strong commitment to principles of sustainable development. Sustainable development is the challenge of meeting current human needs for natural resources, industrial products, energy, food, transportation, shelter, and effective waste management while conserving and, if possible, enhancing Earth's environmental quality, natural resources, ethical, intellectual, working and affectionate capabilities of people and socioeconomic bases, essential for the human needs of future generations. The proper observance to these principles will considerably help to the eradication of the world poverty.
>
> (WFEO, Code of Ethics, 2001)

To enable innovation for sustainability, engineers need to apply some key individual principles and the latter must lie within the overall professional ethics. These key individual principles are as follows:

1. Learn new skills and competencies for sustainable engineering such as systems thinking, life cycle assessments, and multicriteria analysis. Continue to build knowledge on sustainability and resilience throughout the career.

2. Develop soft skills also at the interpersonal level including communication, critical thinking, negotiation, collaboration, consensus building, networking, and leadership. Learn to engage better with communities. Develop these skills as you apply the principles described in Chapter 2 in your practices and also inspire and motivate others.

3. Challenge orthodoxy and encourage change. All engineering is done in teams, within management structures and so this requires the engineer to persuade others to listen, and to agree to try out new ideas. Sustainable practices are ultimately cost-saving, and you must be able to challenge the assumption that sustainability always cost by demonstrating their financial viability and return on investment in the long term.

Individual actions can help to change the system. It is only by embracing the concept that engineers will integrate social, environmental, and economic factors in the decision-making together with the cost, scope, and schedule management. The ability of an engineer to innovate for sustainability depends on his or her individual capabilities and the characteristics of the organization or sector in which he or she is working, which comprises the context for innovation. The engineer's role is to facilitate the emergence of change and act proactively to enable it by first trying out new solutions and then embedding these in standard practice. There will always be a gap in an organization between declared intentions on sustainability and the ground reality. A passion for sustainability will motivate the creativity for the innovation needed to change the system.

10.5 Summary

Sustainability is a key driver of innovation for improving productivity (doing more with less) while significantly reducing the impacts on the environment. In order to achieve sustainability, design and innovation efforts of engineers should intervene at the four different levels of innovation and the higher the levels of innovation, the more significant will be the contribution to "Factor X" reductions in environmental impacts. Sustainability is about a new set of questions to be asked at the right time throughout project delivery and the engineer can specifically act at each stage of project delivery. To enable innovation for sustainability, engineers need to apply some key individual principles and the latter must lie within the overall professional ethics. Professional ethics now require all engineers to take some responsibility for delivering on sustainability. Engineers also need to build the necessary competencies during their education and career to take a leadership role in the whole debate about sustainability.

SUSTAINABLE ENGINEERING IN FOCUS: 5 ESSENTIAL GREEN ENGINEERING INNOVATIONS

The Institute for Sustainable Infrastructure – an organization cofounded by ASCE in 2011 – developed the Envision rating system to measure sustainable infrastructure projects in five categories: quality of life, leadership, natural world, resource allocation, and climate and risk.

Here are five Envision innovations to consider when developing projects:

1. *Develop a long-term sustainable energy plan*:
   The Holland Energy Park project in Holland, MI, became the first power plant and park project to win an Envision Platinum award. The park is part of a 40-year plan for sustainability energy that incorporates natural gas with supplemental purchase power agreements for renewable energy that will provide affordable power for the surrounding community for decades to come. The project uses the latest in combined-cycle, natural-gas-generating technology to produce up to 145 MW of power to meet the needs of a growing community. The project doubles the fuel efficiency of Holland's present power generation and includes a modern building design that creates an eastern gateway to the city. In addition to its power generation capabilities, the park's features have reduced carbon emissions by half and virtually eliminated solid-particle pollutants.

2. *Clear outdated infrastructure and alleviate congested areas*:
   The Port Metro Vancouver's Low Level Road project earned an Envision Platinum award for its ability to enhance rail and port operations while addressing community safety and traffic-congestion challenges. The project involved realigning and elevating 2.6 km of Low Level Road in North Vancouver, British Columbia, opening space for two new rail tracks. The project also eliminated three existing road and rail crossings, thereby providing direct access to major port terminals. The alleviated congestion helped bolster recreation in the area while reducing noise challenges associated with port operations along the Low Level Road. This was accomplished through the reconfiguration of three intersections and improved lanes for cyclists as well as continuation of the Spirit Trail pedestrian walkway.

3. *Strengthen urban centers through integrated transportation systems*:
   The Kansas City Streetcar Project in Kansas City, Missouri, became the first transit and first streetcar project to earn the Envision Platinum award. The streetcar improves local circulation, supports local development goals, strengthens urban centers, and maintains

long-term sustainability standards. The initial planning behind the project developed around four goals – connect, develop, thrive, and sustain. The transportation system provides "last mile" connectivity to other regional transit services and integrates with bicycle and pedestrian facilities.

By investing in the transit system, Kansas City secured transportation for an area that is home to 65,000 employees and 4,600 residents and a host to more than 10 million visitors each year.

4. *Recycle materials and energy*:
    The Detroit Metropolitan Airport's Runway and Associated Taxiways Reconstruction project received the Envision Silver award for providing a safe connection for aircraft from the runway to the passenger terminals. The project construction used several initiatives to reduce resource consumption, including the installation of LED lighting and signage, storm-water reuse for dust control (with over 10 million gallons of recycled water to date), and reuse of excavated materials both on the project and in locations west of the runway to balance the site and minimize maintenance needs. The project contractor also brought a concrete crusher onsite to enable efficient pulverizing of materials for use on other projects. The project went beyond compliance standards by reducing existing impervious surfaces at the site.

5. *Repurpose projects to incorporate artistic elements and recreational capabilities*:
    Atlanta's Historic Fourth Ward Park earned Envision Gold in providing a network of public parks, multi-use trails, and transit along a historic 22-mile railroad corridor circling downtown Atlanta. The centerpiece for the design includes a storm-water retention pond that provides drainage relief within a 300-acre drainage basin. The pond uses artistic elements to aerate and recycle pond water and is surrounded by walking trails, urban plazas, native plantings, and an amphitheater. The success of this project has generated adjacent development, revitalized the surrounding urban area, and even helped preserve the iconic City Hall East building in Atlanta by helping mitigate the flooding that had prevented its restoration.

(Retrieved with permission from Walpole B. 2017. American Society of Civil Engineers (ASCE) News. https://news.asce.org/5-green-engineering-innova tions-to-consider-for-your-next-project/)

References

Ainger, C. and Fenner, R. 2014. *Sustainable Infrastructure: Principles into Practice*, ICE Publishing, London.

Ceschin, F. and Gaziulusoy, I. 2016. Evolution of Design for Sustainability: From Product Design to Design for System Innovations and Transitions. *Design Studies*, 47, 118–163.

Engineers Australia. 2012. Innovation in Engineering Report June 2012.

Frantzeskaki, N. and de Haan, J. 2009. Transitions: Two Steps from Theory to Policy. *Futures*, 41, 593–606.

Geels, F.W. 2005. *Technological Transitions and System Innovations: A Coevolutionary and Socio-technical Analysis*, Edward Elgar, Cheltenham, UK and Northampton, MA.

Hargroves, K. and Smith, M.H. 2005. The Natural Advantage of Nations: Business Opportunities, Innovation and Governance in the 21st Century. In K. Hargroves and M.H. Smith (Eds.), *Chapter 1: Natural Advantage of Nations*. Earthscan, London, 17.

Manzini, E. 2014. Making Things Happen: Social Innovation And Design. *Design Issues*, 30(1), 57–66.

Manzini, E. 2015. *When Everybody Designs. An Introduction to Design for Social Innovation*, The MIT Press Cambridge, London.

Porter, M. and Stern, S. 2003. *Ranking National Innovative Capacity: Findings from the National Innovative Capacity Index, in the Global Competitiveness Report 2003–2004 (Eds.)*, Oxford University Press, New York.

Rahimifard, S. and Trollman, H. 2018. UN Sustainable Development Goals: An Engineering Perspective. *International Journal of Sustainable Engineering*, 11(1), 1–3.

Roorda, N. 2012. *Fundamentals of Sustainable Development*, Routledge, London.

UNEP. 2002. *Product-service Systems and Sustainability. Opportunities for Sustainable Solutions*, United Nations Environment Programme, Paris.

Vezzoli, C., Kohtala, C., Srinivasan, A., Diehl, J.C., Moi Fusakul, S., Xin, L. and Deepta, S. 2014. *Product-Service System Design for Sustainability*, Greenleaf Publishing Limited, Sheffield, UK.

Wilt, C. 1997. Product Stewardship at Xerox Corporation. In G. Davis and C. Wilt (Eds.), *EPR: A New Principle for Product- Oriented Pollution Prevention*. Report, The University of Tennessee –Centre for Clean Products and Clean Technologies, Tennessee, 1–7 and 11.

WFEO. 2001. The WFEO Model Code of Ethics. World Federation of Engineering Organizations, Tunis,Tunisia. www.wfeo.org/wp-content/uploads/code_of_ethics/WFEO_MODEL_CODE_OF_ETHICS.pdf (Accessed 13th November 2019).

Exercises

1. Through a literature review, give some examples of rebound effects following an innovation.

2. Distinguish between the four innovation levels, providing examples for each.

3. Through an internet search, provide 20 examples of sustainability innovations created during the last decade.

4. What are the benefits of products service systems from a sustainability point of view?

5. Provide a checklist for the sustainability requirements of a design project that you can embark during your final year of studies that will develop your competencies in sustainable development.

6. You are a self-employed engineering consultant. You have been employed to produce an environmental impact statement for a new road tunnel on behalf of the construction company proposing the tunnel. It has been made clear to you that the expectation of your client is that the statement will not find significant environmental problems with the project. However, you are concerned that if you produce a report that meets these expectations, it will not fully represent the adverse effects of the project and could lead to the project proceeding even though its benefits do not outweigh the environmental damage it will cause. How should you go about completing the environmental impact statement? Should you aim to meet their expectations, adapting the methodology to get the desired results; warn them that the report may highlight problems; or simply produce the most honest, accurate report that you can? (Source: Royal Academy of Engineering, UK).

Recommended Reading and Websites

- Hargroves, K. and Smith, M. H. (2005) The Natural Advantage of Nations: Business Opportunities, Innovation and Governance in the 21st Century, Earthscan, London.

- UNEP (2002). Product-service systems and sustainability. Opportunities for sustainable solutions. United Nations Environment Programme, Paris.

- Engineering Ethics in Practice Royal Academy of Engineering, UK. https://www.raeng.org.uk/publications/other/engineering-ethics-in-practice-shorter.

Appendix A

State of the World Environment: Key Messages of GEO-6

(Source: UN Environment (2019). Global Environment Outlook – GEO-6: Key Messages. Nairobi. www.unenvironment.org/resources/assessment/geo-6-key-messages)

The United Nations Environment Programme's Global Environment Outlook (GEO) is often referred to as UN Environment's flagship global environmental assessment. UN Environment launched the first GEO in 1997. By bringing together a community of hundreds of scientists, peer reviewers, and collaborating institutions and partners, the GEO reports build on sound scientific knowledge to provide governments, local authorities, businesses, and individual citizens with the information needed to guide societies to a truly sustainable world by 2050. GEO-6 builds on the findings of previous GEO reports, including six regional assessments, and outlines the current state of the environment, illustrates possible future environmental trends, and analyzes the effectiveness of policies. This flagship report shows how governments can put the world on the path to a truly sustainable future. It emphasizes that urgent and inclusive action is needed by decision makers at all levels to achieve a healthy planet with healthy people. The key messages of the report are as follows:

Healthy planet, healthy people: Time to act!

(1) The United Nations Environment Programme's sixth Global Environment Outlook (GEO-6) is the most comprehensive report on the global environment since 2012. It shows that the overall environmental situation globally is deteriorating and the window for action is closing.

(2) GEO-6 shows that a healthy environment is both a prerequisite and a foundation for economic prosperity, human health, and well-being. It addresses the main challenge of the 2030 Agenda for Sustainable Development: that no one should be left behind, and that all should live healthy, fulfilling lives for the full benefit of all, for both present and future generations.

(3)   Unsustainable production and consumption patterns and trends and inequality, when combined with increases in the use of resources that are driven by population growth, put at risk the healthy planet needed to attain sustainable development. Those trends are leading to a deterioration in planetary health at unprecedented rates, with increasingly serious consequences, in particular for poorer people and regions.

(4)   Furthermore, the world is not on track to achieve the environmental dimension of the Sustainable Development Goals or other internationally agreed environmental goals by 2030; nor is it on track to deliver long-term sustainability by 2050. Urgent action and strengthened international cooperation are urgently needed to reverse those negative trends and restore planetary and human health.

(5)   Past and present greenhouse gas emissions have already committed the world to an extended period of climate change with multiple and increasing environmental and society-wide risks.

(6)   Air pollution, currently the cause of between 6 and 7 million premature deaths per year, is projected to continue to have significant negative effects on health, and still cause between 4.5 million and 7 million premature deaths annually by mid-century.

(7)   Biodiversity loss from land-use change, habitat fragmentation, overexploitation and illegal wildlife trade, invasive species, pollution, and climate change is driving a mass extinction of species, including critical ecosystem service providers, such as pollinators. That mass extinction compromises Earth's ecological integrity and its capacity to meet human needs.

(8)   Marine plastic litter, including microplastics, occurs at all levels of the marine ecosystem and also appears in fisheries and shellfish at alarming levels and frequency. The long-term adverse impact of marine microplastics on the marine system is as yet unknown, with potential impact on human health through the consumption of fish and marine products. More research on the magnitude of the problem is still needed.

(9)   Land degradation is an increasing threat for human well-being and ecosystems, especially for those in rural areas who are most dependent on land for their productivity. Land degradation hotspots cover approximately 29% of land globally, where some 3.2 billion people reside.

(10) Natural resources, including fresh water and oceans, are too often overexploited, poorly managed, and polluted. Approximately 1.4 million people die annually from preventable diseases, such as diarrhea and intestinal parasites, that are associated with pathogen-polluted drinking water and inadequate sanitation.

(11) Antibiotic-resistant infections are projected to become a major cause of death worldwide by 2050. Affordable, widely available wastewater treatment technologies to remove antibiotic residues could have huge benefits for all countries. Even greater efforts are needed to control mismanagement of antibacterial drugs at source, in both human and agricultural use.

(12) The harmful impact of inappropriate use of pesticides, heavy metals, plastics and other substances are of significant concern, as such compounds appear in alarmingly high levels in our food supply. They primarily affect vulnerable members of society, such as infants, who are exposed to elevated levels of chemicals. The impact of neurotoxins and endocrine-disrupting chemicals is potentially multigenerational.

Transformative change: A call for systemic and integrated policy action

(13) The social and economic costs of inaction often exceed the costs of action and are inequitably distributed, often being borne by the poorest and most vulnerable in society, including indigenous and local communities, particularly in developing countries.

(14) Current environmental policy alone is not sufficient to address those challenges. Urgent Cross-sectoral policy action, through a whole-of-society approach, is needed to address the challenges of sustainable development.

(15) Achieving internationally agreed environmental goals on pollution control, clean-up, and efficiency improvements is crucial, yet alone is insufficient to achieve the Sustainable Development Goals. Transformative change is needed to enable and combine long-term strategic and integrated policymaking while building bottom-up social, cultural, institutional, and technological innovation.

(16) The key features of effective environmental policies for sustainable development include integrated objectives, science-based targets, economic instruments, regulations, and robust international cooperation.

(17) Transformative change that will enable us to achieve the Sustainable Development Goals and other internationally agreed targets includes a tripling of the current decarbonization rate as we head toward 2050, a 50% increase in food production, and the adoption of healthy and sustainable diets across all global regions.

(18) The transformative changes needed to achieve sustainable development will be most successful when they are just, respect gender

equality, recognize that there will be different impacts on men, women, children and the elderly, and take into account inherent societal risks.

(19) The health co-benefits of reducing greenhouse gas emissions and air pollutants, including short-lived climate pollutants, can together outweigh the costs of mitigation, while achieving climate and air quality targets, increasing agricultural production, and reducing biodiversity loss. Access to safe drinking water and sanitation can also provide environmental and health co-benefits.

(20) Sustainable outcomes can best be achieved by combining objectives for resource-use efficiency, with ecosystem-based management and improved human health, drawing on scientific, indigenous, and local knowledge.

Governance of innovations: Innovations in governance

(21) Food, energy and transport systems, urban planning, and chemical production are primary examples of systems of production and consumption that need innovative, effective, and integrated policies.

(22) Innovations are part of the solution, but can also create new risks and have a negative environmental impact. Where relevant scientific evidence is insufficient to inform decision making, precautionary approaches can reduce threats of serious or irreversible damage.

(23) Innovation in and deployment of technologies to reduce greenhouse gas emissions and increase resource efficiency can strengthen the economic performance of enterprises, municipalities, countries, and other stakeholders.

(24) Agreement on desired pathways for transformative change under conditions of uncertainty can be fostered by coalitions among governments, businesses, researchers, and civil society.

(25) Sustainable development is more likely to be achieved through new modes of governance and adaptive management that give greater priority to the environmental dimension of the Sustainable Development Goals, while promoting gender equality and education for sustainable production and consumption.

Harvest time: Knowledge for sustainability

(26) These new sustainability governance models should also ensure adequate investment in knowledge systems such as data, indicators,

assessments, policy evaluation, and sharing platforms, and act on internationally agreed early signals from science and society to avoid unnecessary environmental impact and costs.

(27) Data from satellites, combined with monitoring on the ground, can enable quicker actions across the world, in response to extreme weather events, for example. Widening access to data, information, and knowledge, and improving the infrastructure and capacity to harness that knowledge will enable those data to be put to the most effective use.

(28) More investment in indicators that integrate different data sources and clearly delineate gender and inequality aspects will facilitate better-designed policy interventions and their evaluation.

(29) Further developments are needed in environmental and natural resource accounting to ensure that environmental costs are internalized into economic decision-making for sustainability.

(30) Harnessing the ongoing data and knowledge revolution and ensuring the authenticity and validity of those data to support sustainable development, combined with international cooperation, could transform capacities to address challenges and accelerate progress toward sustainable development.

(31) Most important of all is the need to take bold, urgent, sustained, inclusive, and transformative action that integrates environmental, economic and social activity to set society on the pathway to achieving the Sustainable Development Goals, multilateral environmental agreements, internationally agreed environmental goals, and other science-based targets.

Appendix B

International Energy Agency: World Energy Outlook 2018. Executive Summary

The world is gradually building a different kind of energy system, but cracks are visible in the key pillars:

- *Affordability*: The costs of solar PV and wind continue to fall, but oil prices climbed above $80/barrel in 2018 for the first time in four years, and hard-earned reforms to fossil fuel consumption subsidies are under threat in some countries.

- *Reliability*: Risks to oil and gas supply remain, as Venezuela's downward spiral shows. One in eight of the world's population has no access to electricity and new challenges are coming into focus in the power sector, from system flexibility to cyber security.

- *Sustainability*: After three flat years, global energy-related carbon dioxide (CO_2) emissions rose by 1.6% in 2017 and the early data suggest continued growth in 2018, far from a trajectory consistent with climate goals. Energy-related air pollution continues to result in millions of premature deaths each year.

Affordability, reliability, and sustainability are closely interlinked: each of them, and the trade-offs between them, requires a comprehensive approach to energy policy. The links between them are constantly evolving. For example, wind and solar photovoltaics (PV) bring a major source of affordable, low-emissions electricity into the picture, but create additional requirements for the reliable operation of power systems. The movement toward a more interconnected global gas market, as a result of growing trade in liquefied natural gas (LNG), intensifies competition among suppliers while changing the way that countries need to think about managing potential shortfalls in supply.

Robust data and well-grounded projections about the future are essential foundations for today's policy choices. This is where the *World Energy Outlook* (*WEO*) comes in. It does not aim to forecast the future, but provides a way of exploring different possible futures, the levers that bring them about and the interactions that arise across a complex energy system. If there is no change in policies from today, as in the *Current Policies*

Scenario, this leads to increasing strains on almost all aspects of energy security. If we broaden the scope to include announced policies and targets, as in our main *New Policies Scenario*, the picture brightens. But the gap between this outcome and the *Sustainable Development Scenario*, in which accelerated clean energy transitions put the world on track to meet goals related to climate change, universal access and clean air, remains huge. None of these potential pathways is preordained; all are possible. The actions taken by governments will be decisive in determining which path we follow.

How Is the World of Energy Changing?

In the New Policies Scenario, rising incomes and an extra 1.7 billion people, mostly added to urban areas in developing economies, push up global energy demand by more than a quarter to 2040. The increase would be around twice as large if it were not for continued improvements in energy efficiency, a powerful policy tool to address energy security and sustainability concerns. All the growth comes from developing economies, led by India. As recently as 2000, Europe and North America accounted for more than 40% of global energy demand and developing economies in Asia for around 20%. By 2040, this situation is completely reversed.

The profound shift in energy consumption to Asia is felt across all fuels and technologies, as well as in energy investment. Asia makes up half of global growth in natural gas, 60% of the rise in wind and solar PV, more than 80% of the increase in oil, and more than 100% of the growth in coal and nuclear (given declines elsewhere). Fifteen years ago, European companies dominated the list of the world's top power companies, measured by installed capacity; now six of the top 10 are Chinese utilities.

The shale revolution continues to shake up oil and gas supply, enabling the United States to pull away from the rest of the field as the world's largest oil and gas producer. In the New Policies Scenario, the United States accounts for more than half of global oil and gas production growth to 2025 (nearly 75% for oil and 40% for gas). By 2025, nearly every fifth barrel of oil and every fourth cubic meter of gas in the world come from the United States.

Shale is adding to the pressure on traditional oil and gas exporters that rely heavily on export revenues to support national development.

The energy world is connecting in different ways because of shifting supply, demand, and technology trends. International energy trade flows are increasingly drawn to Asia from across the Middle East, Russia, Canada, Brazil, and the United States, as Asia's share of global oil and gas trade rises from around half today to more than two-thirds by 2040. But new ways of sourcing energy are also visible at local level, as digitalization and increasingly cost-effective renewable energy technologies enable distributed and community-based models of energy provision to gain ground.

The convergence of cheaper renewable energy technologies, digital applications, and the rising role of electricity is a crucial vector for change, central to the prospects for meeting many of the world's sustainable development goals. This vista is explored in detail in the *WEO-2018* special focus on electricity.

Electricity Is the Star of the Show, but How Bright Will It Shine?

The electricity sector is experiencing its most dramatic transformation since its creation more than a century ago. Electricity is increasingly the "fuel" of choice in economies that are relying more on lighter industrial sectors, services, and digital technologies. Its share in global final consumption is approaching 20% and is set to rise further. Policy support and technology cost reductions are leading to rapid growth in variable renewable sources of generation, putting the power sector in the vanguard of emissions reduction efforts but requiring the entire system to operate differently in order to ensure reliable supply.

In advanced economies, electricity demand growth is modest, but the investment requirement is still huge as the generation mix changes and infrastructure is upgraded.

Today's power market designs are not always up to the task of coping with rapid changes in the generation mix. Revenue from wholesale markets is often insufficient to trigger new investment in firm generation capacity; this could compromise the reliability of supply if not adequately addressed. On the demand side, efficiency gains from more stringent energy performance standards have played a pivotal role in holding back demand: 18 out of the 30 International Energy Agency member economies have seen declines in their electricity use since 2010. Growth prospects depend on how fast electricity can gain ground in providing heat for homes, offices and factories, and power for transportation.

A doubling of electricity demand in developing economies puts cleaner, universally available, and affordable electricity at the center of strategies for economic development and emissions reduction. One in five kilowatt-hours of the rise in global demand comes just from electric motors in China; rising demand for cooling in developing economies provides a similar boost to growth. In the absence of a greater policy focus on energy efficiency, almost one in every three dollars invested in global energy supply, across all areas, goes to electricity generation and networks in developing economies. This investment might not materialize, especially where end-user prices are below cost-recovery levels. But in highly regulated markets, there is also a risk that capacity runs ahead of demand: we estimate that today there are 350 GW of excess capacity in regions including China, India, Southeast Asia, and the Middle East, representing additional costs that the system, and consumers, can ill afford.

Flexibility Is the New Watchword for Power Systems

The increasing competitiveness of solar PV pushes its installed capacity beyond that of wind before 2025, past hydropower around 2030 and past coal before 2040. The majority of this is utility-scale, although investment in distributed solar PV by households and businesses plays a strong supporting role. The *WEO-2018* introduces a new metric to estimate the competitiveness of different generation options, based on evolving technology costs as well as the value that this generation brings to the system at different times. This metric confirms the advantageous position of wind and solar PV in systems with relatively low-cost sources of flexibility. New solar PV is well placed to outcompete new coal almost everywhere, although it struggles in our projections to undercut existing thermal plants without a helping hand from policy. In the New Policies Scenario, renewables and coal switch places in the power mix: the share of generation from renewables rises from 25% today to around 40% in 2040; coal treads the opposite path.

The rise of solar PV and wind power gives unprecedented importance to the flexible operation of power systems in order to keep the lights on. There are few issues at low levels of deployment, but in the New Policies Scenario many countries in Europe, as well as Mexico, India, and China are set to require a degree of flexibility that has never been seen before at such large scale. The cost of battery storage declines fast, and batteries increasingly compete with gas-fired peaking plants to manage short-run fluctuations in supply and demand. However, conventional power plants remain the main source of system flexibility, supported by new interconnections, storage and demand-side response. The European Union's aim to achieve an "Energy Union" illustrates the role that regional integration can play in facilitating the integration of renewables.

The share of generation from nuclear plants – the second-largest source of low-carbon electricity today after hydropower – stays at around 10%, but the geography changes as generation in China overtakes the United States and the European Union before 2030. Some two-thirds of today's nuclear fleet in advanced economies is more than 30 years old. Decisions to extend, or shut down, this capacity will have significant implications for energy security, investment and emissions.

How Much Power Can We Handle?

A much stronger push for electric mobility, electric heating and electricity access could lead to a 90% rise in power demand from today to 2040, compared with 60% in the New Policies Scenario, an additional amount that is nearly twice today's US demand. In the *Future is Electric Scenario*, the share of electricity in final consumption moves up towards one-third, as almost half the car fleet goes electric by 2040 and electricity makes rapid inroads into the residential and industry sectors. However, some significant parts of the energy system, such as long-distance road freight, shipping and aviation, are not "electricready" with today's technologies. Electrification brings benefits, notably by reducing local pollution, but requires additional measures to decarbonize power supply if it is to unlock its full potential as a way to meet climate goals: otherwise, the risk is that CO_2 emissions simply move upstream from the end-use sectors to power generation.

Where Does the Rise of Electricity, Renewables and Efficiency Leave Fossil Fuels?

In the New Policies Scenario, a rising tide of electricity, renewables and efficiency improvements stems growth in coal consumption. Coal use rebounded in 2017 after two years of decline, but final investment decisions in new coal-fired power plants were well below the level seen in recent years. Once the current wave of coal plant projects under construction is over, the flow of new coal projects starting operation slows sharply post-2020. But it is too soon to count coal out of the global power mix: the average age of a coal-fired plant in Asia is less than 15 years, compared with around 40 years in advanced economies. With industrial coal use showing a slight increase to 2040, overall global consumption is flat in the New Policies Scenario, with declines in China, Europe, and North America offset by rises in India and Southeast Asia.

Oil use for cars peaks in the mid-2020s, but petrochemicals, trucks, planes and ships still keep overall oil demand on a rising trend. Improvements in fuel efficiency in the conventional car fleet avoid three-times more in potential demand than the 3 million barrels per day (mb/d) displaced by 300 million electric cars on the road in 2040. But the rapid pace of change in the passenger vehicle segment (a quarter of total oil demand) is not matched elsewhere. Petrochemicals are the largest source of growth in oil use. Even if global recycling rates for plastics were to double, this would cut only around 1.5 mb/d from the projected increase of more than 5 mb/d. Overall growth in oil demand to 106 mb/d in the New Policies Scenario comes entirely from developing economies.

Natural gas overtakes coal in 2030 to become the second-largest fuel in the global energy mix. Industrial consumers make the largest contribution to a 45% increase in worldwide gas use. Trade in LNG more than doubles in response to rising demand from developing economies, led by China. Russia remains the world's largest gas exporter as it opens new routes to Asian markets, but an increasingly integrated European energy market gives buyers more gas-supply options. Higher shares of wind and solar PV in power systems push down the utilization of gas-fired capacity in Europe, and retrofits of existing buildings also help to bring down gas consumption for heating, but gas infrastructure continues to play a vital role, especially in winter, in providing heat and ensuring uninterrupted electricity supply.

Where Are We on Emissions and Access – and Where Do We Want to Be?

The New Policies Scenario puts energy-related CO_2 emissions on a slow upward trend to 2040, a trajectory far out of step with what scientific knowledge says will be required to tackle climate change. Countries are, in aggregate, set to meet the national pledges made as part of the Paris Agreement. But these are insufficient to reach an early peak in global emissions. The projected emissions trend represents a major collective failure to tackle the environmental consequences of energy use. Lower emissions of the main air pollutants in this scenario are not enough to halt an increase in the number of premature deaths from poor air quality.

In 2017, for the first time, the number of people without access to electricity dipped below 1 billion, but trends on energy access likewise fall short of global goals. The New Policies Scenario sees some gains in terms of access, with India to the fore. However, more than 700 million people, predominantly in rural settlements in sub-Saharan Africa, are projected to remain without electricity in 2040, and only slow progress is made in reducing reliance on the traditional use of solid biomass as a cooking fuel.

Our Sustainable Development Scenario provides an integrated strategy to achieve energy access, air quality, and climate goals, with all sectors and low-carbon technologies – including carbon capture, utilization, and storage – contributing to a broad transformation of global energy. In this scenario, the power sector proceeds further and faster with the deployment of low-emissions generation. Renewable energy technologies provide the main pathway to the provision of universal energy access. All economically viable avenues to improve efficiency are pursued, keeping overall demand in 2040 at today's level.

Electrification of end-uses grows strongly, but so too does the direct use of renewables – bioenergy, solar, and geothermal heat – to provide heat and mobility. The share of renewables in the power mix rises from one-quarter today to two-thirds in 2040; in the provision of heat it rises from 10% today to 25% and in transport it rises from 3.5% today to 19% (including both direct use and indirect use, e.g. renewables-based electricity). For the first time, this *WEO* incorporates a water dimension in the Sustainable Development Scenario, illustrating how water constraints can affect fuel and technology choices, and detailing the energy required to provide universal access to clean water and sanitation.

Can Oil and Gas Improve Their Own Environmental Performance?

Natural gas and oil continue to meet a major share of global energy demand in 2040, even in the Sustainable Development Scenario. Not all sources of oil and gas are equal in their environmental impact. Our first comprehensive global estimate of the indirect emissions involved in producing, processing, and transporting oil and gas to consumers suggests that, overall, they account for around 15% of energy sector greenhouse gas emissions (including CO2 and methane). There is a very broad range in emissions intensities between different sources: switching from the highest emissions oil to the lowest would reduce emissions by 25% and doing the same for gas would reduce emissions by 30%.

Much more could be done to reduce the emissions involved in bringing oil and gas to consumers. Many leading companies are taking on commitments in this area that, if widely adopted and implemented, would have a material impact on emissions. Reducing methane emissions and eliminating flaring are two of the most cost-effective approaches. There are also some more "game-changing" options, including the use of CO_2 to support enhanced oil recovery, greater use of low-carbon electricity to support operations, and the potential to convert hydrocarbons to hydrogen (with carbon capture). Many countries, notably Japan, are looking closely at the possibility of expanding the role of zero-emissions hydrogen in the energy system.

Is Investment in Fossil Fuel Supply Out of Step with Consumption Trends?

Today's flow of new upstream projects appears to be geared to the possibility of an imminent slowdown in fossil fuel demand, but in the New Policies Scenario this could well lead to a shortfall in supply and

a further escalation in prices. The risk of a supply crunch looms largest in oil. The average level of new conventional crude oil project approvals over the last three years is only half the amount necessary to balance the market out to 2025, given the demand outlook in the New Policies Scenario. US tight oil is unlikely to pick up the slack on its own. Our projections already incorporate a doubling in US tight oil from today to 2025, but it would need to more than triple in order to offset a continued absence of new conventional projects. In contrast to oil, the risk of an abrupt tightening in LNG markets in the mid-2020s has been eased by major new project announcements, notably in Qatar and Canada.

Government Policies Will Shape the Long-Term Future for Energy

Rapid, least-cost energy transitions require an acceleration of investment in cleaner, smarter and more efficient energy technologies. But policy makers also need to ensure that all key elements of energy supply, including electricity networks, remain reliable and robust. Traditional supply disruption and investment risks on the hydrocarbons side are showing no signs of relenting and indeed may intensify as energy transitions move ahead. The changes underway in the electricity sector require constant vigilance to ensure that market designs are robust even as power systems decarbonize. More than 70% of the $2 trillion required in the world's energy supply investment each year, across all domains, either comes from state-directed entities or responds to a full or partial revenue guarantee established by regulation. Frameworks put in place by the public authorities also shape the pace of energy efficiency improvement and of technology innovation. Government policies and preferences will play a crucial role in shaping where we go from here.

Appendix C

Energy and Climate Change

Historical and current energy systems are dominated by fossil fuels (coal, oil and gas) that produce carbon dioxide (CO_2) and other greenhouse gases – the fundamental driver of global climate change. Table A3.1 summarizes climate facts and figures from latest UN reports on climate. Energy production and use is the single biggest contributor to global warming, accounting for roughly two-thirds of human-induced greenhouse gas emissions. If we are to meet our global climate targets and avoid dangerous climate change, the world urgently needs to use energy efficiently while embracing clean energy sources. Our challenge is to reduce our reliance on fossil fuels to produce electricity and heat and power our transportation systems, while making reliable, clean and affordable energy available to everyone on the planet. This Annex gives some key energy and climate facts and figures. It is based largely on the Advanced Learning Materials on Green Economy published by the UN Partnership for Action on Green Economy (www.un-page.org/resources/green-economy-learning/advanced-learning-materials-green-economy) and on climate and energy facts and figures from the United Nations and from the website of Our World in Data.

Key Energy Facts

Energy Is Essential

- Energy is a key input in the production process for a number of goods in the modern economy.
- Access to affordable energy is a main diver for development: social and economic.
- For least developed countries, transformation of the energy sector would mean reducing energy poverty. It is estimated that 20% of people in the world, approximately 1.3 billion people, lack access to electricity. In least developed countries, this figure is almost 80%.
- A lack of energy services affects health, limits opportunities for education and development, and also the ability to rise out of poverty.

TABLE A3.1

Climate Facts and Figures (Source: www.un.org/sustainabledevelopment/climate-facts-and-figures/)

UN Environment Emissions Gap Report [November 2018]	World Meteorological Organization Greenhouse Gas Bulletin [November 2018]	Intergovernmental Panel on Climate Change Special Report on Global Warming of 1.5°C [October 2018]
• To achieve the goal of limiting climate change to 2°C, countries need to triple the level of their commitments made under the Paris Agreement. • To achieve the goal of limiting climate change to 1.5°C, countries would have to increase their level of ambition by 5x. • Global emissions have reached historic levels at 53.5 GtCO2e, after a three-year period of stabilization, with no signs of peaking. • Only 57 countries (representing 60% of global emissions) are on track to meet their commitments by 2030. • If the emissions gap is not closed by 2030, it is extremely unlikely that the 2°C temperature goal can still be reached.	• Globally, averaged concentrations of CO_2 reached 405.5 parts per million (ppm) in 2017, up from 403.3 ppm in 2016 and 400.1 ppm in 2015. • Concentrations of methane and nitrous oxide also rose, whilst there was a resurgence of a potent greenhouse gas and ozone depleting substance called CFC-11, which is regulated under an international agreement to protect the ozone layer. • Since 1990, there has been a 41% increase in total radiative forcing – the warming effect on the climate – by long-lived greenhouse gases. CO_2 accounts for about 82% of the increase in radiative forcing over the past decade, according to figures from the US National Oceanic and Atmospheric Administration. • The last time Earth experienced a comparable concentration of CO_2 was 3–5 million years ago, when the temperature was 2–3°C warmer and sea level was 10–20 meters higher.	• The report highlights a number of climate change impacts that could be avoided by limiting global warming to 1.5°C compared to 2°C, or more. • By 2100, global sea level rise would be 10 cm lower with global warming of 1.5°C compared with 2°C. • The likelihood of an Arctic Ocean free of sea ice in summer would be once per century with global warming of 1.5°C, compared with at least once per decade with 2°C. • Coral reefs would decline by 70-90% with global warming of 1.5°C, whereas virtually all (> 99%) would be lost with 2°C. • We have the tools we need to address climate change: the world doesn't have to come up with some magic machines to curb climate change – the technology is available to take necessary climate action. • Rapid changes must take place in four key parts of society: energy generation, land use, cities and industry

Energy Demand Is Increasing

- Due to population and economic growth, global energy demand is increasing rapidly. This is especially true in the developing world.

- Figure A3.1 below plots the global energy consumption from 1800 through to 2015. In 2015, the world consumed 146,000 terawatt-hours (TWh) of primary energy – more than 25 times of that in 1800. Renewables account for less than 5%. There is a long way to go if we are to transition from a fossil fuel-dominated energy mix to a low-carbon one.

- By 2035, an estimated 5,900 gigawatts (GW) of new gross capacity are needed with over 90% of this capacity in the developing world. As a comparison, in 2009, global installed power generation capacity was about 5,000 GW.

- Especially middle-income countries are expanding economically, often at high rates of industrialization. At this stage, a green economy strategy can bring essential elements for development

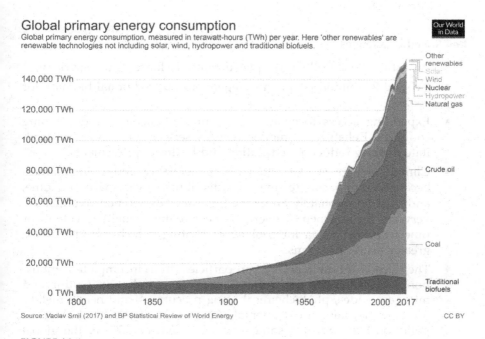

FIGURE A3.1

Global primary energy consumption from 1800 to 2015.

(Retrieved from: https://ourworldindata.org/energy-production-and-changing-energy-sources)

and vast opportunity to invest in large green infrastructure as the energy system expands, avoiding the problem of locking in emissions in the future.

Key Environmental Challenges

- The Intergovernmental Panel on Climate Change's (IPCC) fifth assessment report underscored the importance of mitigating future human-induced climate change – mostly driven by the combustion of fossil fuels – and adapting to the changes that occur.
- Pollution arising from combustion of fossil and traditional fuels is associated with high indirect costs.
- The use of fossil and traditional energy sources in both developed and developing countries also impacts global biodiversity and ecosystems through deforestation, decreased water quality and availability, acidification of water bodies, and increased introduction of hazardous substances into the biosphere. These impacts also reduce the natural capabilities of the planet to respond to climate change.

Key Social Challenges

- An estimated 1.4 billion people currently lack access to electricity.
- Around 2.7 billion people are dependent on traditional biomass for cooking.
- Expanding access to energy is a central challenge for developing countries. Reliable and modern energy services are needed to facilitate poverty reduction, education and health improvements.
- Climate change is likely to worsen inequality and vulnerabilities because its impacts are unevenly distributed over space and time, and disproportionately affect the poor and the women. Women are very much concerned by energy issues are they usually devote labor time to collect wood for cooking or heating, which can also create greater health problems.
- The release of both black carbon particles (from incomplete combustion of fossil fuels) and other forms of air pollution (sulfur and nitrogen oxides, photochemical smog precursors, and heavy metals, for example) have a detrimental effect on public health. Indoor air pollution from burning solid fuel accounted for 2.7% of the global burden of disease in 2000 and is ranked as the largest environmental contributor to health problems after unsafe drinking water and lack of sanitation.

Key Economic Challenges

- The current highly carbon-intensive energy system depends on a finite supply of fossil fuels that are getting harder and more expensive to extract, leading to concerns about national energy security in many countries, that is, the reliability and affordability of national energy supply.
- The current state of the energy sector leaves many countries exposed to large swings in oil import prices and also costs billions in public subsidies.
- The economic cost of fossil fuels amount to billions of USD each year. The annual global costs of adapting to climate change have been estimated by the United Nations Framework on Climate Change Convention to be at least US$ 49, which would rise to US$ 171 billion by 2030. About half of these costs will be borne by developing countries.
- Economic profit can be realized by investing in energy efficiency and renewable energies due to increased productivity and resource efficiency. These potential new economic opportunities are not explored in a business as usual scenario.

Approaches Available for Reducing Global CO_2 Emissions within the Energy Sector

In order to meet the two-degree warming goal as set out in the Paris Agreement, a large-scale transformation of our energy systems is needed, and innovation is central to this transformation. The world has to transition from an energy system dominated by fossil fuels to a low-carbon one. We call this process of transitioning from fossil fuels to low-carbon energy sources "decarbonization". Our progress in decarbonizing our total energy system (including transport, heat, and electricity) has been slow. Fossil fuels are still the dominant energy source.

End-Use Energy Efficiency

- End-use energy efficiency covers technologies that are based on the simple concept of using less energy to provide the same level of output or perform the same task.
- The reason end-use energy efficiency technologies are increasingly being seen as practical solutions is that they avoid the environmental and economic costs of energy supply and distribution, and reducing

the consumption at the end use provides leverage for reducing emissions from generation.

- Some common examples of these technologies are weather stripping buildings, adopting green building principles, and the use of energy control systems.

Renewable Energy

Renewable energy covers a variety of technologies that derive their energy from renewable sources. These include solar, wind, biomass, hydro, and geothermal.

- Hydropower is the largest single renewable electricity source today, providing 16% of world electricity at competitive prices. It dominates the electricity mix in several countries – developed, emerging, or developing.
- Bioenergy is the single largest renewable energy source today, providing 10% of the world's primary energy supply.
- Shifting our energy systems away from fossil fuels toward renewable technologies will require significant financial investment. In Figure A3.2 we see global investments in renewable technologies from 2004 to 2015 (measured in billion USD per year). In 2004, the world invested 47 billion USD. By 2015, this had increased to 286 billion USD, an increase of more than 600%. In 2016, solar and wind energy both received 47% of investment (combining to account for 94% of global finance). These two technologies have been taking an increasing share, especially over the last five years. In 2006, bioenergy (both in the form of biomass and liquid biofuels) took a sizable share of global investment, peaking at 36%. This has dwindled over the last decade, receiving less than 4% in 2016. These trends suggest that investors see solar and wind energy as the dominant renewable technologies of the future.

End-Use Fuel Switching

- End-use fuel switching involves choosing the most appropriate energy carrier to supply a given end use while minimizing primary energy consumption.
- For example, in most contexts, natural gas is more than twice as efficient at providing space heating, domestic hot water and heat for cooking compared to electricity.
- Consequently, switching to more efficient fuels for specific end-uses can increase the overall efficiency of the energy system and reduce the environmental impacts.

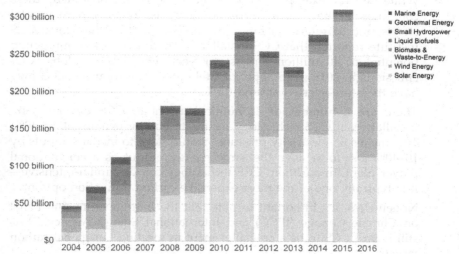

FIGURE A3.2

Investment in renewable energy, by technology.

(Retrieved from: https://ourworldindata.org/energy-production-and-changing-energy-sources)

Power Generation Efficiency

- Power generation efficiency improvements can increase the output of power while reducing, or keeping the same, the quantity of inputs. Fuel switching in power generation means using less carbon-intensive fuel sources such as feedstock and biofuel or using natural gas instead of coal.

Advanced Technologies under Development

- In addition to the technologies areas described above, there are several advanced technologies currently under development that may prove pivotal in reducing the environmental impacts of the energy sector.

- The most well-known of these is carbon capture and storage (CCS). CCS captures carbon dioxide from large point sources and stores it in deep geological formations, in deep ocean masses or in the form of mineral carbonates. This effectively prevents 90% of

CO_2 from being emitted to the atmosphere, meaning that we could avoid the negative CO_2 impacts of fossil fuel combustion

- While several examples of CCS have been demonstrated, there continues to be issues surrounding costs, other environmental effects, and long-term viability. By the end of 2016, the Global CCS Institute reported that CCS installations had a collective mitigation potential of 40 million tonnes per year. For context, global CO_2 emissions in 2016 were approximately 36 billion tonnes – CCS may have therefore averted around 0.1%.
- There are a number of economic barriers to CCS development. Installation of CCS technology incurs an energy penalty of 10–40%. This means an electricity producer would have to increase inputs by 10–40% just to achieve the same energy output as a conventional power plant. In addition, CCS technology can be capital-intensive – it is typically one of the most expensive carbon mitigation options.
- Nonetheless, it's important to note that the Intergovernmental Panel on Climate Change (IPCC) and International Energy Agency (IEA) still regard CCS to be a crucial element in meeting our global carbon targets.

Prices

- Prices can strongly influence our choice of energy sources. In this regard, it is the relative cost between sources which is important. For many countries, increasing the share of the population with access to electricity and energy resources is a key priority, and to do so, low-cost energy is essential.
- The dominant energy source in the transport sector is liquid fuels (diesel and gasoline) for which relative costs are less important than changes in price through time. In the electricity sector, we compare costs based on what we call the "levelized cost of electricity" (LCOE). The LCOE attempts to provide a consistent comparison of electricity costs across sources but taking the full life-cycle costs into account. It is calculated by dividing the average total cost to build then operate (i.e., both capital and operating costs) an energy asset (e.g., a coal-fired power station, a wind farm, or solar panel) by the total energy output of that asset over its lifetime. This gives us a measure of the average total cost per unit of electricity produced. Measuring sources on this consistent basis attempts to account for the fact that resources vary in terms of their capital and operating costs (e.g., solar PV may have higher capital costs, but lower operating costs relative to coal over time). Note that this cost of energy production has an obvious impact on electricity prices for the consumer: the LCOE represents the minimum cost producers would

have to charge consumers in order to break-even over the lifetime of the energy project. To be truly competitive, renewable technologies will have to be cost-competitive with fossil fuel sources. In Figure A3.3, sourced from IRENA's latest Rethinking Energy report, we see the LCOE (measured in 2016 USD per megawatt-hour of electricity produced) across the range of renewable technologies in 2010, and in 2016.

For the global chart, this range of costs is represented as vertical bars for each technology. The white line in each represents the global weighted average cost per technology. The average range of fossil fuel costs is shown below as the grey horizontal block. What we see is that in terms of the 2016 weighted average cost, most renewable technologies are within a competitive range of fossil fuels. The key exception to this is solar thermal which remains about twice as expensive (although is falling). Hydropower, with the exception of traditional biomass, is our oldest and well-established renewable source: this is reflected in its low price (which can undercut even the cheapest fossil fuel sources). Note, however, that although the weighted average of most sources is competitive with the average fossil fuel cost, the wide range of potential costs means that this is not true for all countries. This is why the selection of particular technologies needs to be considered on a local, context-specific basis.

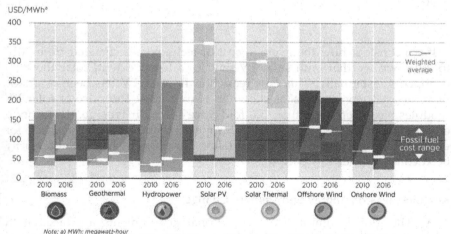

Note: a) MWh: megawatt-hour
b) All costs are in 2016 USD. Weighted Average Cost of Capital is 7.5% for OECD and China and 10% for Rest of World

FIGURE A3.3
Levelized cost of electricity (LCOE) 2010 and 2016.
(Retrieved from: https://ourworldindata.org/energy-production-and-changing-energy-sources)

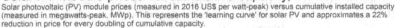

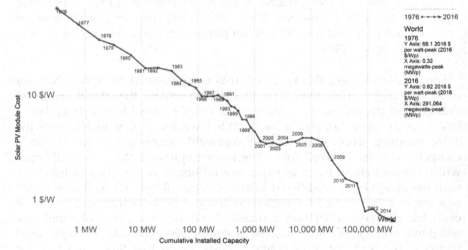

Source: Lafond et al. (2017); IRENA; SolarServer   OurWorldInData.org/renewables • CC BY

FIGURE A3.4

Solar PV module prices vs cumulative capacity, 1976 to 2016.

(Retrieved from: https://ourworldindata.org/energy-production-and-changing-energy-sources)

If we consider how the average cost of technologies changed from 2010 to 2016, we see that both solar PV (and to a lesser extent, solar thermal) dropped substantially. This cost reduction in solar PV has been dramatic over the past few decades, as shown in Figure A3.4. The price of solar PV modules has fallen more than 100-fold since 1976. On average, the technology has had a learning rate of 22%; this means that the cost falls by 22% for every doubling in solar PV capacity (although progress has not necessarily been constant over this period).

Key Messages on Energy and Climate Change

- The global community and national governments are faced with major cross-cutting challenges with respect to the energy sector, such as combating climate change; reducing pollution and public-health hazards; and addressing energy poverty.

- Climate change and the production and consumption of energy from fossil fuels are closely related. In order to reduce CO_2 emissions, it is necessary to change the world's energy model. Fighting climate change requires an urgent transition towards a sustainable development model based on efficiency and equity, as well as a decided commitment regarding renewable energy.

- Greening the energy sector involves a large-scale transformation of the energy systems, using existing and new technologies.
- Options for greening the sector include: improving end-use and power generation energy efficiency, advancing renewable energies, switching to alternative fuels, as well as advancing new technologies (e.g., CCS).
- Climate change is a challenge, but it can be an opportunity to undertake true sustainable development. Measures to save and make energy more efficient, as well as to promote renewable energy, favor genuine development, and reduce external dependence.

Appendix D

Key Environmental Impact Categories

The following is a brief description of the most common environmental impact results reported in LCAs including information about which emissions and mechanisms contribute to the specific environmental degradation.

Acidification

Acidification is the process of changing the pH balance of water and soil, resulting in more acidic or more basic substances. Acidifying substances cause a wide range of impacts on soil, groundwater, surface water, organisms, ecosystems, and materials (buildings). Plants and animals have optimum pH levels within which they thrive and have a limiting range within which they can live. A change in the pH balance reduces the productivity of soil, and in water can lead to a decline in plant and animal health. The emissions that cause acidification can damage exposed building materials and finishes, leading to reduced durability and increased maintenance of facilities. One of several chemical reactions related to acidification is the combination of carbon dioxide, CO_2, with water to form carbonic acid (H_2CO_3). The generation of electricity through fossil fuel combustion is a prime source of CO_2 and also a contributor to sulfur and other gas emissions that result in man-made acidification. Chemicals that contribute to acidification are sulfur oxides (SO_x), nitrogen oxides (NO_x) and ammonia (NH_3). These chemicals can travel long distances as airborne gases before being deposited on the ground or in water as dust or soot or through rain, fog or snow, known as "acid rain." Natural sources including volcanic eruption and decomposition of plants, can also lead to acidification.

Acidification potential (AP) for emissions to air is usually calculated with the RAINS (Regional Acidification INformation and Simulation) computer model describing the fate and deposition of acidifying substances. AP is expressed as kg SO_2 equivalents/kg emission. The time span is eternity and the geographical scale varies between local scale and continental scale. The effect of acidifying emissions depends upon the region.

Climate Change

Climate change refers to the change in global temperature caused via the greenhouse effect by the release of "greenhouse gases" such as carbon dioxide by human activity. There is now scientific consensus that the increase in these emissions is having a noticeable effect on climate such as precipitation patterns. We characterize the potency of a gas for global warming as compared to an equivalent mass of carbon dioxide (CO_2). In addition to CO_2, gases such as methane and nearly 100 other documented gases impact global warming. This impact is reported in units of CO_{2e}, as the Global Warming Potential or GWP or GHG emissions or a carbon footprint or climate change impact. The GHGs absorb energy, trapping the heat close to the earth, functioning like a greenhouse trapping heat within a glass building. Emissions of GHGs from energy generation are a critical component. Some industrial processes such as cement production which release CO_2 and agricultural production which release methane and nitrous oxide emissions also contribute significantly to global GHG emissions.

Climate change can result in adverse effects upon ecosystem health, human health and material welfare. The characterization model as developed by the Intergovernmental Panel on Climate Change (IPCC) is selected for development of characterization factors. Factors are expressed as Global Warming Potential for time horizon 100 years (GWP100), in kg carbon dioxide/kg emission. The geographic scope of this indicator is at global scale.

Eutrophication

Eutrophication (also known as nutrification) includes all impacts due to excessive levels of macro-nutrients in the environment caused by emissions of nutrients to air, water and soil. Algae blooms and dead zones in water are common manifestations of the problems of eutrophication. In freshwater, it is generally phosphate which is the limiting nutrient, while in salt waters it is generally nitrogen which is limiting. Emissions of ammonia, nitrates, nitrogen oxides, and phosphorous to air or water all have an impact on eutrophication . Urban sources of eutrophication include septic field seepage, storm and wastewater runoff – particularly from fertilized landscapes and fossil fuel combustion. Rural eutrophication sources include rainfall runoff after contact with fertilized agriculture and manure and aquaculture. Eutrophication risk has a large variability and impacts are most significant if emissions are released in a region in which the critical load is close to or already exceeded.

Nutrification potential (NP) is based on the work of Heijungs using a stoichiometric procedure and is expressed as kg PO_4 equivalents per kg emission. Fate and exposure is not included, time span is eternity, and the geographical scale varies between local and continental scale.

Ozone Depletion

Ozone (O_3) is a gas that is present in the atmosphere and is in constant flux, undergoing chemical reactions with other elements such as chlorine and oxygen. Ozone high in the stratosphere functions to protect the Earth from the sun's UV rays. Ozone close to the Earth's surface is characterized as smog and its environmental impact is described in the next section. Chlorofluorocarbons (CFCs), Halons and HCFCs are the major causes of ozone depletion. They are refrigerant chemicals or from manufacturing processes that react with the energy of sunlight and release chlorine which then reacts with ozone in a cycle, breaking it down into oxygen. Because of this stratospheric ozone depletion, a larger fraction of UV-B radiation reaches the Earth's surface. This can have harmful effects upon human health, animal health, terrestrial and aquatic ecosystems, biochemical cycles and on materials. The characterization model is developed by the World Meteorological Organization (WMO) and defines ozone depletion potential of different gasses (kg CFC-11 equivalent/kg emission). The geographic scope of this indicator is at global scale. The time span is infinity.

Photo-Oxidant(smog) Formation

In atmospheres containing nitrogen oxides (NO_x, a common pollutant) and volatile organic compounds (VOCs), ozone can be created in the presence of sunlight. Although ozone is critical in the high atmosphere to protect against ultraviolet (UV) light, low level ozone is implicated in impacts as diverse as crop damage and increased incidence of asthma and other respiratory complaints. The major sources of ground level ozone are the S0x and NOx emissions related to fossil fuel consumption and the release of volatile organic compounds (VOCs). Photochemical ozone creation potential (also known as summer smog) for emission of substances to air is calculated with the United Nations Economic Commission for Europe (UNECE) trajectory model (including fate) and expressed using the reference unit, kg ethene (C_2H_4) equivalent.

Like for acidification, the location of the emission and the local conditions are essential to understanding the environmental impact of smog formation.

Human Toxicity

This category concerns effects of toxic substances on the human environment. Health risks of exposure in the working environment are not included. Chemical and particulate matter emissions can have wide-ranging impacts on human health. Other measured environmental impacts have a direct and indirect impact on human health. Correlating between emissions and human health requires integration of models that can predict the fate, that can predict exposure dosage to humans as well as model the relationship between dosage and human response. Environmental and biological scientists continue to advance methods to quantify these connections.

Characterization factors, human toxicity potentials (HTPs), are calculated with USES-LCA, describing fate, exposure and effects of toxic substances for an infinite time horizon. For each toxic substance, HTPs are expressed as 1,4-dichlorobenzene equivalents/kg emission. The geographic scope of this indicator determines on the fate of a substance and can vary between the local and global scale.

Eco-Toxicity

Environmental toxicity is measured as two separate impact categories which examine freshwater and land, respectively. The emission of some substances, such as heavy metals, can have impacts on the ecosystem. Assessment of toxicity has been based on maximum tolerable concentrations in water for ecosystems. Ecotoxicity potentials are calculated with the USES-LCA, which is based on EUSES, the EU's toxicity model. This provides a method for describing fate, exposure and the effects of toxic substances on the environment. Characterization factors are expressed using the reference unit, kg 1,4-dichlorobenzene equivalent (1,4-DB), and are measured separately for impacts of toxic substances on fresh-water aquatic ecosystems and terrestrial ecosystems.

Fresh-water aquatic eco-toxicity: This category indicator refers to the impact on fresh water ecosystems, as a result of emissions of toxic substances to air, water and soil. Eco-toxicity Potential (FAETP) is calculated with USES-LCA. The time horizon is infinite. Characterization factors are expressed as 1,4-dichlorobenzene equivalents/kg emission. The indicator applies at global/continental/regional and local scale.

Marine ecotoxicity: Marine eco-toxicity refers to impacts of toxic substances on marine ecosystems (see description fresh water toxicity).

Terrestrial ecotoxicity: This category refers to impacts of toxic substances on terrestrial ecosystems (see description fresh water toxicity).

Resource Use and Depletion

Resource use and depletion is of environmental significance given the broad environmental, social and economic impacts of declining resources. Nonrenewable resources, such as fossil fuels, do not renew themselves within a human time scale. Typically, the use of material and energy resources are converted into units of energy, often termed "embodied energy" representing both the energy available within a product (what could be attained if the product were incinerated) and the energy used as a fuel to manufacture the product. Alternatively, we can look at resource depletion and at the relative impact of consumption of different resources can be characterized and reported as a weighted impact. This impact takes into account both the quantities of resources consumed and the impacts of their consumption. The impacts are typically reported for both abiotic elements and abiotic resources.

The Abiotic Depletion Factor (ADF) is determined for each extraction of minerals and fossil fuels (kg antimony equivalents/kg extraction) based on concentration reserves and rate of de-accumulation. The geographic scope of this indicator is at global scale.

Consumption of Freshwater

Water is a critical resource upon which all social and economic activities and ecosystems functions depend. Around the world, water is becoming an increasingly scarce resource, due to increased demand, and changes in patterns of rainfall. To recognize the value of water as a resource, the damage that over-extraction from rivers and aquifers can cause water consumption can be tracked as an inventory item within an LCA or developed independently as a water footprint. Water footprints divide water consumption into blue (fresh surface or groundwater), green (precipitation on land that stays on surface/in soil), and gray (volume required to dilute pollutants). Typically, LCAs report blue water consumption. In both of methods, the severity of the impact is not accounted for, as local conditions of scarcity are not included in the assessment. The interpretation of results must thus integrate knowledge of regional water scarcity issues.

Waste Generation

The quantity of waste and materials for reuse and recycling is often reported as inventory items in LCA. These inventory items are reported in kilogram of material. LCA typically classifies waste as hazardous, nonhazardous, or radioactive. The definition of the waste types depends upon the region.

Material ReUse and Recycling

Material available for reuse and recycling (kg of total material) is an inventory item that can be reported in LCA. These quantities attempt to capture the efficiency of the manufacturing or complete life cycle of products and can be compared with the quantity of waste materials generated.

The following table gives an overview of the relevant impact categories, the typical unit in which they are measured in and what they describe.

Environmental Impacts

Impact category/ Indicator	Unit	Description
Global warming	kg CO_2-eq	Indicator of potential global warming due to emissions of greenhouse gases to air
Ozone depletion	kg CFC-11-eq	Indicator of emissions to air that cause the destruction of the stratospheric ozone layer
Acidification of soil and water	kg SO_2-eq	Indicator of the potential acidification of soils and water due to the release of gases such as nitrogen oxides and sulphur oxides
Eutrophication	kg PO_4^{3-}-eq	indicator of the enrichment of the aquatic ecosystem with nutritional elements, due to the emission of nitrogen or phosphor containing compounds
Photochemical ozone creation	kg ethene-eq	Indicator of emissions of gases that affect the creation of photochemical ozone in the lower atmosphere (smog) catalyzed by sunlight.
Depletion of abiotic resources – elements	kg Sb-eq	Indicator of the depletion of natural nonfossil resources
Depletion of abiotic resources – fossil fuels	MJ	Indicator of the depletion of natural fossil fuel resources
Human toxicity	1,4-DCB-eq	Impact on humans of toxic substances emitted to the environment
Fresh water aquatic ecotoxicity	1,4-DCB-eq	Impact on freshwater organisms of toxic substances emitted to the environment
Marine aquatic ecotoxicity	1,4-DCB-eq	Impact on sea water organisms of toxic substances emitted to the environment
Terrestrial ecotoxicity	1,4-DCB-eq	Impact on land organisms of toxic substances emitted to the environment

Resources Used

Impact category/Indicator	Unit	Description
Primary renewable energy (materials)	MJ	Use of renewable primary energy resources as raw materials
Primary renewable energy (energy)	MJ	Use of renewable primary energy, excluding renewable primary energy resources used as raw materials
Primary renewable energy (total)	MJ	Sum of the two values above
Primary nonrenewable energy (materials)	MJ	Use of nonrenewable primary energy resources as raw materials
Primary nonrenewable energy (energy)	MJ	Use of nonrenewable primary energy, excluding renewable primary energy resources used as raw materials
Primary nonrenewable energy (total)	MJ	Sum of the two values above
Use of secondary material	kg	Material recovered from previous use or from waste which substitutes primary materials
Use of fresh water	m^3	Freshwater use
Use of renewable secondary fuels	MJ	Renewable fuel recovered from previous use or from waste which substitutes primary fuels
Use of nonrenewable secondary fuels	MJ	Nonrenewable fuel recovered from previous use or from waste which substitutes primary fuels

Waste Type

Impact category/ Indicator	Unit	Description
Hazardous waste disposed	kg	Hazardous waste has a certain degree of toxicity that necessitates special treatment
Nonhazardous waste disposed	kg	Nonhazardous waste is nontoxic and similar to household waste. It consists of inert waste and ordinary household waste
Radioactive waste disposed	kg	Radioactive waste mainly originates from nuclear energy reactors

Output Flows

Impact category/ Indicator	Unit	Description
Components for reuse	kg	Material or components leaving the modelled system boundary which is destined for reuse
Materials for recycling	kg	Material leaving the modelled system boundary which is destined for recycling
Materials for energy recovery	kg	Material leaving the modelled system boundary which is destined for use in power stations using secondary fuels
Energy production	J	Energy exported from waste incineration and landfill

Appendix E

Key Standards That Support Sustainable Development

Voluntary sustainability standards are an important means of providing assurance that products and materials traded in complex, global supply chains have been produced in an ethical and environmentally friendly way. This appendix provides an overview of some of the key standards that are used in organizations to support sustainable development.

ISO 26000, Guidance on Social Responsibility: The extent to which an organization contributes to sustainable development and its impact on society and the environment is known as "social responsibility." It is increasingly becoming a critical measure of performance, with it influencing everything from an organization's reputation to its ability to attract high-caliber employees. The standard provides guidance – and not certification – on how businesses and organizations can operate in a socially responsible way, which includes encompassing the principles of non-discrimination and equal opportunities. The core subjects and issues defined by the standard comprise human rights, labor practices, the environment, fair operating practices, consumer issues, and community involvement.

The Global Reporting Initiative (GRI): a non-governmental organization (NGO) founded to develop and manage a sustainability reporting framework, has published a guidance document called GRI G4. The document provides assistance to organizations who wish to use GRI guidelines as the reporting framework for their implementations of ISO 26000.

The ISO 14000: This family of standards details practical tools for companies and organizations of all kinds to manage the impact of their activities on the environment. This suite of standards covers overall frameworks, audits, communications, labeling, life cycle analysis, and methods to mitigate and adapt to climate change.

ISO 14001:2015 Environmental management systems – Requirements with guidance for use: This is one of ISO's most widely used standards. It sets out the criteria for an environmental management system and can be certified to. It maps out a framework that a company or organization can follow to set up an effective environmental management system. It can be used by any organization regardless of its activity or sector. Using ISO 14001:2015 can provide assurance to company management and employees as well as external stakeholders that environmental impact is being measured and improved.

The ISO 14040 series: This series of standards relate to Life Cycle Assessment (LCA). The following standards are included under this series.

- ISO 14040 – General Principles and Framework
- ISO 14041 – Goal and Scope Definition and Inventory Analysis
- ISO 14042 – Life Cycle Impact Assessment (LCIA)
- ISO 14043 – Life Cycle Interpretation
- ISO 14044 – Requirements and guidelines

ISO 14024 (Type I Label): This is a voluntary, multiple-criteria-based, third-party program developed for a specific product or products. Examples include the EU eco-label.

ISO 14021 (Type II Label): This is for any written or spoken environmental statement or claim. There are no fixed criteria, and manufacturers simply declare the information they wish to communicate about the environmental attributes of their products.

ISO 14025 (Type III Label): This is based on ISO 14040/44 and introduces two concepts: product category rules (PCRs) and environmental product declarations (EPDs). PCRs are specific guidelines for the calculation of the environmental impact of products with similar characteristics. By following the requirements in the PCR, a company can develop an EPD, which is a concise document containing relevant environmental information about a product. PCRs are subject to administration of a program operator.

ISO 14064: This standard provides governments, businesses, regions and other organizations with a complimentary set of tools for programs to quantify, monitor, report and verify greenhouse gas emissions. It specifies principles and requirements for the reporting of greenhouse gas (GHG) emissions at both the organization and more specifically at the project level. Part 1 focuses on the organization level and part 2 at the project level. Part 3 provides guidance for those conducting or managing the validation and/or verification of GHG assertions.

ISO 14067: Carbon Footprint of Product: This standard is based on ISO 14040/44 and ISO 14025, but focuses on climate change only. Quantification of the carbon footprint of a product (CFP) is largely based on ISO 14040/44, but includes requirements on specific issues relevant for carbon footprints, including land-use change, carbon uptake, biogenic carbon emissions and soil carbon change. The standard also provides specific requirements on communication with or without the intention to be publicly available.

GHG Protocol Product Standard: Product Life Cycle Accounting and Reporting Standard: The GHG Protocol Product Standard has been developed by the World Resources Institute (WRI) and the World Business Council on Sustainable Development (WBCSD). Like ISO 14067, this standard is largely in compliance with ISO 14040/44, but is specifically focused on greenhouse gas accounting. It quantifies the GHG inventories of products and also requirements for public reporting.

PAS 2050: UK's Product Carbon Footprint Standard: This standard was published by British Standards Institution (BSI) in 2008 and revised in October 2011. PAS 2050 was the first carbon footprint standard and has been applied by many companies worldwide. The 2011 revision resulted in a standard that is largely aligned with the GHG Protocol product standard.

ISO 14046: Water Footprinting: This is a standard which demonstrates an organization's leadership in environmental protection and helps manage and reduce water consumption.

ISO 14051: Material Flow Cost Accounting (MFCA): This is a management process tool standard that helps trace all materials through production and measures the output in finished products or waste material. It provides a framework which helps develop an integrated approach to optimize the use of materials.

ISO 31000: 2018 – Risk Management: This is a standard that supports businesses to embed risk management into every aspect of their organization, systems, and processes.

ISO 45001: 2018, Occupational Health and Safety Management Systems – Requirements with Guidance for Use: This is designed to help companies and organizations worldwide protect the health and safety of the people who work for them.

ISO 50001: 2018 – Energy Management: This standard supports businesses to improve energy performance, efficiency, use, and consumption.

ISO 20400: Sustainable procurement – Guidance: This helps organizations develop sustainable and ethical purchasing practices that also benefit the societies in which they operate. It includes guidelines for implementing ethical processes throughout the supply chain

ISO 15392: Sustainability in building construction – General principles: This identifies and establishes general principles for sustainability in buildings and other construction works throughout their whole life cycle, from inception to end of life.

SA 8000: Social Accountability: This is a widely recognized global standard for managing human rights and provides a framework for organizations to develop, maintain, and apply socially acceptable practices in the workplace.

Index

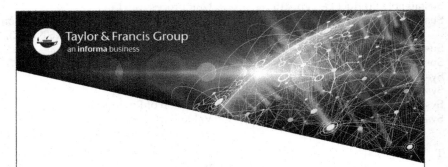